Streamflow Generation

Streamflow Generation

Processes and Perceptual Models

Jeffrey J. McDonnell
University of Saskatchewan

OXFORD
UNIVERSITY PRESS

OXFORD
UNIVERSITY PRESS

Great Clarendon Street, Oxford, OX2 6DP,
United Kingdom

Oxford University Press is a department of the University of Oxford. It furthers the University's objective of excellence in research, scholarship, and education by publishing worldwide. Oxford is a registered trade mark of Oxford University Press in the UK and in certain other countries.

Published in the United States of America by Oxford University Press
546 Fifth Avenue, New York, NY 10036, United States of America

British Library Cataloguing in Publication Data
Data available

Library of Congress Control Number: 2026942777

ISBN 9780192862693

DOI: 10.1093/9780191953736.001.0001

Printed and bound by
CPI Group (UK) Ltd., Croydon, CR0 4YY

The manufacturer's authorized representative in the EU for product safety is Oxford University Press España S.A. of Parque Empresarial San Fernando de Henares, Avenida de Castilla, 2 – 28830 Madrid (www.oup.es/en or product.safety@oup.com). OUP España S.A. also acts as importer into Spain of products made by the manufacturer.

Cover image: Juergen Schacke

MIX
Paper | Supporting responsible forestry
FSC® C013604
www.fsc.org

For my students' students

Contents

Dedication

[Books] are the voices of the distant and the dead

—William Ellery Channing

This book is dedicated to John Hewlett (1922–2004). His name is often lost in the annals of hydrological history. No awards or induction as Fellow into any professional society. Nothing named after him. But Hewlett radically altered our understanding of streamflow generation. He disrupted thinking in a way unsurpassed by anyone before or after him. Like many great scientists, he thought like an artist and worked like an accountant. He did important field work and built on the work of his predecessor, Charles Hursh. Hewlett ushered in radical shifts in thinking at the iconic Coweeta Catchment in North Carolina, USA that reverberated worldwide. The dedication is especially apt, as Hewlett struggled to change the engineering paradigm of his day and felt, allegedly, that his work did not have the impact or recognition it deserved. Yet, with the fullness of time, it is clear that Hewlett's thinking on hillslope hydrology, catchment rainfall–runoff processes and perceptual model integration with numerical models was decades ahead of its time: combining bottom-up and top-down approaches; his use of physical models in the field and lab; the dialog he started between field experimentalists and modelers; his fill-and-spill bucket model; and thinking about what criteria would best test a model's robustness. Beyond his papers, Hewlett's short text with Wade Nutter *An Outline of Forest Hydrology* (later updated as *Principles of Forest Hydrology*, by Hewlett and UGA Press) was (and still is) a gem—a book that outlined the essentials of the field in a readable, compelling and compact way. A colleague of mine was one of the last undergraduates to be taught by Hewlett. She still recalls his clear, concise and systematic teaching descriptions of catchment hydrology. My hope is that my small effort with this book can be a modern update to Hewlett's text and do justice to his early, important contributions and continued legacy.

Preface

In vain have you acquired knowledge if you do not impart knowledge to others

—Deuteronomy Rabbah

I am a field hydrologist. And my experiences in the field color my writing of this book. But over the past decades I've seen field work diminished in value in our model-obsessed hydrology world. Therefore, this book is about field-based studies linked to streamflow generation and the process understanding and perceptual models that can come from field-based inquiry.

To paraphrase perhaps the greatest-ever field scientist, Alexander von Humboldt, hydrologists only ever possess the imperfect knowledge of their time. My time in hydrology began in the mid-1980s in the Harp Lake research catchment on the Precambrian Shield of Canada. During five months of daily visits to the slopes as snowmelt played out, I witnessed striking spatial and temporal patterns of streamflow generation; largely slow, directional wetting-up through six weeks of melt followed by days to weeks of moderate hillslope drainage to the streams. My PhD field work commenced a few years later at a different site: the Maimai catchment in New Zealand. There, the rains were largely low intensity but frequent. The hillslopes were short and very steep. During the relentless back-to-back storms that I monitored, I encountered streamflow responses that were sudden and comparatively violent; where the chart recorder next to the v-notch weir rose *and fell* almost vertically. This was a different form of streamflow generation.

Since those early graduate days, I've seen many more headwater streams and catchments—North American field sites at well-known catchments like the HJ Andrews catchment in Oregon, the Panola catchment in Georgia, the Savannah River Site in South Carolina, Sleepers River in Vermont, Reynolds Creek in Idaho and the Swift Current research hillslope in Saskatchewan where I did my own research—but also those of collaborators where I've spent time doing field work in Europe and the UK, Mexico, SE Asia, South America and Australia. Beyond the sites used for my formal research, I've observed the hydrology of hillslopes and catchments in dozens of other countries on research trips, university lectures where hosts showed me their sites, during conference field trips, and family adventures. Across these many different places, I've seen similarities and differences. This book tries to summarize what I have read and seen and learned—to hand off to the next generation of catchment hydrologists a narrative of streamflow generation processes and perceptual models (and how they have evolved over time—both at and between sites).

This book also tries to revive something now rarely discussed in hydrology: the development of process knowledge; often for knowledge's sake. And to re-invigorate discussion of discovery science. I came of age scientifically learning from books by Chorley like "Water, Earth and Man"; Kirkby's edited volume "Hillslope Hydrology" and other collections of process understanding. With the advent of modeling—a topic I'll have more to say about in the first and last chapters—we've somehow lost what I think is the most important thing to pay attention to: how things work. We have over-indexed our models of today and underestimated our model structural uncertainty.

In other words, we've been distracted by decades of model calibration and "validation" without paying attention to the most important piece of models: their structure, informed by process understanding and descriptions which themselves are based on field observations and measurement. And then testing them with metrics that go beyond hydrograph shape and timing matching.

This book is a personal synthesis—meaning, my attempt at a readable narrative; an attempt to distill the perceptual models of how catchments store and release water to help a beginning graduate student to know how certain ideas came about. What we know, what we think we know and what we need to know. Because what we know now is tiny compared to what we need to know about hydrological processes linked to streamflow generation; to understand the effects of climate change, land use change and our ever-more altered natural world.

Lastly, this book has no equations. My experience as an educator over the past many decades is that while it can often feel easier to teach with equations—given how an equation can encode complex ideas compactly; and that they can feel precise with relations between hydrological variables that are unambiguous—many students struggle with such equations, that to them, seem unrelatable. They often want to understand what lies behind these equations. My goal is to separate the math from the explanation; to complement our many textbooks that lead from a purely quantitative perspective. Clearly both approaches are ultimately needed. But I hope this book can offer that balance, so lacking in hydrological teaching today.

Jeffrey J. McDonnell
Vernon, British Columbia, Canada
Corvallis, Oregon, USA
September, 2025

CHAPTER 1

Introduction

1.1 Streamflow generation

I am among those who think that science has great beauty

—Marie Curie

Streamflow generation is fundamental to the science of hydrology. The most important questions related to streamflow generation—how catchments store and release water—are simple and timeless: where does water go when it rains? What flowpath does the water take to the stream? How long does the water reside in the catchment? How does vegetation compete for water with streamflow? They are at the core of many of today's most pressing water sustainability problems of how climate change impacts floods and drought and how landuse change impacts streamwater quantity and quality.

Figure 1.1 shows an image of a stream draining a small catchment—in Scotland in the drizzling rain. My hope is that as you read this book, you will get curious about how catchments work—be it the catchment that you live in[1] or a research catchment where you may be working; and to start thinking in questions. To go beyond the undergraduate-level question of how infiltration, evaporation and groundwater recharge "work," toward graduate-level thinking about: how different topographic positions in the catchment contribute water to the stream? How old the water is in the stream and what is its age distribution? What are the precipitation thresholds for determining how and when hillslopes connect and disconnect to the stream? How deep does infiltrating rain or meltwater go on these slopes? Does the stream discharge increase linearly downstream? How do rainfall intensity and antecedent wetness conditions combine to affect the stream response? Where does the vegetation growing in the catchment get its water from and how might the transpired water be linked to the water flowpaths to the stream?

But rather than thinking curiously about how streamflow generation works in a setting like the one in Figure 1.1, we usually turn much of this incredible beauty and complexity and curiosity into a hydrological "problem"; one that we "solve" with computer models. Likely most students reading this book will have seen or used a hydrological model—they are now the lingua franca of our science. Rather than thinking about the dominant processes of streamflow generation at a site, the model approach is an office-based mechanical exercise, dealing with shape files, look-up tables of rainfall excess and the like. There are 100s of models out there. The model you may have used could be a statistical model, a conceptual bucket model, an integrated physically based runoff model, or perhaps a groundwater–surface water model. Examples, of some of the most popular ones are SWAT, WEAP, HEC-HMS, PRMS, VIC, HBV and MIKE SHE, to name a few.

Of course, we need hydrological models. We use them to forecast and predict. Models help us with "what if" questions in terms of a planned and anticipated change. They can be an aid to understanding: to help summarize and test knowledge about a catchment system. The best book out there on catchment-scale hydrological models is Rainfall–Runoff Modeling: The Primer by Keith Beven. In it, Keith describes the modeling process: where for a new model, one begins one's perception of a

[1] We all live in a catchment!

Streamflow Generation. Jeffrey J. McDonnell, Oxford University Press. © Jeffrey J. McDonnell (2026). DOI: 10.1093/9780191953736.003.0001

Figure 1.1 A quintessential headwater stream and its catchment during light rainfall, Scottish Highlands.
Source: Michal Klajba, Wikipedia.com

catchment (i.e., the (subjective) qualitative mental model that this book focuses on). This sets up the flowchart of steps in the modeling process to the conceptual model stage where one decides on the equations to use, and then what is called the procedural model where one gets the code to run on the computer; followed by model calibration and finally, model validation.

The problem now though is that with so many models already developed, most beginning graduate students simply do the last two steps—calibration and validation—of *existing* models. And often with the assumption that the available models and their model structures (derived initially from field-based process studies and a perceptual model or set of equations based on our few laws in hydrology) are applicable anywhere and everywhere. This sets up a dangerous cycle where fewer and fewer students go into the field to develop their *own* perceptual model for the site that they study or ever question whether or not the simplified numerical model usefully describes the field behavior.

Without seeing first-hand the complexity of water flow in a soil pit or trenched hillslope, or in the tracer response in a stream, students begin to believe that models represent some sort of hydrological reality.[2] We get stuck—as we are now, I think—in "model land," where modeling is now akin to an all-explaining ideology. This so-called epistemic[3] arrogance is for me, the defining issue of our time and now openly discussed in hydrology (Gleason and Brown, 2025). I hope this book exposes something about how much we understand and don't understand streamflow generation. Because this epistemic arrogance can have a double effect. First, it means that we overestimate what we know. And second, we underestimate uncertainty by limiting the range of possible streamflow generation processes and perceptual models by artificially reducing the space of the unknown in a model. We need to go out into the field—ideally during rainfall or snowmelt events—and observe hydrological

[2] Nassim Nicholas Taleb's (2007) book "The Black Swan" discusses epistemic arrogance in great detail in relation to scientific understanding in general.

[3] Fancy word for "knowledge."

processes (van Stan et al., 2023). In other words, we need to escape model land.

We need to escape model land because our lack of understanding of the hydrological processes within catchments mocks the precision[4] of any one of our hundreds of hydrological models. You may have heard the common phrase "all models are wrong; some are useful."[5] But to say that all models are wrong, but some are useful misses the point about delusive accuracy and why hydrologists, like environmental scientists in general, cannot predict the future.[6] This is because our catchment systems that generate streamflow can be diabolically complex. To escape model land, we need to assess what we know, what we think we know and what we need to know; to move our understanding of catchment hydrological processes forward. We need to appreciate the limits of our understanding—that what we know is infinitesimally small compared to the process complexity of how catchments store and release water. And perhaps most worrying now, as we move into an AI world, that the success of machine learning for hydrological modeling could mean that processes knowledge of streamflow generation becomes fully decoupled from modeling (Nearing et al., 2021).

1.2 Headwater streams and catchments

"...the truth is, the science of Nature has been already too long made only a work of the brain and the fancy. It is now high time that we should return to the plainness and soundness of observations"

—Robert Hooke

Before we get too far along, we must first remind ourselves that to understand streamflow generation—what Cook (1946) called "the central problem" of hydrology—is to understand the land-area generating that streamflow: the catchment. Different languages and scientific cultures have different words to define the same thing: catchment, basin, watershed, drainage basin. In this book, we will stick with the word "catchment," with a good technical definition being: "the entire area providing runoff to and sustaining part or all of the streamflow of, the main stream and its tributaries" (as noted by Gregory and Walling, 1973, in their textbook "Drainage Basin Form and Process"). The term itself dates back to the late 1700s in a memoir by Philippe Bauche presented to the French Academy of Science.[7] Since that time, the catchment has been our hydrological north star; the fundamental landscape unit: the place where we compute water balances, link internal processes to streamflow generation and integrate water, energy, sediment and biogeochemical budgets. The catchment then is the place where we imagine the storage and release of water; its age, origin and pathway to the stream.

This book is mostly about small headwater catchments like the one in Figure 1.1. This is because headwater catchments are where most of the streamflow is generated. The term "first-order stream" or headwater was first coined by Strahler (1952) (following the original work of Horton—and others—who we will discuss at length in Chapter 2)—as streams that lack any tributary. In the Strahler (1957) system now in use today, a second-order stream forms when two first-order streams come together. These first- and second-order streams will be the focus of this book. And recent analysis shows that they account for ~80% of the total length of flow waters in the USA (Golden et al., 2025), and likely globally.[8] Beyond this scale (usually ~0.1–1.0 km^2) we often see a shift from colluvial to alluvial geomorphic processes (Montgomery and Foufoula-Georgiou, 1993), meaning from hillslope dominance to stream dominance. It is the former that we will focus on in this book.

[4] Another great book is Erica Thompson's (2022) "Escape From Model Land" with the subtitle "how mathematical models can lead us astray and what we can do about it." While not geared to the hydrologist, many of the points she makes summarize our hydrological state of affairs.

[5] There is an old aphorism by the British statistician George Box.

[6] Finally, another great book is "Useless Arithmetic: Why Environmental Scientists Can't Predict The Future" by Pilkey and Pilkey-Jarvis, 2007. It exposes many predictive failures with examples across the environmental sciences.

[7] As noted by Smith (1969).

[8] No one has yet computed or calculated that as far as I know.

Figure 1.2 Catchments of the USA showing the grain of the hydrological landscape.
Source: Robert Szucs and used with permission.

We can see how the headwaters link to their higher-order downstream counterparts in Figure 1.2. This is a visualization of Strahler's stream classes for the 18 major drainages of the USA.[9] We can see how the grain of the hydrological landscape is dominated by streams draining headwater catchments. Like Russian nesting dolls, there are catchments within catchments. The headwaters are usually the steepest zones in the catchment, where the most precipitation falls (due to orographic uplift, snow accumulation, etc.). But these recognizable and repeatable hillslope forms are also a result of fascinating feedbacks—where certain areas are steep because they are dissected by the water—all in a hydrogeomorphic dance of feedbacks, thresholds and non-linearities. There are "rules" for scaling of catchment steepness, numbering, drainage density, stream length, etc. in the geomorphology literature.

In process hydrology, headwater catchments are also the scale where we can make field measurements; dig holes, install instruments and see connections between our point measurements, the hillslopes and catchment response. But while headwaters are the dominant catchment unit and are accessible to the field hydrologist, they are not where streamflow measurements from government gauging networks are made. Figure 1.3 shows that almost all of the faint lines that dominate most of Figure 1.2 (and similarly for most of the world) are *ungauged;* not to mention the much larger number of intermittent headwater streams that are smaller than these channel tips and have flowing water for only part of the year (all a function of map scale of course[10]). It shows, starkly, how extraordinarily data-limited we are for getting to grips with the variability of the natural world vis-à-vis streamflow

[9] Note that in the map we can see no headwater streams per se: they are simply too small and too numerous to show on a map at this scale.

[10] As mentioned in footnote 9, headwater streams are too small to be seen at a national scale. And related to gauging, I'd wager that well less than 1% of the world's headwater streams are gauged.

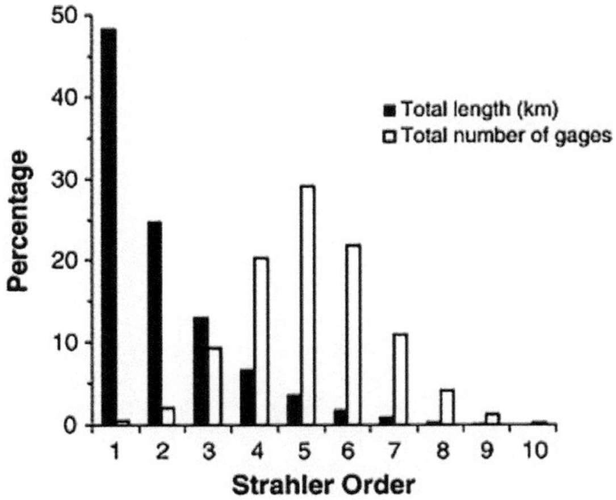

Figure 1.3 Distributions of stream length and total number of USGS stream gauges (conterminous USA).

Source: Poff et al. (2006).

generation. That is why this book will rely largely on case studies from experimental headwater catchments. They are our experimental windows into how streamflow generation works since they represent the very few examples with field data necessary to obtain process understanding.

1.3 Processes and perceptual models

Imagination is more important than knowledge. Knowledge is limited. Imagination encircles the world.

—Albert Einstein

This book will discuss processes and perceptual models of streamflow generation. The word process is likely familiar to you. In the context of streamflow generation, a process might be infiltration, evaporation or transpiration, like you learned in your undergraduate class. A perceptual model is a mental model of a site that places these processes in a wider context. First introduced by Beven (1987), the term perceptual model is a description or a collection of hydrological processes—in whole or in part—often in a highly detailed, yet qualitative manner. So, a perceptual model is ultimately a creative image of something that can incorporate far more complexity than can be described mathematically. It is a mind-drawing of sorts. And if Picasso was correct that "art is the lie that helps us see the truth," then a perceptual model is the "lie" that helps us understand how the many key hydrological processes operate and interact. It is a caricature. Sometimes it is a first guess of how you think a catchment works. But one that can be refined and improved with repeat field studies and new methods; using the old perceptual model as null hypothesis to try and reject and then to propose a new and better one. The act of bringing measured fragments of a catchment into a mental model is what this book is about. But to be clear, unlike art, a perceptual model is informed by data and measurements. Therefore, with the same data and measurements, two different field hydrologists should come up with the same perceptual model.[11] This is different to art where the point of view is unique to each artist—for example without someone like Edvard Munch, the painting "The Scream" would never have been painted.[12]

So perceptual models are about the aggregate behavior leading to streamflow generation. We will explore how we go from measurements and some vital signs of catchment into a hydrological diagnosis: a picture of how the catchment "works." New perceptual thinking too can unfold as you see more and more places and form process connections between them. Measurements are how we probe the catchment, to force it to unlock its secrets. Data collection is usually the easy part (although field work can be extraordinarily challenging at times). But it is the *interpretation* and assembly of that data into a comprehensive picture or mental model that is the challenge for catchment hydrologists—framing field observations in an "understandable context." Perceptual models are then a way to instantiate our process understanding and the relationship between the stream and the headwater catchment—and indeed the scaffolding for future theories (and models) of streamflow generation.

[11] That said, Beven notes that one's training and background may influence one's perceptual model. Indeed, often our field data is not so clear, such that what we make of it can depend on the conditions, places and processes that we have observed before.

[12] This analogy is modified from a story often told by the American astrophysicist and writer, Neil deGrasse Tyson.

1.4 The teaching philosophy of this book

Real knowledge is to know the extent of one's ignorance

—Confucius

Teaching catchment hydrology has always had a creative tension between the engineering-based idea that there are laws and generalities that are independent of place and time—and the field-based paradigm of the earth scientists and foresters who argue that place, and the history of that place, matter. Most textbooks in hydrology have leaned heavily toward the engineering approach. And what is taught usually assumes that local hillslope and catchment details do not matter (that much). The descriptive approaches from the field hydrology community are sometimes mentioned in textbooks but they have played second fiddle to the engineering approach. Since the 1970s when computer models came on the scene, field hydrology has diminished year on year as hydrologists realized the risks, challenges and citation differential between modeling (a profitable indoor sport, as noted by some) and field work. Many have lamented over the past several decades that with the rise of computing power and the expense and time needed for field work, "there is a trend away from field work and towards more complete dependence on simulation" (Kirkby, 2004, p. 16). The situation gets worse every year (Burt and McDonnell, 2015). Because models are needed for solving many important and pressing problems, we have a current state where "the pragmatics of applications [are] outweighing the information from direct observations" (Beven 2021, p. 854). In other words, we've become detached from process realities in our model obsessed world. Whither field hydrology. Model über alles.

The philosophy of this book is that streamflow generation is not some problem to be solved (with a computer model) but rather first, a field of basic research to understand how nature works (Figure 1.4). We'll talk about the needed dialog between experimentalist and modeler at the end of the book. But first I want to take you on a model-free (and equation-free!) journey through "boots on the ground, hands in the dirt, streamflow generation studies."[13] We will discuss, among other things, the age, origin and pathway of water from the hillslope to the stream and from the soil to plant transpiration. And my hope is that one day, you can add to this narrative with your own field work.

Figure 1.4 The importance of field-based inquiry. Curiosity. Chin scratching. Asking oneself: hmm, I wonder how/when/where this does that? Even better: I wonder why this does that?
Source: Louise Arnal.

It has been said that writing about music is like dancing about architecture; so too with process hydrology. It would be infinitely easier and more effective for us to stand and discuss these processes together in the field (e.g., at the site in Figure 1.1) with a shovel in hand as raindrops fall; where we might dig a hole together and add some blue dye to trace the water flowpaths; or to examine the stable isotope dynamics in streamflow through a storm event and to then age-date that water and compare it to the water in storage. Field work is the best teacher. But alas, we cannot do this. So, let's all keep in mind our local field images from our own experiences as we go through the book and explore what we know and how we arrived at this knowledge, what we think we know and what we need to know.

[13] To paraphrase Grant and Dietrich (2017)

To remember that hydrology is largely a historical geoscience and that understanding streamflow generation depends on qualitative and quantitative observations recorded in the past and the experimental testing of hypotheses formed from these observations. In this way, catchment hydrology is also an inductive science, as noted by Beven et al. (2025), in that it relies on inferences made from those observations in different ways to make statements about places where there are no observations available. A key philosophical pillar of this book is that understanding how our knowledge has changed over time is crucial since these changes inform how we got to our current state of knowledge.

1.5 Why should a beginning graduate student read this book?

Pay attention to the important things
—Solon of Athens

Entry into the field of catchment hydrology is now *much* more daunting than it was when I was a graduate student. Back then, in a relatively short time, one could read the literature, be caught up and be waiting in eager anticipation for the next issue of *Journal of Hydrology* or *Water Resources Research* to arrive (in hard copy) to the university library. Back then, people read more and published less. Two papers per year was seen as exceptionally productive; achieved by few. The warp and weft of the field is tough to keep up with now—with >5× the number of papers published each year compared to the year I finished my PhD. One can't easily follow a single thread of ideas anymore. The poor beginning graduate student is overwhelmed with articles—often of exquisite narrowness. It is like drinking through a fire hose of information. And the question is often "Where to start"? This book will help with that.

But beyond where to start, it is also difficult to follow the threads of ideas when ideas have bounced around so much from place to place for several decades. A student entering the field of catchment hydrology expects to see a clear path from A to B where the field has progressed in new places, with new techniques and new ideas in an easily decipherable linear fashion. But this is often not that way for streamflow generation research (nor for most other fields of research). Figure 1.5 shows that often the path of progress from A to B—especially when looked at over the span of decades and the full arc of processes and perceptual models—is more a squiggle than line! The goal of this book, therefore, is to take a new graduate student by the hand from A to B—something that can seem insurmountable for today's students encountering the journal literature. I will try my best to lead you through the squiggle. It may be tortuous and confusing at times. But it will be fun. Remember, it is a mystery novel. And we're channeling our inner Agatha Christie, on a long train journey with many characters. You'll meet all of them—or at least the notable ones. And by the end, you will be like Christie's famous detective Hercule Poirot—able to understand all the pieces and to go on to solve your own cases at your own research catchments.

Figure 1.6 shows what most undergraduate textbooks present for the basic process descriptions of runoff generation: infiltration excess overland

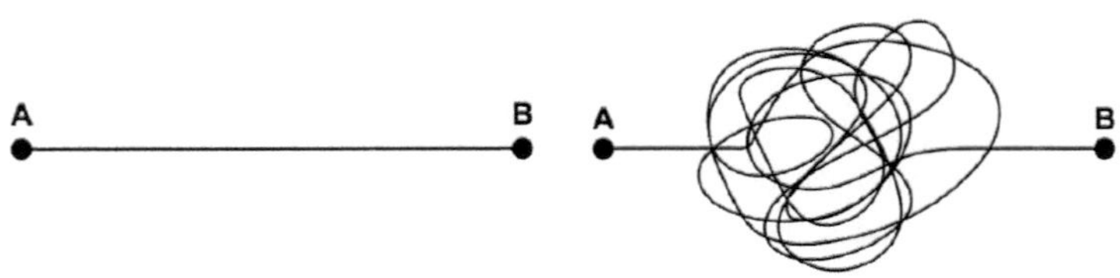

Figure 1.5 The path of learning in process hydrology: more squiggle than line. This is the case for the field of hydrology more generally and most graduate theses. In fact, this is a terrific video on the pathway of learning and how an "option C" may indeed be our real path: https://www.ted.com/talks/uri_alon_why_science_demands_a_leap_into_the_unknown?utm_campaign=tedspread&utm_medium=referral&utm_source=tedcomshare

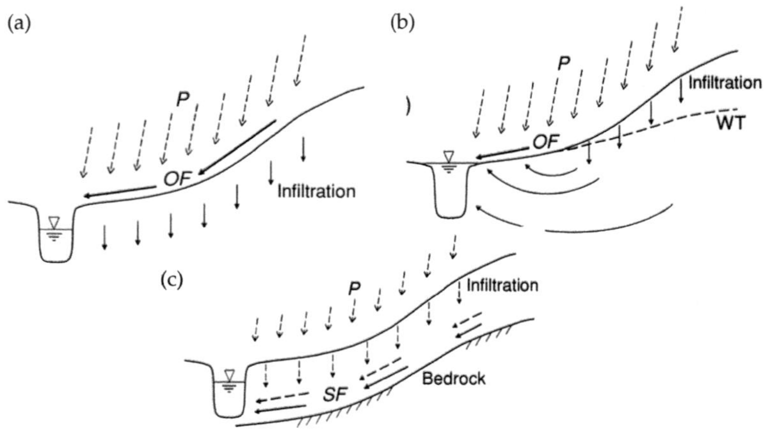

Figure 1.6 The most popular sort of diagram found in most undergraduate hydrological textbooks on runoff generation. In fact, while a chapter might be devoted to infiltration or groundwater, only a few pages in our hydrology texts are ever devoted to runoff processes.
Source: Brutsaert (2005).

flow (Figure 1.6a) where the rain intensity exceeds the soil's ability to absorb it (i.e., infiltration rate); saturation excess overland flow (Figure 1.6b), where the water table rises to intersect the soil surface during a rain or snowmelt event; and subsurface stormflow (Figure 1.6c), where rainfall is delivered to the stream via lateral flow through the soil profile. While not wrong, these descriptions and diagrams—like my taste in music—are stuck in the 1970s. They represent our process understanding from that era and not our present understanding.

If it were all as simple as depicted in Figure 1.6 and if it were clear how, where and why each of these mechanisms are activated, then perhaps our current hydrological models *would* be enough. For a catchment with a fully impervious surface, and with some careful measurement of precipitation inputs, we probably could get by with a model-only approach using say, the shallow wave equation. Or if streamflow was fed exclusively by perfectly homogeneous and saturated porous media with known lower boundary conditions, then a 3D Darcy–Richards model *could* probably work, and we could set aside our shovels and thinking caps.

The problem is that most catchments are intermediate between these two simplistic extremes. With anything but uniform pore sizes and soil depths. Our catchments are variably saturated; riddled with preferential flowpaths, covered with pesky vegetation that does not obey simplistic engineering notions of plant water uptake; with incredibly flexible responses of their root distributions to soil moisture and rock moisture distributions. These things matter for streamflow generation. And understanding where in the landscape stream water originates—in both time and space (as depicted in Figure 1.7) is still a major challenge. This is something that few models are able to predict, especially if the question goes beyond the basic form and shape of the hydrograph to questions of what parcels of rainfall or snowmelt make up the hydrograph.

I will assume that anyone picking up this book will have had some sort of undergraduate course in hydrology. So, you will know Darcy's law; the water balance equation; the pieces of the soil moisture release curve, what a hydrograph is, etc. The goal of this book is to simplify and make readable the narrative of processes and perceptual models of streamflow generation. It is important to note that there has been a new and renewed interest in perceptual model construction for mathematical models (Wagener et al., 2021; Beven and Chappell, 2021) and summarizing some of the perceptual model similarities and differences (McMillan et al., 2022; 2025). There have also been some useful recent summaries of process advancements in

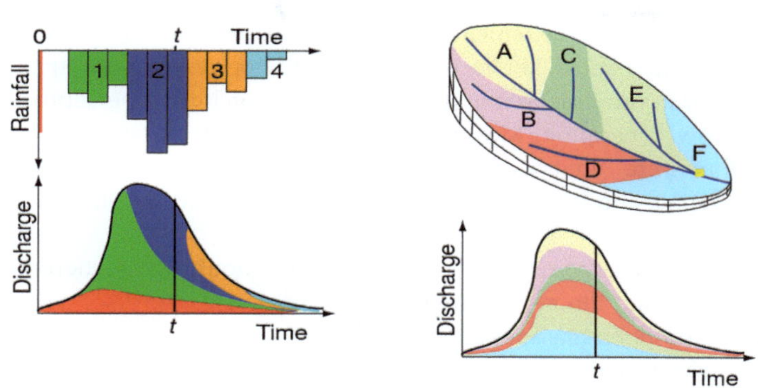

Figure 1.7 Our challenge remains: where does water come from at the catchment scale—in time and space.
Source: Sayama and McDonnell (2009).

catchment hydrology (Pfister and Kirchner, 2017; Kirchner et al., 2023; McGuire et al., 2024).

But still there is the question of why has there been so little uptake of perceptual models in our teaching? When I ask this to my engineering hydrology colleagues in universities around the world, the answers are usually two-fold: (1) equations are easier to teach and (2) the messy world of perceptual models is simply too complicated and "all-over-the-place" to convey to students. This is a sad state since we need to systematize field-based discovery science to take us to the next level in catchment hydrology and improve hydrological models. Like Kirchner (2006) has noted, it is not that we do not want our models to be physics-based, we simply do not know what physics to use. I hope that this book will begin a discussion on the physics we should start to use based on what field hydrology has taught us over the past months, years and decades.

1.6 Outline of the book

Learn as much by writing as by reading
—Lord Acton

This book is divided into three parts: what we know (Part 1), what we think we know (Part 2) and what we need to know (Part 3). The underlying thread of the book is to show the chronology of how key disruptive papers[14] played a critical role in advancing our understanding of streamflow generation processes. And how creativity and originality are central to the hydrological sciences. The main chapters all have dates associated with them. That is not to say that the time period of applicability of those ideas is restricted to these dates—rather when the ideas came about. They are ideas that we are still building upon. For the new graduate student, it is important to understand that hydrology, like science in general, is a process of dissent with modification. And in hydrology, at least in the last 100 years or so, there have been no major shake-ups[15] that have demolished the foundations of what we view as knowledge (Kuhn, 1997).[16] Therefore, knowing the how, when and where of key perceptual models is essential before one can do new research. That is, to know the status quo and what is wrong with the status quo, before you can do research that goes beyond the status quo.

Going back to the original papers in Parts 1 and 2 will help you build the capacity and skill sets that you will need to become a better hydrologist: To know what benchmark science is by seeing what has stood the test of time. And like we remember Sir Edmund Hillary and Tenzing Norgay because they

[14] And especially the so-called persistently disruptive papers, as defined by Deng et al. (2025) as ones that stand the test of time.

[15] Perhaps the biggest shake-up—one we will explore—is the use of stable isotope tracers. But even that knowledge gained with a new tool did not demolish what came before; it simply clarified key differences between pressure response and particle response (as we will see in Chapters 5 and 7).

[16] And as Park et al. (2023, p. 138) note that there is a decline in disruptiveness due to "a narrowing in the use of previous knowledge."

were the first to summit Mt Everest, we remember the authors who first summited some intellectual mountain in hydrology—or in our case, some new discovery about streamflow generation. Meaning, we remember papers that were "firsts" and that have had longstanding impact and influence. And for this, to read old papers is useful.[17] The first few chapters of this book will use a lot of quotes from the older papers in the hope that the new student will see first, how many questions have remained the same in streamflow generation research. And secondly, that some of the things that we think are recent discoveries were already thought of or mentioned quite some time ago.

Part 1 covers the only tectonic shift we have had in catchment hydrology—the first catchment water balance and the rejection of ancient theories of the origins of streams and springs. It then covers the first process descriptions of rainfall–runoff that moved us beyond the simple water balance and how ideas on the soil infiltration capacity related to storm runoff mesmerized the field. We will also explore in Part 1 the parallel history that explores the subsurface stormflow view of the world—the details of which we are still learning about today. We'll cover the stunning advancements made during the International Hydrological Decade (1965–1974) the early days of stable isotope tracing.

Part 2 covers our modern perceptions of streamflow generation; less related to rainfall–runoff (or as many numerical models would call it—"rainfall excess"—to what is best described as storage excess. It focuses on different scales given the extreme scale dependency that we now understand to occur. Part 2 is place- and idea-driven. It draws partly from my own body of research from the many catchments where I have worked and covers what we think we know as a community in catchment hydrology. It is comparative in nature and addresses different scales as shown in Figure 1.8. Part 2 also shows how the combination of bottom-up and top-down field approaches can help pin down a catchment perceptual model with a focus on catchment storage and streamflow age. And how plant transpiration links to water flowpaths and streamflow generation, during and between events.

Throughout each section of Parts 1 and 2, I will try to describe who discovered "what" and "where" linked to certain processes at first; then the first perceptual models of the newly discovered behaviors and then potentially widely applicable concepts. Like other areas of science, we will see ample evidence of Alfred North Whitehead's adage "Everything of importance has been said before by somebody who did not discover it." But we will also see how the discoveries of the past relate to today's pressing research questions in streamflow generation research: the old influencing the new; and the new helping to contextualize the old.

Part 3 is a look to the future: to future classic problems; to opportunities and challenges in catchment hydrology. My crystal ball is not very clear, so it is the immediate future that we will focus on—related to the urgency for perceptual models of streamflow generation in a changing climate. We'll discuss how can we develop a better dialog between experimentalists and modelers and, for those doing field work and developing their own perceptual models: what we should measure; in what order; and why?

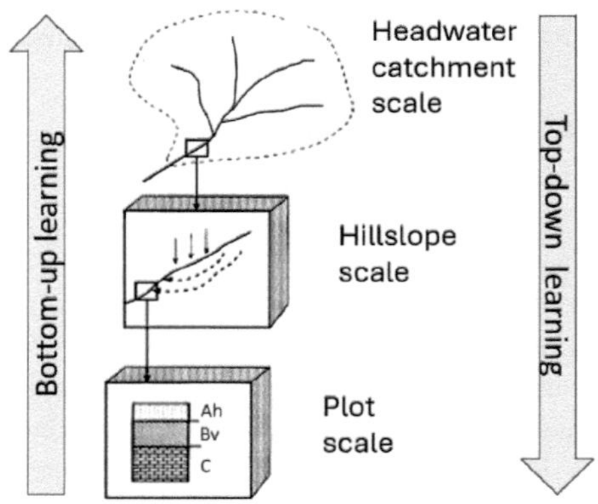

Figure 1.8 The scales that we will address in this book—in both a bottom-up and top-down manner.

[17] I would say that reading the right old papers in the right way is better than reading *only* new papers. To be clear, I am not saying that you should read *only* old papers. Or that old papers are better because they are old. Or that you should never read any new papers or that new papers are worthless. Only that you should read and learn how to read some old papers.

Finally, as a new student in catchment hydrology with a field interest for process research, one often feels that they have arrived late to the party. In other words, it is easy to think that "all the big stuff has been discovered already, so what's the point of more field investigation?"[18] I hope to show with this book that this is most certainly *not* the case. And that our steady arc of learning over the past century is one that builds on the discoveries of the past *and* continues today and for another century into the future. There is *so much* to still be discovered. I hope you will know and embrace that. I hope that this book will inspire you to launch your own field discoveries. And that the small quotes sprinkled throughout—from scientists, writers and artists[19] —inspire you along this journey, as they do me.

[18] A comment passed to me by Tom Dunne (pers. comm.) who, like me, has heard this same lament from students in recent years

[19] These quotes are less gender-mixed than originally laid out. Sadly, quotes used in the way (called epigraphs) require permission to use; just like figures. But unlike figures, such permissions are trickier to obtain. Therefore, in the end, most are old enough not to require permission, but as a consequence much more male-dominated.

PART 1

What Do We Know?

CHAPTER 2

Streamflow: The first process discoveries (1674–1960)

2.1 Streamflow generation: The beginnings

> *If there is but little water in the stream, it is the fault, not of the channel, but of the source*
>
> ***—St Jerome***

We will start with the first thinking on streamflow—linked to its origins in the context of the water cycle.[1] Most readers of this book will have seen the water cycle. It would be like studying physics and never having heard of gravity; it is *that* fundamental to catchment hydrology and how we understand streamflow generation. The water cycle in a headwater catchment is quantified with the water balance—described in undergraduate texts as input (precipitation) minus output (streamflow and evapotranspiration) equals the change in storage Usually, we do not equate a perceptual model to this rhythmic cycle of the catchment water balance since it is a *known* process.[2]

But it was not always that way.

The progression of ideas leading up to what we now know as the origin of streams in the context of the water cycle is shown in Figure 2.1. The big question was: why do streams flow when it is not raining? That question was pondered by early thinkers that you may have heard of, like Aristotle (384–322 BCE), Pliny the Elder (23–79 CE) and Seneca (4 BC–65 CE). There is much scholarly writing on this topic (Tuan, 1968; Biswas, 1970) and more recent work summarized in a 2025 Special Issue on the History of Hydrology in the *Hydrological Sciences Journal* (Beven et al., 2025).

Figure 2.1a shows the earliest idea—unsupported by any field evidence of course—of a "reverse hydrological cycle." This is the idea that streamflow is formed by a subsurface connection to the ocean. Imagine you are in Greece or southern Italy and seeing water in a stream during an otherwise rain-free period. You too might have been tempted to think that there was some connection to the permanent water in the sea since—without maps of catchment areas and understanding of what a catchment source area is, it would have seemed that rainfall was too little to possibly create permanent flow (or even seasonally ephemeral flow) in streams or springs. This is what early Greek and Roman scholars thought, where Aristotle noted that "it is unreasonable for anyone to refuse to admit that air becomes water in the earth for the same reason that it does [from] above it" (Aristotle, as quoted in Adams (1938), p. 427).

Both Pliny the Elder and Seneca were proponents of the idea outlined in Figure 2.1a where the sea is the source of streams where subterranean air is "unchanging and constant" (Duffy, 2017) and could continuously supply water for streamflow.

[1] This chapter and a few that follow will include a number of important quotes. Don't be put off by them. My goal is to give you snippets of classic papers to show (a) that some things we think of now as new were pondered already some time ago, and (b) that some things were known and debated already a lot earlier than we may think.

[2] Although I'll have more to say on this at the end of this chapter and in Chapters 8 and 9.

Streamflow Generation. Jeffrey J. McDonnell, Oxford University Press. © Jeffrey J. McDonnell (2026). DOI: 10.1093/9780191953736.003.0002

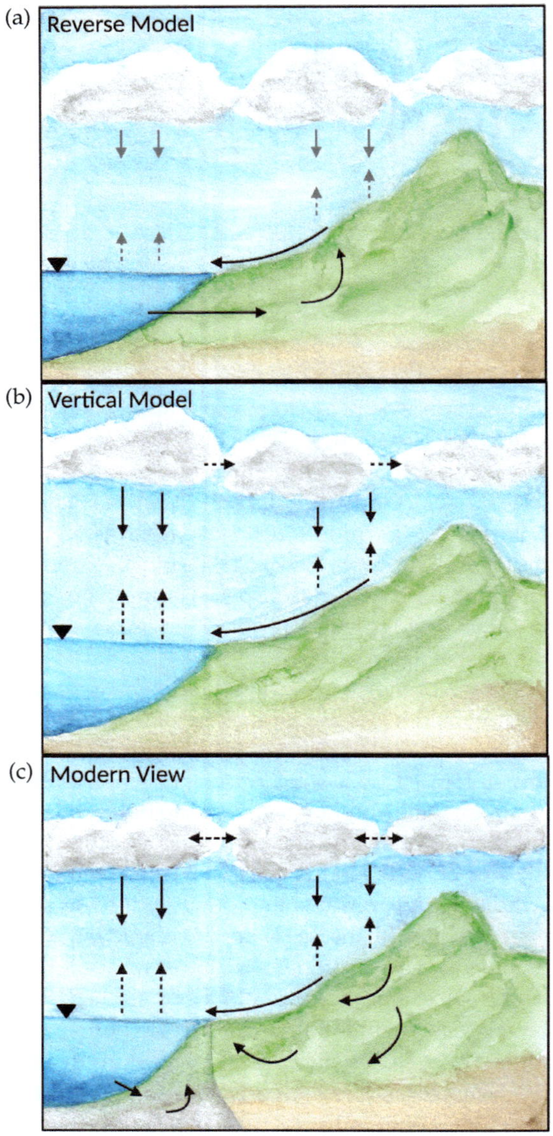

Figure 2.1 Evolving perceptual models of the terrestrial water balance. (a) A subsurface origin of springs and rivers, (b) a vertical model where precipitation and dew are the source of springs and rivers (as discovered by Perrault, 1674) and (c) the modern view—one that also recognizes that precipitation is the origin of springs and river but includes flowpaths into and out of the subsurface that are deeper than Perrault imagined (modified from Duffy, 2017 by Harvinder Sidhu).

Seneca[3] rejected the idea that rainfall could supply streamflow:

> . . . as a diligent digger of vines I can affirm from observation that no rain is ever so heavy as to wet the ground to a depth of more than 10 feet. . .how then can rain, . . . store up a supply sufficient for rivers? (Seneca, as quoted in Tuan, 1968)

This is a very reasonable point if you ponder it. Indeed, if I were able to dig a hole this deep in many of the catchments where I have worked, I might have come to the same conclusion had I not already known of the Richards equation.[4]

What might be the first perceptual model of the water cycle derived from data? That is shown in Figure 2.1b. It highlights the recognition that rainfall and snowmelt are sufficient to sustain streamflow. The first field evidence of this came from a headwater catchment in the upper Seine River basin in France. It was published in 1674 by Pierre Perrault. Figure 2.2 shows the Perrault publication[5] and a photo of his site. Perrault measured rainfall for several years and estimated the flow in the small stream to utter these immortal (translated) words: "Only about one-sixth part of the rain and snow water that falls is therefore needed to cause this river to flow continually for one year." This was pretty much the first time that data[6] was brought to the question of streamflow generation. As such, Perrault might be considered the first process hydrologist. He understood the value of field work and field experiments as he noted that "it is to experiments that we owe the finest knowledge we now have concerning the things of nature." Tres bien Pierre!

[3] I love Seneca's writings and stoic thinking. Completely unrelated to hydrology—but perhaps *very* related to "launching your field hydrology career" that we will discuss at the end of Chapter 9—is his book "On the Shortness of Life." That monograph is highly recommended reading—words as true now as they were 2000 years ago.

[4] The Richards equation (Richards, 1931) describes water movement in unsaturated soils—it combines Darcy's Law with the continuity equation to describe capillary-driven flow and gravity-driven flow.

[5] A first edition of which I feel very fortunate to have on my shelf!

[6] Although as we note in McDonnell et al. (2025) in a re-visiting of Perrault, his perceptual model and his peculiar catchment, streamflow was not measured but was a guesstimate.

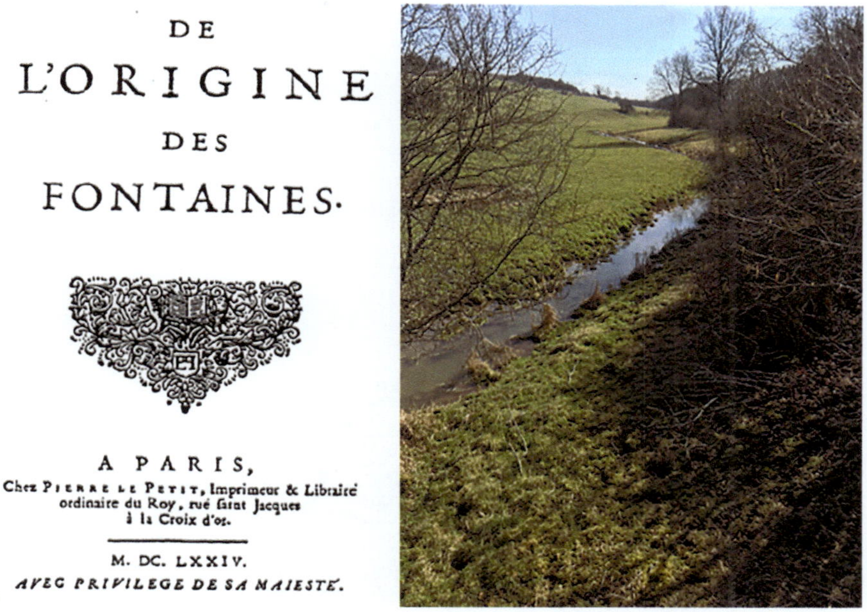
DE
L'ORIGINE
DES
FONTAINES.

A PARIS,
Chez PIERRE LE PETIT, Imprimeur & Libraire ordinaire du Roy, rue saint Jacques à la Croix d'or.

M. DC. LXXIV.
AVEC PRIVILEGE DE SA MAIESTE'.

Figure 2.2 The actual stream in the upper Seine near where Perrault made his measurements. Note the small channel draining a catchment and that he did not actually measure the streamflow but "guestimated" it.
Source: The author.

But Perrault had a few things not quite right. For instance, he did not get the area of his headwater catchment sorted out—perhaps understandable for the mid-1600s when there were no detailed topographic elevation maps (or digital elevation models!). He also did not think that rain infiltrated into the soil deeply enough to be the cause of springs (again, he lived ~250 years before Richards). Nor would there be enough rain falling on high ground to supply springs when nearly all that water, he thought, was absorbed strongly by the soil and later lost by evaporation. Nevertheless, the totality of Perrault's work and contribution (i.e., his many accompanying controlled experiments, his measurements, and his theorizing) marks the start of catchment hydrology and field-based streamflow generation research (McDonnell et al., 2025).

But while Perrault showed that precipitation was sufficient to explain annual streamflow, it was not until Dalton (1802), over a century later, that marked "the last time that any other 'opinions' on the origins of springs were brought up in the scientific literature" (Brutsaert, 2005, p. 584); meaning, that the uppermost panel of Figure 2.1 was officially obsolete. John Dalton[7] did extensive gauging and measurements in the UK and noted that:

> . . .we can fairly conclude that the rain and dew of this country are equivalent to the quantity of water carried off by evaporation and by the rivers (Dalton, 1802, p. 365)

While catchments had been delineated and mapped before him, Figure 2.3 shows Dalton's so-called "districts" draining the main rivers of the UK. Dalton was very clear on the relation of streamflow to catchment area where he noted that:

> All other circumstances being the same, the quantity of water carried down by any river should be as the area of the ground from which the water is derived, and on this account, the Humber ought to exceed the Thames (Dalton, 1798, p. 356)

[7] Dalton (1798) is something to have a quick read on if you have time and it is freely downloadable at: https://www.biodiversitylibrary.org/item/52767. While the Perrault book is in French and even in translated form (by La Rocque, 1967) is a rather difficult bit of text to navigate and follow, Dalton (1798) is clear, lucid and a rollicking good read!

Figure 2.3 Dalton's sketch of his so-called "rivers in England and Wales divided into districts."
Source: Dalton (1798).

So, the first perceptual models of streamflow generation were all in the context of the catchment water balance. But it is important to recognize that the catchment water balance itself is a black box (Figure 2.4)—even with modern, correct perceptions of deeper flowpaths and groundwater contributions to streamflow (Figure 2.1c). In other words, measuring and quantifying a catchment water balance on an annual basis does not equate to understanding the processes of streamflow generation during

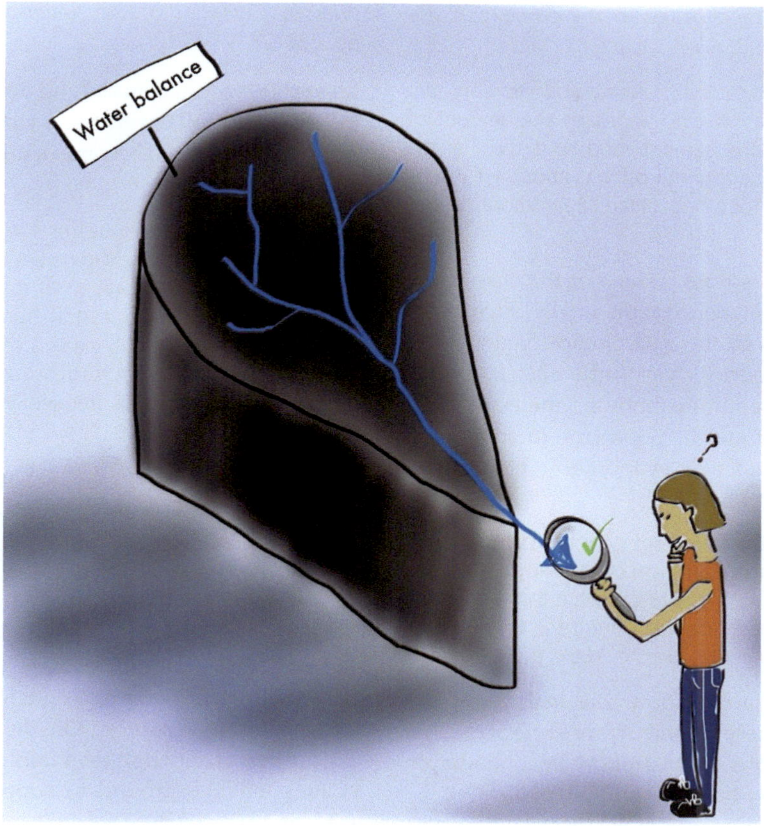

Figure 2.4 The black box accounting model of streamflow generation in the context of the catchment water balance.
Source: Louise Arnal.

or between events. Streamflow in the water balance context is simply a percentage of the precipitation that falls in a given year and runs off. It says nothing about flowpaths, flow sources and flow ages of water. When I reflect on the history of hydrology, I wonder how the field might have evolved had Perrault known that his stream water during low flow was ~50 years old (as we have determined recently (McDonnell et al., 2025))?! Might that have influenced him and others who followed to think more about "inventory" (i.e., water in storage) in the catchment water balance accounting model, and to re-direct the focus from exclusively flow-related quantities to storage-related measurements? But I get ahead of myself with these thoughts, and this is something we will come back to in Chapter 7.

2.2 The earliest thinking on rainfall–runoff processes

> *There is nothing new, except what has been forgotten*
>
> ***—Marie Antoinette***

Process descriptions of streamflow generation did not evolve much after Perrault and Dalton until the mid-1800s. And even then—and up to ~1920—there were no descriptions of "processes" as we might know them today, as we saw from Figure 1.6 in Chapter 1. The first real hydrology textbook after the consensus on the catchment water balance and precipitation as the source of streamflow was Beardmore's 1862 *Manual of Hydrology*. It shows early rudimentary ideas on how streamflow is affected by catchment water storage in the context of the catchment water balance:

> . . .in November, December, January and February, the tendency of percolation is to approach the amount of rain fallen during the same period, but that in the eight months the quantity run off was about 4.7 inches out of 10 inches that had fallen (Beardmore, 1862, p. 148)

Here, Beardmore is noting some of the changes in where water goes when it rains in UK catchments at different times of the hydrological year: in the winter months, there is very little abstraction by evaporation; in the summer months, there are much lower amounts of storm runoff due to competing demand on the water by evaporation. He goes on to say that:

> This result is tolerably consistent with the Dalton gauge experiments, for the quantity run off is nearly that due to the year of low rainfall; and all experience shows that very little more would have filtrated during the four months of June to September. (Beardmore, 1862, p. 148)

So Beardmore is noting how seasonality impacts streamflow and these comments are some of the first process descriptions of streamflow beyond simply enumerating water balance components.

In Meyer's *The Elements of Hydrology*, an important text from 1917, further insights into the mechanisms of streamflow generation are made. Meyer (1917, p. 279) notes that "runoff constitutes the residual of precipitation after evaporation, transpiration, and deep seepage losses have been supplied." But he goes on to say that runoff, being a residual, means that it cannot be a percentage rainfall and that "the runoff must be determined by deducting losses from precipitation and not by taking a percentage of the precipitation" (Meyer, 1917, p. 279). So, with this statement about deducting losses, Meyer is starting to really go beyond the water balance and to get at processes. He links this to the notion of detention storage, stating:

> A watershed, impervious on account of outcropping rocks, nevertheless absorbs a moderate amount of water at the beginning of a rainstorm on account of the moss and humus found in the crevices of the rocks. After such storage capacity is exhausted, however, substantially all of the succeeding rainfall will run off into the water courses and into lakes, ponds and marshes and swamps common to such watersheds (Meyer, 1917, pp. 280–281)

Meyer's perceptual model and description of runoff in areas with soil cover is one where surface runoff occurs when the soil becomes saturated. As such, he goes on to say:

> When heavy rains continue for some time, all but the most sandy and gravelly watersheds become temporarily impervious through saturation of the soil. For any given rainfall, the total surface runoff from a pervious, sandy watershed will necessarily be less than that from other watersheds, by the amount of water required for saturation. Sandy watersheds frequently exhibit no signs of surface runoff (Meyer, 1917, p. 281)

Meyer even goes on to describe some of the effects of topography on runoff amounts, the effect of tile and ditch drainage on runoff and "seepage flow," defined by him as "water contributed to streams as groundwater supplied by percolation" (Meyer, 1917, p. 284). Meyer summarized knowledge of the day that was surprisingly accurate vis-à-vis snowmelt inputs where: ". . .melting snow permits the greatest percolation. On the whole, a greater proportion of snowfall than rainfall eventually percolates into the ground to supply seepage flow" (Meyer, 1917, p. 286). This anticipates the first global analysis by Jasechko et al. (2014) that confirms this statement some 100 years later!

Following Meyer, one of the earliest known field studies that explored the factors influencing rainfall–runoff relations was for a set of seven headwater catchments ranging from ~1–112 acres (~0.4–45 ha) near Jackson, Tennessee, USA (Figure 2.5). In that work, Ramser (1927, p. 799) noted that the rate of runoff from these catchments depended on a long list of factors, including the area and shape of the catchment, its slope angle, vegetation cover, soil permeability and the antecedent wetness (that he described as the "saturated condition of the soils due to previous rains"). He also noted the impact of what he called the "arrangement and character of drainage channels," the "duration and intensity of the rainfall" and the "catchment's storage and underground conditions." He noted that "of these factors affecting the rate of run-off, evaporation is practically negligible for small watersheds." This last statement portends

Figure 2.5 Photos of two of the 7 headwater catchments studied by Ramser (1927).

many decades of similar thought vis-à-vis the role of evaporation (and evapotranspiration) on runoff process generally—something we will come back to in Chapter 8 when we explore current work on plant transpiration effects on streamflow.

So, while not getting at processes per se, Ramser understood the importance of saturation, rain intensity and storage as key factors for runoff generation. But he was also perplexed by what he called the "unaccountable variations in the run-off coefficients" and the "many interdependent factors entering into the relation between rainfall and run-off." Ramser lamented that it was "practically impossible to evaluate all of them accurately." That said, he comes very close in 1927 to articulating a clear perceptual model of runoff generation when he noted the importance of antecedent wetness:

If the maximum rate of rainfall occurs at the beginning of the rain, before the surface of the ground has been thoroughly wetted, the percentage of run-off is less than when the maximum rate occurs some time after the beginning of the rain (Ramser, 1927, p. 819)

In a way, he is getting at what we will examine in Chapter 4 as saturation overland flow. And Ramser's early thinking portends the clearer articulation of infiltration excess overland flow that we will explore a little later in this chapter, where he mentions the importance of rainfall intensity on runoff response:

a greater percentage of run-off also occurs for heavy than for light rains, an illustration of which is the rain of April 19 on watershed No. 5, when the percentage of run-off was 85.2 as compared with the next highest percentage of 51.5 for the comparatively light rain on June 6 on this watershed (Ramser, 1927, p. 819)

A few years after the Ramser work, Horton (1931) would summarize the pre-process era of catchment hydrology as follows:

many attempts have been made to correlate runoff with rainfall and other factors, principally temperature and evaporation. Even with the best available data an estimate of seasonal runoff, year in and year out, if based on these factors alone, is likely to be considerably in error. All existing formulas become still more unreliable when applied to a single day or to the fortuitous rainfall and still more casual runoff of arid regions (Horton, 1931, p. 199)

Horton is saying that hydrologists at the time did not know what were the factors that affect runoff on timescales of events. While he may have been aware of the Ramser (1927) work, who lists some of these factors, he did not cite it. And like many studies around this time, they did not *build* upon each other—rather they developed independently; largely in geographical and intellectual isolation from one another.

2.3 The first process descriptions of subsurface stormflow

Indeed, history is nothing more than a tableau of crimes and misfortunes

—Voltaire

Geographical and intellectual isolation was certainly evident in the first process studies of subsurface stormflow; today defined as "lateral subsurface flow above a restricting layer of lower hydraulic conductivity that occurs during and following storm events" (McGuire et al., 2024, p. 1). One might argue that subsurface stormflow was the first true runoff generation process to be identified from field observations.[8] In humid catchment settings with steep slopes and permeable soils, subsurface stormflow can be the main runoff generating mechanism—either when a restrictive layer impedes the downward percolation of water and forces it laterally or if a (temporary) water table develops and rises into higher conductivity soils near the soil surface.

Figure 2.6 shows what may be the first perceptual model of subsurface stormflow—from Arnold Engler at ETH Zurich in Switzerland. It describes subsurface stormflow on a steep forested slope at the Emmental catchment in Switzerland. Engler imagined that flow through "uncountable veins" was responsible for storm-induced downslope flow of water at his forested hillslopes. Engler was a forester and his perceptual model was based on 15 years of runoff measurements. A key feature of Engler's work was the recognition of the importance of the forest and soil factors on subsurface stormflow response.

It is clear from Figure 2.6 that preferential flow through soil cracks and macropores was a key element of Engler's thinking. Ideas on preferential flow had been around since Schumacher (1864),[9] who discussed preferential flow in the context of infiltration. Early work by Fletcher et al. (1954) noted five conditions for piping (one particular form of preferential flow—usually with larger pore spaces oriented downslope) to take place: (1) a source of water, (2) soil infiltrabilities greater than the permeability of some subsurface layer, (3) an erodible layer immediately above the impeding layer, (4) water above the impeding layer under a hydraulic gradient and (5) the presence of an outlet for the lateral flow.

[8] Note I say "might" as the history of hydrology is still being written and many early manuscripts from the Russian literature—and from other languages—are yet to be translated into English and fully digested in the historical context.

[9] A tome written in High German and with book font to match!

Figure 2.6 A hillslope section from Engler (1919) showing his perceptual model of subsurface stormflow where hashed areas denote vertical infiltration through the Dammerde (humus) and Rodboden (soil); deeper in the profile water is flowing in "veins." Nagelfuh is a conglomerate and Mergel is marl.

Source: Weiler et al. (2005).

Elsewhere, there was early recognition of subsurface stormflow and its control of storm runoff in the channel. For instance, Snyder (1939, p. 738) noted that subsurface stormflow "consists of the discharge in excess of a base or ground-water flow which passes a gaging station within a rational period of time subsequent to the storm causing the rise." Lowdermilk (1934, p. 534) had a perceptual model of storm runoff that recognized subsurface contributions, including "shallow seepage or discharge of wet weather springs." This was echoed by Hoyt a couple of year later where he noted that:

> after prolonged rain when natural storage has already become utilized, subsurface storm-flow may under some conditions reach the stream-channels nearly as promptly as overland storm-runoff (Hoyt, 1936, p. 245)

A few years later Hoyt and Langbein (1939) would observe that:

> there is a rapid passage of a part of the infiltrated water into stream channels, either through the relatively shallow earth-mantel or through the upper parts of the shattered bedrock (Hoyt and Langbein, 1939, p. 174)

Such comments, almost a century ago, highlight some of the most active areas of field research today, namely what is now called "the critical zone"; essentially all the subsurface materials down to unweathered bedrock.[10] Much current work explores the role of deeper flowpaths through weathered bedrock and we will explore these modern concepts and descriptions in Chapter 6.

There is much Russian literature on subsurface stormflow, summarized by Smathkin (2002) and listed in Table 2.1. These studies each examined aspects of subsurface stormflow processes. These papers—and others—are still largely unexplored in the hydrological literature and warrant more investigation linked to the history of process thinking on subsurface stormflow. They are the only ones that I include in the book that I have not read myself—but with the help of AI translation, I hope that these can be "read" and interpreted and contextualized in terms of the arc of learning about subsurface stormflow.

[10] Technically, also including also the above-ground vegetation—we will discuss this in detail in Chapters 7 and 8.

Table 2.1 Some key early Russian literature (~1930–1960) on subsurface stormflow. All are direct quotes from Smathkin (2002).

Authors (Date)	Main Findings/Contribution
Rodnikov (1935–1936)	Found a reduction in porosity at the humus–illuvial horizon boundary, with lateral subsurface flow occurring when soils reached field capacity.
Vasiliev (1948)	Subsurface flow volume during storms exceeded overland flow by 7–8 times. Flow followed the pattern of antecedent moisture, especially in upper soil layers (0–40 cm).
Rutkovskij (1940) and Sozykin (1940)	Demonstrated that subsurface flow dominated in forested areas, while overland flow was prevalent in non-forested areas.
Berg (1945) and Reshetnikov (1945)	Proposed that storm hydrographs were composed of both surface and subsurface flow, but their interpretation was questioned later.
Voskresenskii (1948)	Suggested that multiple semi-pervious layers within soils could contribute to subsurface flow with different response times.
Daraselija (1949)	Showed that subsurface flow was the primary stormflow component unless disrupted by ploughing or soil disturbance.
Burykin (1950–1953)	Found that storm water percolated into soil and re-emerged at slope bases as ephemeral springs. Subsurface flow increased with slope, forest cover and stone content in soil.
Volftsun (1954)	Questioned earlier findings by suggesting subsurface flow may only be part of shallow groundwater flow.
Rode (1954)	Provided strong support for subsurface flow as a separate hydrological process, citing its role in stormflow generation in certain physiographic zones.

Of all the early voices and studies of subsurface stormflow it was the work of Charles Hursh who advanced process understanding of subsurface stormflow the most. Hursh was an ecologist who started the Coweeta Hydrological Laboratory in North Carolina (Figure 2.7), arguably the most famous catchment of all time for streamflow generation research.

Hursh helped place subsurface stormflow into a larger catchment hydrology context by selecting Coweeta so that "comprehensive studies of forest influence could be made," as Burt and Swank (2010, p. 719) note in their historical review of Hursh's work. Hursh's site criteria were the following: "a complete headwater drainage basin with well-developed stream networks, preferential flow, high rainfall, deep soils and complete forest cover." Hursh's motivation with the work at Coweeta was to explore runoff behavior in forested catchments where infiltration rates were high, noting that:

Interpretation of surface-runoff phenomena, and more particularly the movement of water after it has reached river-channels, has been the subject of much excellent research by Horton, Sherman, and others. Interpretation of flood-hydrographs from areas of high soil-infiltration where there is little, if any, surface- or overland-flow, has received less attention (Hursh, 1944, p. 743)

The first formal definition of subsurface stormflow comes from Hursh (1936). By 1944 he had a clear and concise description of its occurrence that distinguished it from groundwater flow:

the dynamic type of storm-water that enters the soil but moves away from the area through upper soil-horizons at a rate much in excess of normal ground-water seepage (Hursh, 1944, p. 744)

He, furthermore, noted that "... .certain soil-profiles are neither true overland storm-runoff nor true ground-water discharge" (Hursh, 1944, p. 744). Hursh and Fletcher (1942) noted the importance of preferential flow for both storage and transmission of gravitational water but:

the exact nature of this macro-pore space occurring in different horizons of the soil profile has yet to be described. It includes all large underground channels formed from decayed roots, fractured rock, insect and animal burrows, and large spaces that may exist . . . In like manner, it is conceivable that a few continuous void spaces may give rise to rapid discharge of groundwater through a soil profile, which when viewed as a uniformly pervious medium,

Figure 2.7 The Coweeta entrance sign circa 1942.
Source: USDA.

would be expected to transmit water slowly (Hursh and Fletcher, 1942, p. 485)

Here, Hursh and Fletcher (1942) recognize the outsized impact of large pores and are stating, effectively, Hagen-Poiseille's Law from a hundred years earlier, that describes that flow in pipes is proportional to the fourth power of their pore radius. In other words, at or near saturation, it only takes a small volume of soil pipes and void space to transmit almost all the flow; something we will return to in Chapters 4 and 6. Later, Hursh would credit Robert Horton[11] as laying the foundation for subsurface stormflow understanding:

In his theory of maximum ground-water levels, Horton (1936) has introduced an essential assumption to the concept of subsurface storm-flow. His assumption is that topographic conditions such as gullies and natural drainage-lines may create temporary outlets for the ground-water aquifer at the upper limits of the rise of the water-table. (Hursh, 1944, p. 744)

These statements would foreshadow work some 30 years later by Anderson and Burt (1977) that we will examine in Chapter 4. Figure 2.8 shows Hursh's subsurface stormflow process description in the context of a perceptual model of the hillslope system in at his site at Coweeta. You can see his thinking of the effect of forest on the runoff response with the comparison of the two slope sections. If you look closely, you will also see his clear perception of groundwater mimicking the surface topography but well below the zone producing subsurface stormflow.[12]

[11] This quote regarding Horton is somewhat ironic given how much Horton seemed to downplay anything other than infiltration excess overland flow, as we will explore in the next section.

[12] Note that Figure 2.8 shows a view of topographic controls on groundwater 25 years before the famous

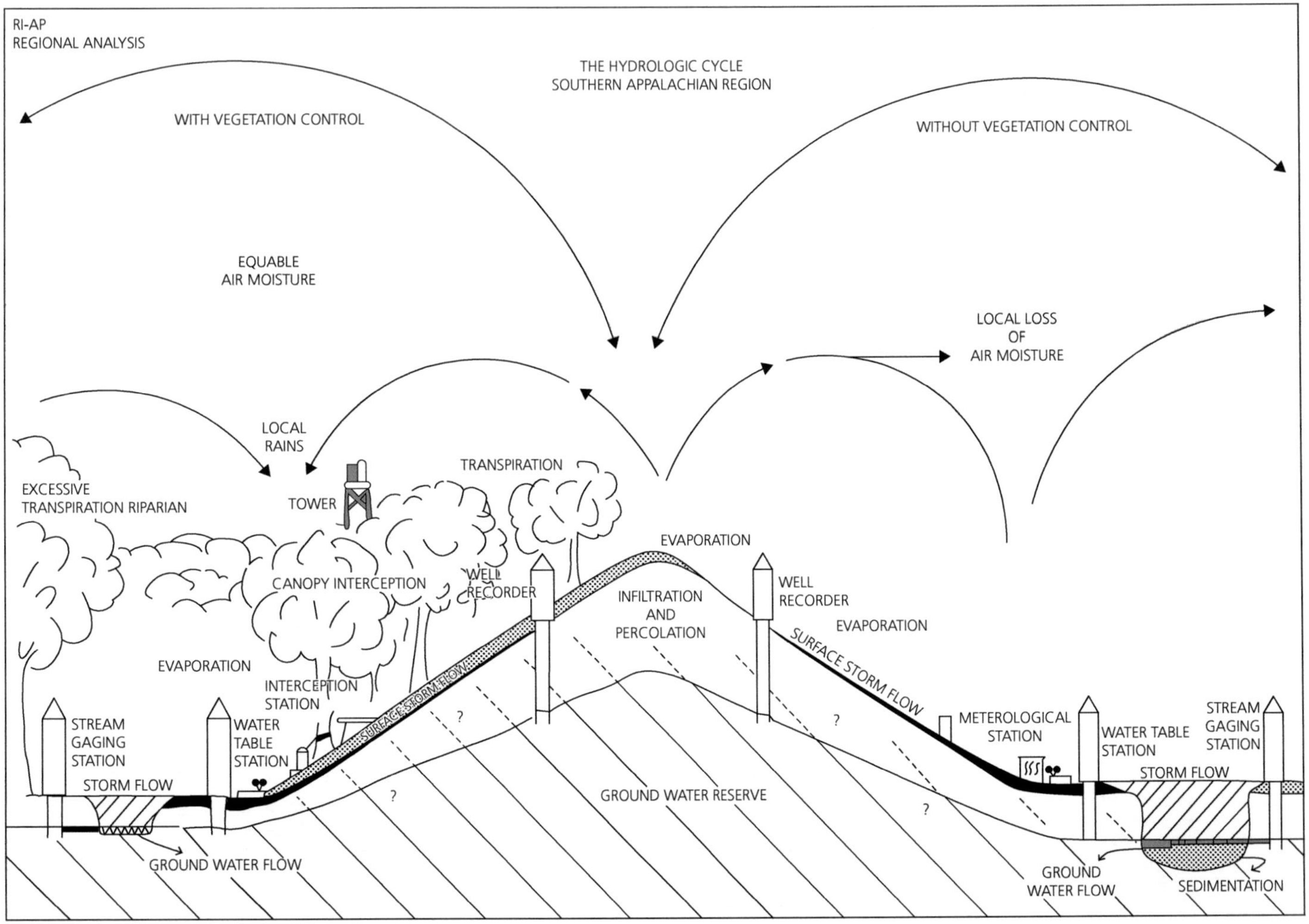

Figure 2.8 Subsurface stormflow in the wider context of the hillslope hydrological cycle driven by subsurface stormflow, by Hursh in 1938.

Source: Reproduced with permission from the Coweeta Hydrologic Laboratory archive.

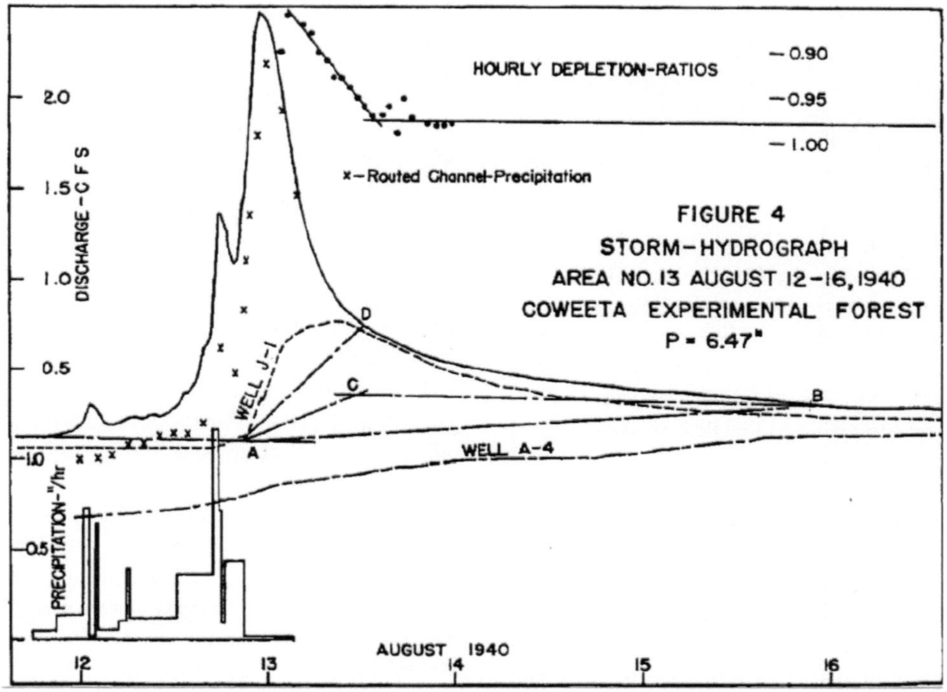

Figure 2.9 Stream hydrograph and groundwater hydrograph analysis[a] of the 12–16 August 1940 storm, Watershed 13, Coweeta. Note that there is early thinking on hydrograph separation—something we will discuss in Chapter 3—where, in the figure, different elements of subsurface stormflow response are shown in lines AC and AD.

Source: Hursh and Brater (1941)

Speaking of groundwater—and anticipating much later work we will examine in Chapter 5—Hursh and Brater (1941) examined data from the different well positions (shown in Figure 2.8). Figure 2.9 shows a set of these well hydrographs for an event in December 1942. They examined the role of groundwater and differentiate two types of hillslope sources: (1) "in close enough proximity to the stream to contribute a significant amount of percolated water during the period of the storm" and (2) "draining from further upslope that takes longer to reach the stream and so supports the gradual recession in stream discharge after the storm hydrograph" (as quoted in Burt and Swank (2010, pp. 4–5 in their retrospective analysis of Hursh and Brater). Burt and Swank (2010) also note that Hursh and Brater (1941) ranked the sequence of top three factors contributing water to streamflow as channel precipitation, shallow water table in close proximity to the stream (as shown for well J1 in Figure 2.9) and subsurface stormflow through "porous soil horizons close to the surface and under high hydraulic gradient." All this would later lay the foundation for follow-on work at Coweeta by Hursh's successor—John Hewlett—and ideas that we will explore in Chapter 3 related to the variable source area concept.

2.4 The first process descriptions of overland flow

Water is the one substance from which the earth can conceal nothing; it sucks out its innermost secrets and brings them to our very lips

—Rudolf Steiner

work of Tóth (1963)—a benchmark paper on groundwater flow in small basins.

The first description of infiltration excess overland flow comes from Hall (1918)[13] who noted that:

Surface runoff will occur when and where the rate of rainfall exceeds the rate of absorption, until the surface soil is saturated then it depends on the difference between rates of rainfall and percolation (Hall, 1918, p. 97)

Unlike subsurface stormflow, overland flow is a visible runoff process. And by the early 1930s, specific reference to the controls on overland flow were identified by Robert Horton—who Beven (pers comm) notes was clearly aware of, and impacted by, the early Hall (1918) paper. Horton was an engineer and interested in prediction. What we call today "infiltration excess overland flow" or sometimes "Horton overland flow" (a term first coined by Kirkby and Chorley, 1967) is a runoff generation process that often occurs on disturbed surfaces where the infiltration capacity has been reduced (overgrazed lands, logging roads etc). But it also occurs in areas of naturally low soil infiltrability and where rainfall intensities can be exceptionally high (e.g., semi-arid and arid catchments).

The process description of infiltration excess overland flow was described by Horton (1933) in his classic paper titled "On the role of infiltration in the hydrologic cycle" as:

The surface of a permeable soil acts like a diverting dam and head-gate in a stream. . .with varying rain-intensity, all of the rain is absorbed for intensities not exceeding the infiltration capacity, while for excess rainfall there is a constant rate of absorption as long as the infiltration capacity is unchanged. (Horton 1933, p. 446)

Horton defined the infiltration capacity, as "the maximum rate at which a given soil can absorb rainfall when the soil is in a specified condition." Being an engineer, he continues with his dam analogy where he notes that:

As in the case of the dam and head-gate, there is usually some pondage which remains to be disposed of after the supply to the stream is cut off, so in the case of infiltration, surface-detention remains after rain ends. Infiltration divides rainfall into two parts, which thereafter pursue different courses through the hydrological cycle. One part goes via overland flow and stream-channels to the sea as surface runoff; the other goes initially into the soil and thence through the groundwater flow again to the stream or else is returned to the air by evaporative processes. (Horton 1933, pp. 446–447)

In this way, Horton is equating the soil surface as a "separating surface" and he believed that "various hydrologic problems are simplified by starting at this surface and pursuing the subsequent course of each part of the rainfall as so divided, separately" (Horton 1933, p. 447).

Horton published over a dozen papers with infiltration in the title (Vimal et al., 2025). His detailed views on infiltration and infiltration capacity all helped sharpen our understanding of the process elements of infiltration excess overland flow. Figure 2.10 shows his thinking on how all this was linked to what he called "rainfall excess." This figure from the Horton Archives shows how important an accurate estimate of rainfall would be to determine rainfall excess—essentially any rainfall about the line marked a to b. Again, Horton was interested in prediction and you can see how infiltration excess could be a very simple process to calculate in a pre-computer age.

But Horton was also acutely aware of how losses along the overland flowpath occur. Figure 2.11 shows several steps that define the progression from the onset of rain to the peak of flow. This was

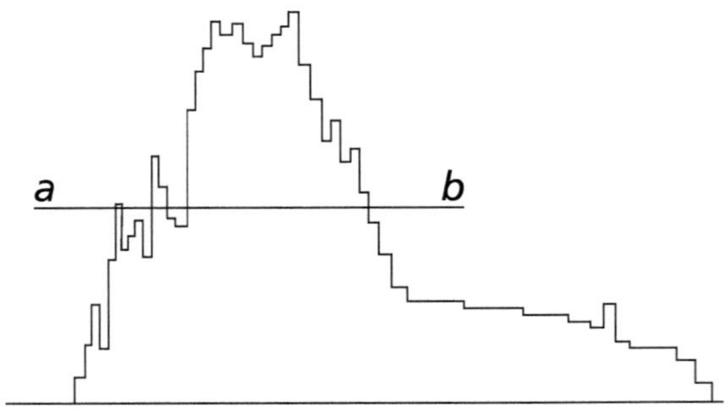

Figure 2.10 Hand-drawn image by Horton of his ideas on how the fine-scale time dynamics of rainfall intensity are related to the production of infiltration excess overland flow.
Source: Horton Archives.

[13] This paper was re-discovered and described recently in Beven (2025).

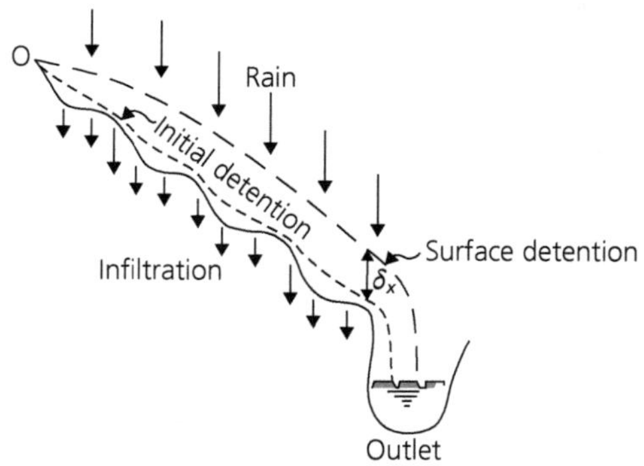

Figure 2.11 How detention storage impacts the transmission of water downslope.
Source: Horton (1935).

summarized by Horton's protégé, Howard Cook about a decade later where he notes the sequence:

1. A thin water layer forms on the surface and downslope surface flow is initiated
2. The flowing water accumulates in surface depressions
3. When full, these depressions begin to overflow
4. Overland flow enters microchannels, which coalesce to form rills, which combine to form rivulets, which in turn, discharge into small gullies, this being continued until discharge into major channels occurs
5. Along each collecting channel, lateral inflow from the land surface takes place (Cook, 1946, p. 729)

Following Horton's death in 1945, more and more research hydrologists started to question the widespread occurrence of infiltration excess overland flow, even in places where rainfall intensity exceeded the soil infiltrability (like arid regions, disturbed lands, etc.). The thinking was that even if infiltration excess overland flow was produced, it would be restricted to certain areas or hot spots in catchments. The first articulation of this idea was in French in Cappus (1960) who said, effectively, that only certain areas would contribute runoff.[14]

[14] Stop the presses! Keith Beven, knowing I am finishing up this book, sent me an e-mail on February 27, 2025 noting (for the first time) that "Horton (1937) includes the treatment of "partial-area flood" contributions" whereas Keith notes "albeit that in Horton's case each area is that of a Thiessen polygon and an average infiltration capacity is calculated for the whole basin. Horton also does not mention the possibility that on the areas where he determines no rainfall-excess there might have been stream gauges showing storm hydrographs." The history of hydrology continues to be written. . .

The Cappus (1960) work also pre-supposes developments that would only occur a decade later with field-based process work on overland flow on saturated areas that we will discuss in the next chapter.

Beven (2006) in his historical review of the Cappus work noted that Cappus:

differentiates between a 'runoff area' (zone de ruissellement) and an 'infiltration area'. He defines the runoff area consisting of impermeable areas (roads, paved tracks, soils compacted due to the effects of man or stock, rock surface, etc.) but also areas of permeable soil that are saturated. He notes that any rainfall falling on these areas will become runoff, either by surface routes (ruissellement superficiel) or near-surface pathways (ruissellement hypodermique) (Beven, 2006, p. 49)

These fixed area hot spots ultimately became known as "partial areas." These partial areas were seen as the only places where infiltration excess could be produced in a catchment. And it was a paper that asked the rhetorical question in the title—"What is watershed runoff?" (Betson, 1964) that helped anchor this partial area idea. Betson defined the so-called partial area concept as a "function of apparent watershed infiltration capacity." Betson (1964) begins with the following statement:

Infiltration, by definition, is the movement of water through the soil surface into the ground. The concept of infiltration is easily understood. Measuring and understanding the physical factors controlling the infiltration process, however, is exceedingly difficult (Betson, 1964, p. 1541)

Betson goes on to develop a mathematical function that links infiltration excess to only a portion of the runoff in the stream where he notes that his function characterizes "the infiltration capacity of that portion contributing to runoff" and that such a function "should prove to be a useful infiltration capacity index with which watersheds can be compared" (Betson, 1964, p. 1541). Figure 2.12 shows a diagram of Betson's partial area concept. It illustrates, in an abstract way, how fixed zones of reduced permeability would be a hot spot for surface runoff generation.

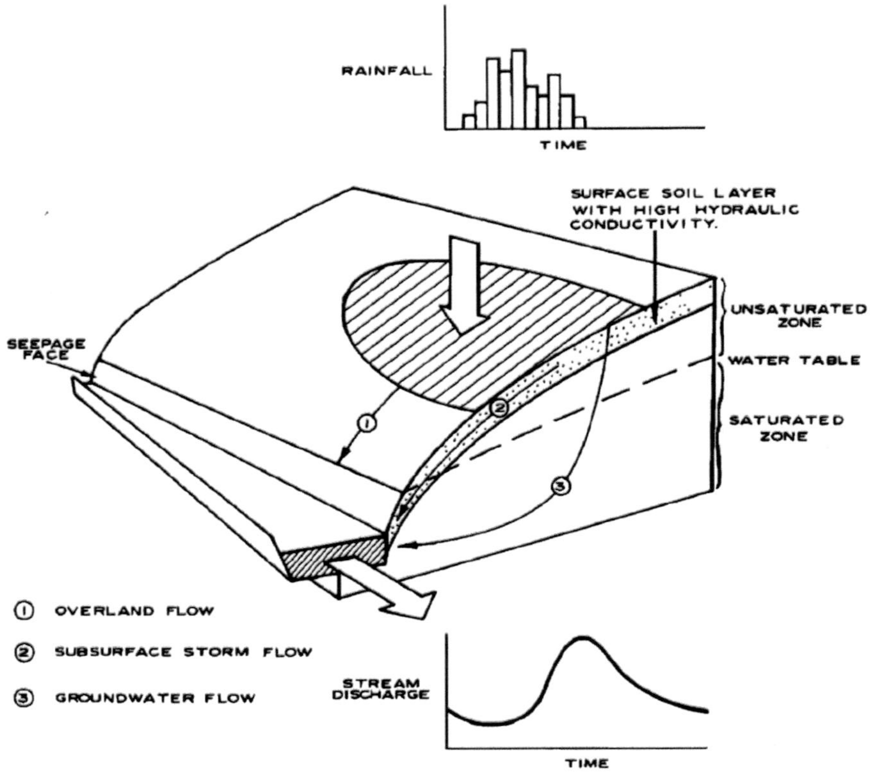

Figure 2.12 A diagram—the only one I am aware of—for the partial area concept of Betson (1964). It shows too other forms of runoff but illustrates nicely the idea that infiltration excess is restricted spatially.
Source: Freeze (1974).

2.5 Reconciling subsurface stormflow and infiltration excess overland flow

> *History never repeats itself, but the kaleidoscopic combinations of the pictured present often seem to be constructed out of the broken fragments of antique legends*
>
> ***—Mark Twain and Dudley Warner***[15]

One might ask how Horton viewed subsurface stormflow? Did he acknowledge its existence? Indeed, he was aware of subsurface stormflow. He called it "concealed-surface runoff":

> surface-runoff may take place in such a manner as not to be visible, as, for example, where it occurs through a layer of coarse material, sometimes through a thick matting of grass or mulch-cover; through a layer of plant roots close to the soil-surface and under forest-litter; or even, in some cases, through a network of sun-cracks in the soil-surface (Horton, 1942, p. 481)

He goes on to say that:

> This has sometimes been called "subsurface-runoff," sometimes "ground-water flow." The term "subsurface-runoff" would not be objectionable were it not for the fact that it is likely to be confused with true ground-water flow (Horton, 1942, p. 481)

Horton is correct, in many ways, in that it is difficult to distinguish subsurface stormflow from ground-water flow. He goes on to say that subsurface

[15] The quotation comes from Chapter 47 of The Gilded Age: A tale of today, by Mark Twain and Charles Dudley Warner; often wrongly quoted elsewhere as "History never repeats itself, but it rhymes."

stormflow (what he terms "flow occurring close to the surface"):

has little in common with true ground-water flow. It is mostly turbulent flow, while true ground-water flow is mostly laminar. It persists only during rainfall-excess or for a short time thereafter, measured in hours or at the most in days, whereas ground-water flow persists on perennial streams at all times (Horton, 1942, p. 481)

Horton (1943, p. 475) did recognize that some soils were capable of absorbing all the water that fell on them. For the Little Tallahatchie catchment in Mississippi, USA, he noted that "owing to somewhat unusual conditions, surface runoff rarely occurs from soil well protected by forest cover." His protégé Cook began to recognize the importance of subsurface stormflow in a 1946 paper where he notes that in addition to overland flow, "in areas of low storage capacity (such as thin-soiled basins) a large proportion of the flood water may consist of subsurface runoff" (Cook, 1946, p. 728).

Nevertheless, some have been harsh in their assessment of Horton, where:

although Horton placed such emphasis on overland flow (rainfall intensity minus infiltration rate) as the origin of storm hydrograph rates and as the motor of surface erosion, he was always strangely defensive regarding one's inability to commonly observe this phenomenon, particularly on vegetated and soil-covered slopes (Chorley, 1978, p. 12)

Beven's multiple retrospective pieces on Horton's work (summarized most fully in Beven, 2021) is more charitable in its assessment where he notes:

I am not sure he [Horton] would have totally approved of such widespread use of the infiltration excess concept. Although he frequently used the infiltration excess concept as a way of calculating the volume of runoff production from a particular rainfall (e.g., Horton, 1933), he also had a hydrological laboratory in his wooded back garden in Voorheesville, New York State (Horton, 1936) where he would surely not have observed infiltration excess overland flow very often (Beven, 2001, pp. 10–12)

Finally, Tom Dunne—who we will hear about in Chapter 3—in correspondence with me during the writing of this book notes that:

it became obvious from Vermont that when Horton described overland flow in Voorheesville, NY he was almost certainly seeing saturation overland flow. But, there was no point in ragging on him for misinterpreting the field observation when the model he conceived of was still valuable somewhere else. (Dunne, pers. comm., 2021)

This is interesting as Beven (2004) has also noted that some of the runoff responses Horton observed at his site in Upstate New York were very likely overland flow due to saturation excess (i.e., due to a rising groundwater table—something we will explore in the next chapter) and *not* infiltration excess.

2.6 Summary and looking ahead

Some people will never learn anything,
for this, reason,
Because they understand everything too soon
—Alexander Pope

This chapter has tackled quite a long time arc in streamflow generation thinking; from the early Greeks and Romans to ~1960. What we see is that the early thinking on streamflow generation was linked to the catchment water balance and streamflow on an annual basis vis-à-vis annual precipitation—at least up until the early 1900s. Work on subsurface stormflow developed largely independent of the work on infiltration excess overland flow. Of course this is partly a matter of place. For instance, in forested and humid catchments, subsurface stormflow may be the dominant storm runoff mechanism for many of the larger events. In arid or semi-arid systems where rainfall is infrequent but very intense when it does come, then infiltration excess overland flow will likely be the dominant process for such events.

While it is interesting that the subsurface stormflow work commenced ahead of the overland flow work, the infiltration theory based overland flow ideas hugely dominated the period ~1930–1960. Cook (1946) called this an "era of infiltration," where here he noted that:

the pre-occupation of contemporary hydrologists with the 'infiltration theory of runoff', and the vast amount of energy they have expended in an effort to turn this concept to practical account, will certainly be put down as a distinctive feature of our times (Cook, 1946, p. 726)

REPORT OF COMMITTEE ON RUNOFF--1944-1945

R. W. Davenport, Chairman

W. C. Ackermann	W. G. Hoyt	Charles C. McDonald
Bertram Barnes	C. R. Hursh	F. T. Mavis
Merrill Bernard	Carl F. Izzard	C. E. Ramser
Bruce R. Colby	Karl Jetter	L. K. Sherman
Howard L. Cook	Phillip Light	Waldo E. Smith
E. S. Cullings	Norbert H. Leupold	F. F. Snyder
G. A. Hathaway		H. G. Wilm

Figure 2.13 The AGU Committee on Runoff for 1944–1945, from *Trans. American Geophysical Union*, Vol 27, Number 1, Published Feb 1946. Note the names we have discussed in this chapter who were part of this committee: Cook, Hoyt, Hursh, Ramser, Snyder. Horton died in April 1945 (at 69 years old).

Beven (2021) has written the definitive review of this era and reflected deeply on the history of how hydrological process thinking developed during 1930–1960. But why was it an era of infiltration when it could have been an era of subsurface stormflow? Of course, much comes down to the fact that overland flow was a convenient process to see and calculate in the era before computers. And while mechanistically, both subsurface stormflow and infiltration excess overland flow made sense for different settings, Horton's ideas "won out" in the court of public opinion.

Take a moment to reflect on subsurface stormflow and infiltration excess overland flow. What is the big difference in how you might imagine these processes to occur at a hillslope near where you live? One can be seen with the naked eye, the other cannot. Therefore, overland flow has observable boundary conditions. And, even better, you can go out with an infiltration ring device and get a sense of whether or not rainfall events in your area might exceed the local infiltrability of the soil. Now ponder subsurface stormflow for a moment. The soil infiltrability is likely high. But other than not having a "barrier" at the soil surface, one can not really make any real inference about subsurface stormflow until one measures soil depth, and the permeability of whatever is below the soil (e.g., rock, saprolite, till) that otherwise causes a break in permeability within the soil column that could re-direct water laterally.

So, overland flow is easy to see, conceptualize and model compared to subsurface stormflow. It was true in the period 1930–1960 and it is still the case today! Read any engineering textbook from that period until recently and overland flow is *the* focus—largely because it offers an easy path to modeling success linked to simple infiltration theory or some simple loss function in the landscape.

A new student may still wonder why the separate overland flow vs. subsurface stormflow histories from ~1930–1960 and why people were not quoting each other? Certainly there were differences in training and backgrounds (e.g., Horton and most of his colleagues were engineers, while Hursh and many of his collaborators were foresters and forest hydrologists). But perhaps the answer is as simple as communication. Before World War 2, people were not communicating like they started to do after the war, when many key hydrological voices were at last talking and interacting. Figure 2.13 shows the membership of the AGU "Committee on Runoff" for the term 1944–1945. In the list, Hursh is there—who also Chaired a "sub-committee on subsurface flow," along with names like Ramser, Barnes, Cook, Hoyt, Sherman and Snyder, who we have already heard from. So at least they were all talking to each other at their annual meeting!

Finally, before we move on to the next chapter, a student may ask if there are any open questions linked to this early history?[16] Indeed, we'll explore

[16] The history of hydrology is still being written. There is still much to come from the Horton boxes. Elsewhere, countries like China are having their process hydrological history as it is related to streamflow generation research written only recently (e.g., Nan et al., 2024). While other national histories are coming out with the History of Hydrology Working group (Beven et al., 2025) and their many new national hydrological

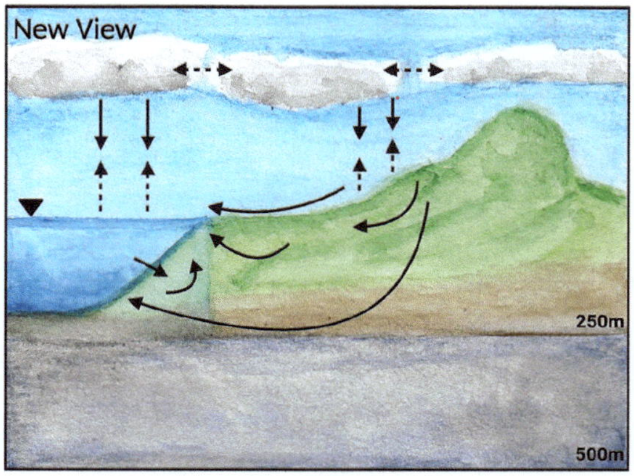

Figure 2.14 A revised view on the catchment water cycle in the context of the bottom of the catchment—where new work suggests that less than 0.1% of groundwater below 500 m ever returns to form streamflow—suggesting a more compartmentalized water cycle than previously thought since most groundwater lies below 500 m.
Source: Drawn by Harvinder Sidhu.

a multitude of open questions related to the processes of overland flow and subsurface stormflow in the following chapters. But, perhaps surprisingly, the water cycle diagram that we started this chapter with is now being re-imaged with new data on the water age. Figure 2.14 shows perhaps a fourth updated water cycle diagram that captures what we know today. It shows, diagrammatically, the recent finding that 2/3 of global groundwater below 250 m is 12,000 years old or older (Jasechko et al., 2017). In other words, fossil water—not recharged on human timescales. So very old water may be only a small part of the water cycle and recent work has suggested that below 500 m, less than 0.1% of global groundwater below 500 m ever returns to form streamflow (Ferguson et al., 2024). So, this could be seen as a global "cut off" for the water cycle. This is a big deal since there is a vast body of largely saline groundwater that exists down to at least 10 km. It has been quantified for the first time by Ferguson et al. (2021, p. 1) where they note that "subtracting off the 0–500 m interval from the total volumes, there is 34.4 km^3 of groundwater between 500 and 10,000 m. That's 78% of the total volume in the crust beneath the continents." So our perceptual model of the global water balance is still changing![17] We will address current thinking on catchment storage and streamflow age in Chapters 7 and 9 and where the bottom of the catchment is in different regions with different rock type, different permeability architectures and topographic conditions.

history efforts for Croatia (Pevlic et al., 2025), Slovakia (Holko et al., 2025), Japan (Nakamura et al., 2025) and the development of runoff generation models in Russia and the former USSR (Kuchment and Gelfan, 2024). And the role of key historical figures on the development of catchment hydrology; for early contributions to runoff monitoring in the UK, e.g., W. N. McClean (Werritty and Black, 2025) and for early contributions to experimental hydrology in China, e.g., Wei-Zu Gu (McDonnell, 2023).

[17] Note also the new USGS water cycle diagram that also includes human activities: https://www.usgs.gov/special-topics/water-science-school/science/water-cycle-diagrams

CHAPTER 3

The variable source area concept (1960–1970)

3.1 Dynamic and variable source areas

> *An important scientific innovation rarely makes its way by gradually winning over and converting its opponents. . .what does happen is that its opponents gradually die out, and that the growing generation is familiarized with the ideas from the beginning*
>
> ***—Max Planck***

Most undergraduate texts describe the variable source area concept[1] and note its importance for runoff generation. But what is it exactly? And why is it perhaps the most revolutionary idea, ever, in rainfall–runoff research? If you reflect on what we discussed in Chapter 2, the discovery of subsurface stormflow and infiltration excess overland flow were all effectively point-scale process discoveries. Starting in the early 1960s with work by Hewlett (1961), a dynamic, whole-catchment, system-based view emerged. In that paper—a Southeast Forest Experiment Station Annual Report for 1961—Hewlett wrote a general description of the "expanding and shrinking source area" concept.[2] The best definition for the variable source area concept was in the paper by Hewett that presented the concept in detail for the first time:

> . . .when the subsurface flow of water from upslope exceeds the capacity of the soil profile to transmit it, the water will come to the surface and the channel length will grow. That in essence is the variable source area concept (Hewlett and Hibbert, 1967, p. 279)

Let's now look at it diagrammatically in Figure 3.1. The variable source area concept is a synthetic, systems view of a catchment and how it processes rainfall; a perceptual model incorporating subsurface stormflow catchment-wide. Hewlett and Hibbert (1967) tried to counter what had come to dominate thinking in the field up until their paper: that infiltration excess overland flow and the ideas of Horton that we explored in Chapter 2 dominated most places. Hewlett and Hibbert (1967, p. 277) described the situation within the community—that runoff was generally assumed to be a "product of overland flow and that other types of flow were mere exceptions to that general rule." But their perceptual model based on the observations at Coweeta during rainfall events was that "the opposite approach is more logical in the case of forest land; that is, to begin with the assumption that all flow is subsurface flow until there is evidence otherwise" (Hewlett and Hibbert, 1967, p. 277). This is one of the most clear and pivotal statements in streamflow generation research up until that time! Hewlett and his co-author were not only the first to forcefully and clearly counter the

[1] This chapter uses the word "concept." Up until now, we have discussed *processes* and *perceptual models*. Why do we use the term "concept" to denote variable source area ideas? What differentiates a concept from a perceptual model is generality. A concept in the context of variable source areas is an overarching framework that can describe a set of behaviors from many places around the world at the catchment scale. We will come back to the notion of concept—and indeed a follow-on concept from the variable source area concept—in Chapter 6.

[2] This was picked up ~1962 by the Hydraulics Branch of Tennessee Valley Authority but without credit to Hewlett. It was one of the TVA anonymous reports which first gave it the name "variable source area concept." Based on letters I have read of Hewlett (see later in this chapter), he noted that he liked the term and stuck with it but, in his words, the phrase variable source area concept "has no particular historical priority."

Streamflow Generation. Jeffrey J. McDonnell, Oxford University Press. © Jeffrey J. McDonnell (2026). DOI: 10.1093/9780191953736.003.0003

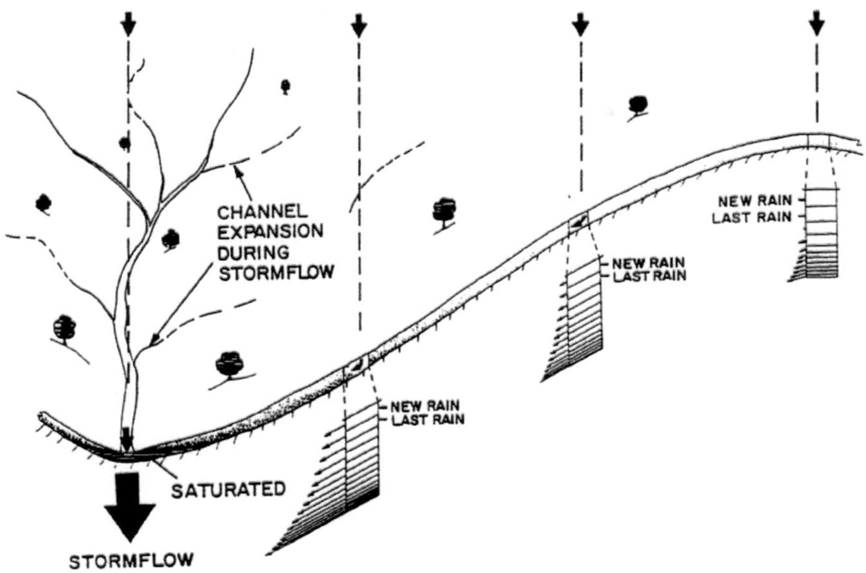

Figure 3.1 The variable source area concept of Hewlett and Hibbert (1967). Note the interesting combination of 2D and plan-view display. *Source*: McDonnell (2009).

widespread adoption of the Horton view but were a match to Horton's eloquence![3]

Hewlett and Hibbert (1967) was one of the first papers to explicitly show how pressure and particles propagate through a catchment during a rainfall event. Indeed, how hillslopes store water for months to years and then release it in minutes to hours, which remains the source of considerable debate and discussion even today (Kirchner, 2003; McDonnell and Beven, 2014)—something we will get to in Chapters 5 and 7. Hewlett and Hibbert were exceptionally clear-thinking on this topic of flow and transport and their paper presented a thoroughgoing analysis of the passage of rainfall to the stream:

> each unit of rainfall contributes more to temporary storage and less to direct runoff. However, of the parts contributed to direct runoff, a fraction will be some of the actual raindrops that fell during the storm – that is, some new rain – and the other fraction will be flow produced by a process of displacement (Hewlett and Hibbert, 1967, p. 279)

They go on to say that the effusion of stored water will be "released in large quantities only when the soil is within the field capacity range or wetter" (Hewlett and Hibbert, 1967, p. 279). They called this process "translatory flow" which can be defined as piston flow displacement. However, Hewlett and Hibbert's translatory flow lacked any explicit treatment of the preferential flowpaths that Hursh had described at Coweeta decades earlier. Nor did Hewlett and Hibbert note an enhanced flow and transport at the soil–bedrock interface—key elements that many subsequent papers would describe as necessary for rapid transmission of soil water to the channel during an event. Nevertheless, Hewlett and Hibbert (1967) helped the hydrologic community begin to understand differences in source areas and timing of water delivery in a radically new way—one that integrated scales.

Hewlett and Hibbert (1967) also spoke to themes that only decades later would emerge in the mainstream hydrological literature. They described how "a topographic pattern of soil water storage and

[3] Horton was perhaps the best scientific writer of his generation of hydrologists—his writing abilities were noted even when he was an undergraduate at Albion College in the USA (see the history of hydrology wiki site at http://www.history-of-hydrology.net/mediawiki/index.php?title=Main_Page for more information if you are interested).

availability is already well established in soil survey, site and vegetation studies" and that there would be "some advantage in bringing hydrologic theory in line with these views" (Hewlett and Hibbert, 1967, p. 280).[4] As the Preface to this book will attest, I am an admitted worshiper of Hewlett and all that he accomplished. But don't just take my word for it. Bonell (1993) in his definitive review of twentieth century hydrology said that:

> By hydrological standards, the introduction of the variable source area concept was close to being a 'Copernican revolution' in terms of how the runoff component of the water balance equation was viewed (Bonell, 1993, p. 218)

3.2 Behind the scenes: Creation of the variable source area concept

> *One cannot accomplish anything without fanaticism*
>
> ***—Eva Peron***

We have already seen a direct and unmistakable line that can be drawn from Hewlett back to his predecessor at Coweeta, Robert Hursh. Hewlett built on Hursh's ideas. But he also built upon the monitoring infrastructure at Coweeta. There were already some 28 groundwater wells at Coweeta with a 20–25-year record when Hewlett started working there. Hewlett expanded the soil moisture monitoring efforts at Coweeta and by 1961 he was able to report that the:

> soil moisture regime at Coweeta are now being effectively monitored to depth of 15 to 18 feet despite rocks and other difficulties". . .including "an analysis of some 14,000 soil moisture determinations over a 6-year period (Hewlett, 1961, p. 61)

But there were two *other* key pieces of work that undergirded the variable source area concept: the so-called "sloping block" work of Hewlett and Hibbert (1963) and the tritium tracer work of Horton and Hawkins (1965) that Hewlett and Hibbert made clear use of (note that the Horton just mentioned is a different Horton to the one we discussed in relation to infiltration excess overland flow in Chapter 2). Let's now look at both of these additional pieces of work in detail as we reflect on the variable source area concept further.

A look at the Hewlett sloping soil block in the present day (Figure 3.2a) does not necessarily suggest a shrine to hydrology; but a shrine to hydrology it is! It's a bit worse for wear compared to its original construction (Figure 3.2b): a tree has grown inside it—now cut down—and some of the walls are now cracked. But let's go back in time to when it looked newly minted and a young team posed studiously on the slope with their chart recorders. Hewlett and Hibbert (1963) focused on one of many such poured-concrete slope sections at Coweeta. They noted that the outflow pipe visible in Figure 3.2:

> established a free water surface which was used as a zero datum for all measurements. The soil under the free water surface was graded to sand, gravel and rock to simulate stream bank conditions and to allow free drainage (Hewlett and Hibbert, 1963, p. 1083)

The artificial hillslope also included 17 access tubes for neutron probe monitoring for soil moisture—the preferred method of the day—and 1 piezometer and 4 tensiometers. More than the site looking quite grand back in the day, it connoted control, accuracy and precision. It also destroyed any soil structure and placed an artificial bottom on the slope that conditioned all the findings. But we'll not nitpick for now and come back to this topic in Chapter 7.

Hewlett and Hibbert (1963) measured outflow from the base of the slope shown for 145 days and recorded matric potential (i.e., the soil water tension) and soil water content within the block. Figure 3.3 shows their main finding: on a log–log plot, a major break in the flow recession follows the stoppage of irrigation to the hillslope block. They explained it as follows:

> the larger pores were substantially emptied in 1.5 days, during which outflow occurred according to the expression, $Q = a_1 T^{-b_1}$, where T is time in days since day zero and a_1 and b_1 are constants. After a 5-day transition period, the logarithm of drainage for the next 80 days was again linearly related to the logarithm of time, but with new constants, a_2 and b_2, expressing unsaturated flow from the entire soil mass. An additional 60 days of

[4] Hewlett and Hibbert are acknowledging here ideas going back to Hans Jenny and his 1941 landmark book "Factors of Soil Formation" and ideas on soil catena (although that book is not cited in their paper).

Figure 3.2 (a) The Hewlett slope in the early 1960s. The block was "3×3×45′ and set on a 40 degree slope." And that a "60–18–22 mix of sand–silt–clay was packed to a bulk density of 1.3 with a porosity of 50% by volume" (from Hewlett and Hibbert, 1963, p. 1082).
(b) The sloping soil block in 2013.

Source: Photo on right taken by the author.

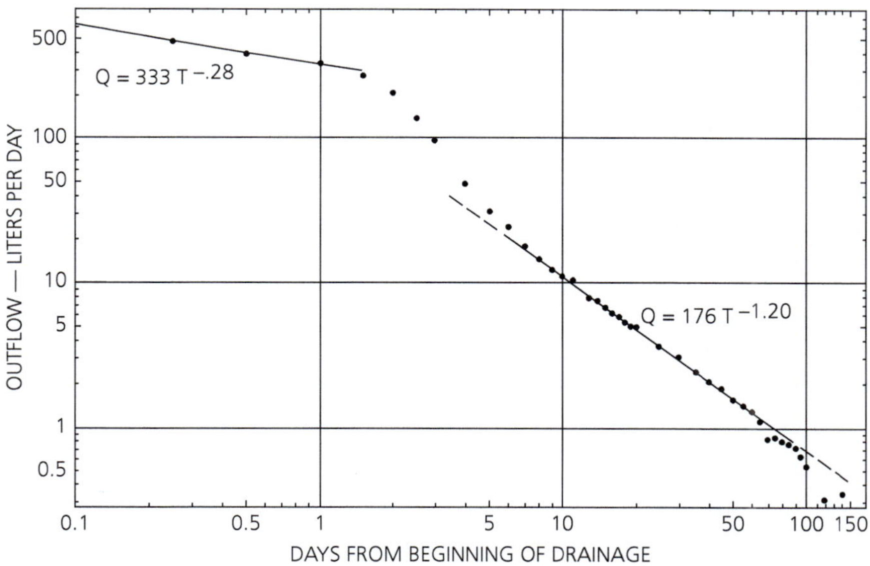

Figure 3.3 Hillslope block outflow as a function of time since the beginning of drainage for the Hewlett and Hibbert (1963) experiment. Note the log–log axes.

Source: Hewlett and Hibbert (1963).

flow occurred, deviating more and more from the log–log relation as the limits of drainage were approached. (Hewlett and Hibbert, 1963, p. 1081)

What they are getting at here is the transition from fast to slow recession and their perceptual model for baseflow in a headwater stream: drainage of soil water from the side slopes, which they note "may well be a primary mechanism for sustaining baseflow" (Hewlett and Hibbert, 1963, p. 1081). By way of anecdote, this log–log recession approach became all the rage in hydrology in the years that followed and two papers that we will explore in the next chapter (e.g., Harr, 1977; Mosley, 1979) made use of this same type of plot. The idea was that for any headwater catchment, the break in the log–log recession denoted a change in drainage characteristics, interpreted sometimes as the transition from macropore to matrix flow drainage conditions or from transient saturation to unsaturated drainage conditions.

The other key influence on Hewlett's variable source area concept was the study of Horton and Hawkins (1965). The Horton[5] and Hawkins work was done nearby at the Savannah River Site in South Carolina and the results heavily influenced some of Hewlett's thinking on translatory flow. Horton and Hawkins traced the "downward displacement of soil moisture by infiltrating rain" using an applied tritium tracer. Their experiment packed 1.2 m soil columns with local sand and their experimental results showed a clear displacement effect with considerable dispersion. So, what Hewlett and Hibbert (1967) did was to modify these 1D column experiment results to buttress their idea of piston flow displacement of soil water *laterally* downslope on the Coweeta slopes (Figure 3.1). Figure 3.4 shows the results from the original column experiments of Horton and Hawkins—interestingly, showing more dispersion[6] than the clean piston flow imagined in the classic Hewlett and Hibbert (1967) diagram in Figure 3.1!

Essentially, the results of these column experiments were that stored water was displaced vertically downwards through the 1.2 m soil columns by the invading new water. The important point here—and the assumption that Hewlett and Hibbert are making—is that slow water displacement from columns is analogous to the very rapid displacement during a storm event—in other words, displacement holds whether it is fast or slow. I will have more to say on this topic in Chapter 5. But for now, let's read a more recent retrospective discussion of Hewlett and Hibbert (below called "HH") by Buttle (2020), focused on this translatory flow idea:

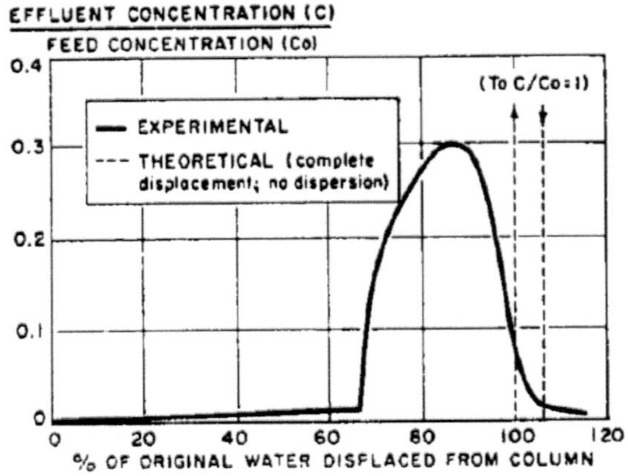

Figure 3.4 Results from the Horton and Hawkins work showing displacement of water "held at field capacity"
Source: Horton and Hawkins (1965)

> Horton and Hawkins concluded that infiltrated rainwater moves primarily through capillary pores, avoiding the larger ones, and that each rain tends to displace most of the water ahead of it" (page 280). The qualifying words "primarily," "tends," and most" are important: HH did not suggest that non-capillary flow could not occur or that complete displacement of pre-event soil water necessarily had to occur. HH tilted Horton and Hawkins's soil column from the vertical. They argued that a similar displacement or translation of stored soil water occurs on hillslopes where vertical infiltration of newer precipitation leads both to downward and downslope translation of older soil water. (Buttle, 2020, p. 5686).

Here, Buttle is making the point that the analogy to the Horton and Hawkins work is a general one and perhaps we should not take the details of displacement too literally.

Many have since commented on these experiments and how Hewlett and Hibbert's work embraced Horton and Hawkins' findings. Beven (2006) makes an important point in his retrospective on Hewlett and Hibbert's use of the tracer work where he notes:

[5] Just another reminder that this is a different Horton from Chapter 2, i.e., not Robert E Horton of infiltration excess overland flow fame.

[6] We will discuss much further the notion of dispersion in Chapters 5 and 7.

What Hewlett and Hibbert do not record, however, is that the time scale for displacement in the Horton and Hawkins experiments was of the order of three weeks, somewhat longer than the timescale of the storm runoff hydrograph. (Beven, 2006, p. 114)

3.2.1 Quickflow separation

A background piece to the development of the variable source area concept was the quickflow hydrograph separation method developed by Hewlett and Hibbert (1967) (Figure 3.5). While graphical hydrograph separation is no longer *de rigueur* (something we now realize is akin to trying to unscramble an omelet!), the power of Hewlett and Hibbert's approach was in their intercomparison of 15 small catchments (including Coweeta) across the Appalachian Piedmont region. Hewlett and Hibbert examined 200 years of record for the 15 small forested catchments and decided that a line projected from the beginning of any stream rise at a slope of 0.05 cubic feet per second per square mile per hour until it intersected the falling limb side of the hydrograph would be a simple and satisfactory method for separating streamflow into "quick flow" and "delayed flow" (Figure 3.5). The Dictionary of Geology now defines quickflow as "a mixture of overland flow and subsurface stormflow" and delayed flow, that they term "delayed runoff" as "runoff that penetrates the subsurface before discharge."

Hewlett and Hibbert (1967) found that none of the forested catchments yielded more than one-third of their total yield as quickflow. They began assembling features that might explain the variations in quickflow and grouping of catchments in terms of their quickflow production, noting that things like topography, soil depth, climate and land use all played a role. In one of the clearest expositions of runoff generation ever written, Hewlett and Hibbert listed the order of importance of these factors based on the process-based findings at Coweeta and their comparison with these other study sites. Foremost among the four factors (as noted on their page 288) was "the average soil mantle depth or depth to a relatively impermeable layer." Second was the "average land slope and its effect on the average length of slope from the channel to the water divide." Third was "the average size and number of the larger storms or the average annual storm."

They noted that:

these influence response by their obvious effect on rates of infiltration" and stated that "average annual

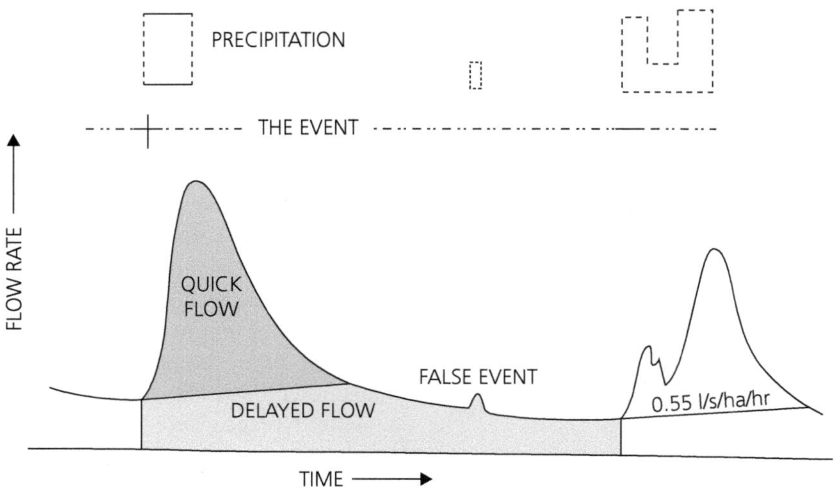

Figure 3.5 Illustration of the quickflow separation method where a line of 0.05 cubic feet per second per square mile per hour is used to separate the hydrograph into quickflow and delayed flow.

Source: Hewlett and Hibbert (1967) redrawn in McDonnell (2009)

precipitation and its distribution may be included here, as it influences response through control of antecedent wetness conditions and variable source areas contributing subsurface flow (Hewlett and Hibbert, 1967, p. 288)

Land use was listed last, as Hewlett and Hibbert noted (also on p. 288), "not because it is minor, but because its effects on the time distribution of flow are superimposed on the effects of the other factors."

Finally and somewhat curiously, the variable source area concept as outlined by Hewlett and Hibbert (1967) said little about rain falling (or snow melting) onto the expanding saturated area adjacent to the channel—even though the famous diagram in Figure 3.1 showed expanding channel areas and saturation around the channel margins. This is perhaps not surprising because that work was done in the unglaciated Southern Appalachian Mountains of North Carolina, USA where the steep terrain with incised valleys had little in the way of extensive flat zones around the stream channels. Nevertheless, we saw in Chapter 2 how Hursh's early work at Coweeta had already shown diagrammatically how his groundwater wells near the stream responded. But conversations with one of Hewlett's collaborators, Wade Nutter, in the preparation of this book, suggested that part of the silence on near stream runoff contributions was driven partly by their intent to counter the engineering hydrology beliefs at the time on the role of overland flow (due to infiltration excess) as a chief runoff-producing mechanism. In other words, infiltration excess overland was not seen in upland forests and the variable source area was presented as a viable (but invisible) alternative to the infiltration excess overland flow paradigm.

3.3 Saturation excess overland flow

Truth is ever to be found in simplicity; and not in the multiplicity and confusion of things

—Isaac Newton

Despite all the effort by Hewlett and colleagues to try and downplay the role of overland flow, runoff *is* produced on the saturated areas near the stream channel. It often stares us in the face. We now call this mechanism saturation excess overland flow or sometimes just saturation overland flow. Kearey's (2001) "Dictionary of Geology" defines it as "runoff from an area of soil on a shallow slope saturated by throughflow[7] which accumulates at the base of the slope." This definition is consistent with its first field-based discovery of saturation overland flow by Tom Dunne at the Sleepers River catchment in Northern Vermont, USA. Unlike Coweeta in North Carolina, USA where Hewlett worked, Sleepers River is a low angled, gently rolling glaciated region with short slopes. And even before the work of Dunne, Kirkby and Chorley (1967)—who were aware of Hewlett's work—synthesized what they saw as factors for overland flow that was not infiltration excess:

areas of flow convergence, surface concavity and thinner or less permeable soil are shown to have higher moisture contents per unit volume, greater flows per unit cross section and, consequently, overland flow at lower rainfall intensities and durations than elsewhere (Kirkby and Chorley, 1967, p. 6)

But even before Kirkby and Chorley's work, there was recognition of saturated areas and their importance for runoff generation in Cappus (1960), a paper we examined in Chapter 2. Figure 3.6 shows the catchment where Cappus worked: the Alrance research catchment in the central Massif of France. Cappus (1960) noted that runoff can be explained by rain falling on saturated areas that are made "impervious" due to a rising groundwater table, and not rain falling on areas made impervious due to their limited infiltration capacity.

In between the work of Cappus and Dunne, there was also the work of Zhao and Zhuang (1963) who proposed a mechanism of saturation excess runoff as the dominant control on runoff generation based on the discharge and precipitation data at 21 stations in southern China. They found that when the precipitation amount exceeded a threshold value, the slope of the rainfall–runoff relation approximated a 1:1 line. Zhao and Zhuang (1963) proposed a "water storage curve" to characterize these saturated areas contributing to storm runoff (as translated and reviewed recently in Nan et al., 2024).

[7] Remember that throughflow is the same as subsurface stormflow.

Figure 3.6 A photo of the research catchment of Alrance where Cappus worked showing the middle part of the catchment. The gauging station is located about 400 m downstream. The site is near the hamlet of Boussinesq. (Incidentally, the name Boussinesq is also a huge one in hydrology: Joseph Valentin Boussinesq, who you may have heard about in your groundwater hydrology classes and known for the Boussinesq Equation.)
Source: Bruno Ambroise.

Now back to Dunne's work in Vermont, USA.

Like the work of Hewlett, Dunne's work was groundbreaking and continues to influence the field. Figure 3.7 shows how Dunne's work influenced thinking on the variable source areas concept. The work reported in Dunne and Black (1970a, b) and the papers that followed, are really what introduced the foundational process understanding of what goes on near the stream channel and the process now known as saturation excess overland flow. Getting back to the definition of saturation overland flow,[8] a more formal definition was offered by Dunne in a 1980s reflection piece where he noted:

Saturation overland flow develops where the soil becomes saturated by the perennial groundwater table rising to the surface or by lateral or vertical percolation above an impeding horizon (Dunne and Black, 1970a, b; Bonell and Gilmour, 1978). Some of the water that has been moving slowly through the topsoil emerges and flows over the ground surface as return flow (Musgrave and Holtan, 1964) and is augmented by direct precipitation onto the saturated area, which expands and contracts seasonally and during rainstorms. (Dunne, 1983, pp. 28–29)

Dunne and Black (1970a, b) showed several new features that had not been previously

[8] Note that the literature—and likely me in this book!—will use interchangeably "saturation overland flow" and "saturation excess overland flow." I like the added word "excess" for this and for infiltration excess. That is because when we get deeper into subsurface stormflow, you will see that it too is an "excess"-type of mechanism, related to *storage excess*. Chapter 6 will explore this in detail and show that ultimately, all runoff processes—at or below the soil surface—are each some form of storage excess.

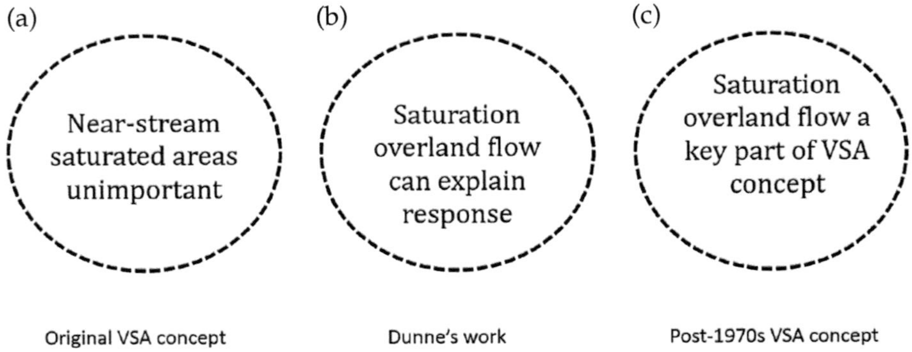

Figure 3.7 The pre-Dunne and post-Dunne view on variable source area theory

described: that saturation overland flow could be the main contributor to headwater runoff, that the saturated areas themselves could expand and contract—seasonally and within an event. Dunne and Black used the term "return flow"—a term first used by Musgrave and Holtan (1964) to define water returning to the ground surface from beneath the soil. Today, we would call this groundwater exfiltration. But this combination of return flow plus the direct precipitation onto saturated areas was what the Dunne work showed to be important. And in the Northern Vermont context, as in many similar rolling upland humid catchments—especially with gentle slopes and wide valley bottoms—that saturation overland flow can be the main generator of storm and snowmelt runoff. Figure 3.8 shows a saturated area linked to a rising water table for one of the headwater forested catchments Sleepers River, a few catchments over from the agricultural catchment where Dunne did his pioneering work.

Dunne and Black (1970a, b) mapped the expansion and contraction of a headwater agricultural catchment at Sleepers River. I've walked this catchment[9]—or what's left of it following highway construction that now bisects it. It is gently rolling as Figure 3.9a attests. The beauty of the site (called WC-4) is that unlike neighboring forested systems as shown in Figure 3.8, it is was open, clear and with visible boundary conditions. Hence, the ability to walk the catchment; to see the saturated area of water exfiltrating to the surface; to map these zones with one's own feet and to see with one's eyes how they change dynamically through and between events.

[9] And indeed most of the sites I describe in this book.

Dunne and Black (1970a) noted (and Figure 3.9b shows), a perceptual model where:

> Flow generated by the mechanism described by Horton did not occur. Although soils and topography were those generally thought to be conductive to subsurface stormflow, the runoff produced by this mechanism was too small, too late, and too insensitive to fluctuations of rainfall intensity to add significantly to stormflow in the channel at the base of the hillside. When the water table rose to the surface of the ground, however, overland flow was generated on small areas of the hillside. Only when this overland flow occurred were significant amounts of stormflow contributed to the channel by the hillside. The return periods of storms that would produce such overland flow were found to be very large (Dunne and Black, 1970a, p. 478)

For natural and artificial sprinkling events, Dunne and Black (1970b) described the runoff from a short slope section above their trench as seen in the photo above. Figure 3.10b shows the hillslope flow from the trench and then streamflow response to one event; a 46 mm rain storm that lasted 2 hours. Their data shows how rain falling onto saturated areas dominates the storm hydrograph and how subsurface stormflow through the sandy loam soil of the overlying glacial till is a much smaller component of flow—at their two measured trough depths. Figure 3.10a shows the catchment's saturated area expansion for the event where the saturated area shown in black is the seepage zone around the channel prior to the event. The shaded gray zone is

Figure 3.8 Near stream saturated area from the forested W9 catchment near the original Dunne agricultural catchment. Photo taken during snowmelt and showing often the smearing between what is channel and what is saturated area; this section with <10 ha upslope contributing area.

Source: Photo by the author.

the expansion through the event—where the peak expansion equates to the peak runoff. Unlike the Hewlett perception of "channel expansions" coming from subsurface stormflow *exceeding the capacity of the soil profile to transmit it*, this saturated area at the low relief, glaciated, WC-4 catchment, with its short slopes, showed a new behavior not quantified before—actual flow over the soil surface.

In follow-up work at Sleepers and other sites in Ontario and Quebec, Dunne and his PhD students Colin Taylor and Tim Moore showed how the dynamics of saturated areas changed based on pre-event wetness and storm size. They developed relations between catchment saturated areas and discharge (Dunne et al., 1975). This relation would be followed-up by many authors in the decade

(a)

(b)

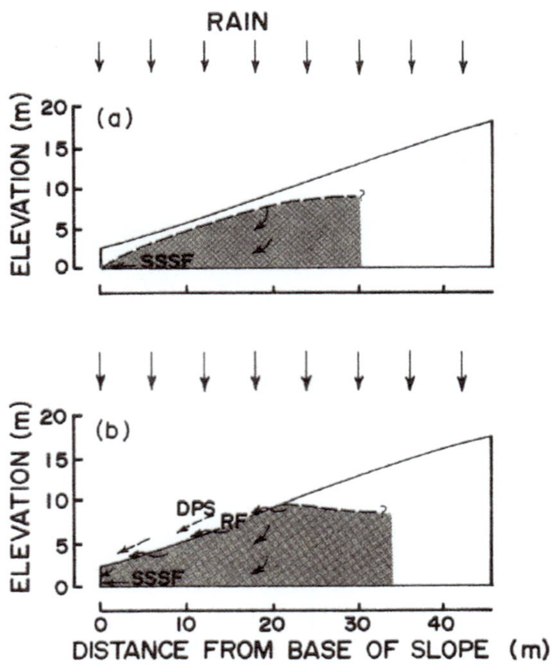

Figure 3.9 The Dunne site. (a) Very different slope lengths and drainage density to the Coweeta slopes and catchments; not to mention that the Sleepers site was glaciated and Coweeta was not. (b) Components of saturation excess overland flow.

Source: (a) Photo from Jamie Shanley, (b) from Dunne (1978)

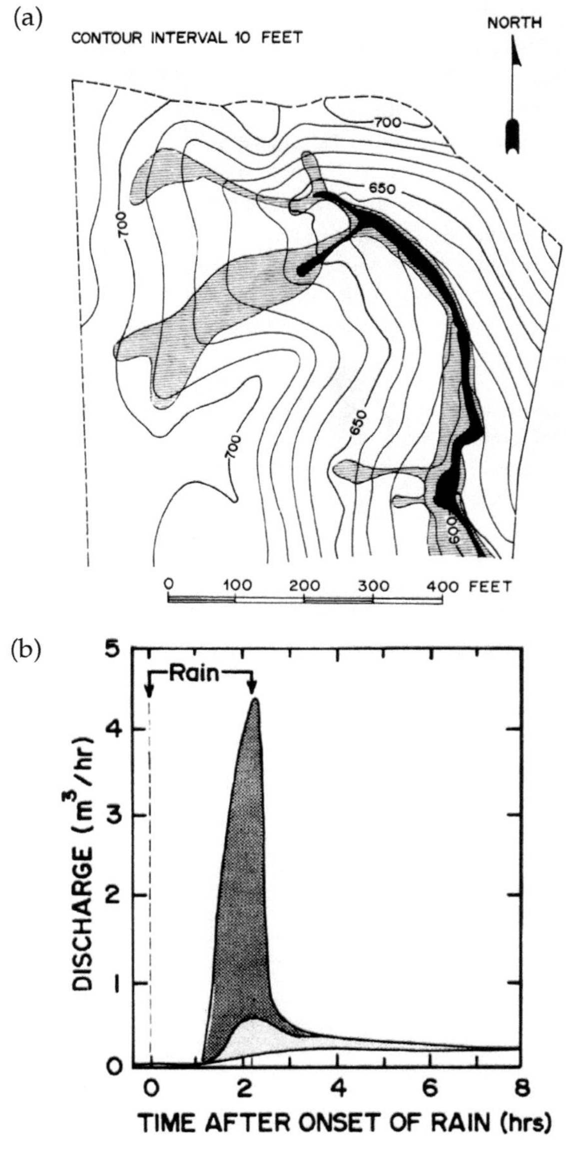

Figure 3.10 (a) Saturated area expansion during a 46 mm storm event for the Dunne and Black (1970b) WC-4 catchment at Sleepers River. The black area shows the pre-storm saturated area, the stippled area the storm-induced expansion and peak extent of saturation at the peak of the storm hydrograph. (b) A 44 mm rain event monitored by Dunne and Black (1970b) showing components of flow linked to saturated area expansion.

Source: Both figures from Dunne (1978)

following Dunne et al. (1975) (e.g., McDonnell and Taylor, 1987; DeWalle et al., 1988)—all using the *high tech* "squishy boot technique," whereby one walks the perimeter of the saturated area to delineate the line that separates surface saturation (from groundwater at the soil surface) and the unsaturated soil surface immediately upslope.

But it wasn't only Dunne and Black (1970a, b) who were mapping changes in near-stream contributing areas. Dickinson and Whiteley (1970) were doing important work in southern Ontario, Canada—another glaciated, low relief landscape—around this same time. They noted the importance of antecedent wetness for defining saturated area expansion and contraction and developed what they called a "basin moisture index." That work would anticipate additional important work in the late 1990s by Western and Grayson in Australia who measured soil moisture across a catchment and linked it with topographic position and threshold increases in runoff (as we will discuss in Chapter 6).

One extension of the largely North American saturation overland flow work of Dunne and his students (and earlier work by Cappus in France) is the work in tropical catchments that followed into the 1980s. In these environments—with extremes in rainfall amount and intensity—saturation excess overland flow was also reported on hillslopes; well upslope from the valley bottom sections seen in the more benign temperate upland environments. These tropical catchment studies and examples of a more hillslope expression of saturation excess overland flow would come from Northeast Queensland, Australia (e.g., Bonell et al., 1981), Panama (e.g., Dietrich, 1982), Western Amazonia (e.g., Elsenbeer and Cassel, 1990) and beyond. Bonell (1993, p. 234) summarized much of this in a landmark review paper, where: these "forested environments where an impeding soil layer occurs at shallow depth below highly transmissive surface layers" are a recipe for surface saturation.[10]

Recognition of the importance of saturation overland flow also continued in wetland-dominated environments with early work on *widespread surface saturation* in forested wetland-dominated

[10] A great review of the tropical hydrology literature some time later was by Elsenbeer (2001).

catchments in northern Minnesota (e.g., Verry and Boelter, 1975; Boelter and Verry, 1977), headwater swamps in tropical Africa (e.g., Balek and Perry, 1973) and small wetland-dominated catchments in eastern Massachusetts, USA (e.g., O'Brien, 1977). Taylor (1982, p. 166) noted extreme seasonality in such wetland environments—"when water levels are high, considerable quickflow can be produced, but during the dry seasons when water levels are low, there may be no response at all"—all linked to the presence or absence of surface saturation near the stream channel.

We've discussed the saturated areas and rain falling or snow melting onto them and running off. But what about the groundwater below and *its* contribution to the stream?

Another landmark study was made in Vermont around the time of Dunne's work; by Ragan (1968) ~100 km away from Sleeper River in Burlington, Vermont, USA. Ragan made careful and unusually detailed measurements along a 188 m length section of a second-order stream draining a 46 ha portion of forested catchment. The glacial deposits at Ragan's site were quite different from Dunne's glacial deposits at Sleepers; rather than glacial till with mineral soils above, the Ragan (1968) site included a 24 m deep deposit of uniform glacial outwash. The part of the Ragan (1968) finding that was so important was the fact that the stream flow response was dictated by a very small area around the channel margin—largely fixed spatially and not expanding and contracting as Dunne had shown at his site. Why might this be, you ask? Groundwater would be the answer.

Ragan (1968) took note of the area of the catchment generating runoff and reported that:

> after subtracting the rainfall interception and measuring the changes in soil moisture storage, it was found that the contributing area ranged between 1.2 and approximately 3% of the total watershed area (Ragan, 1968, p. 246)

This was different to the Sleepers River site (and the other examples above) where saturated areas would grow, at least for Sleepers, up to 50% of the catchment area. Ragan found that water tables in areas within ~2 m of the stream channel responded very quickly to storm rainfall. He would go on to explain this mechanistically (and show it diagrammatically in Figure 3.11b) where:

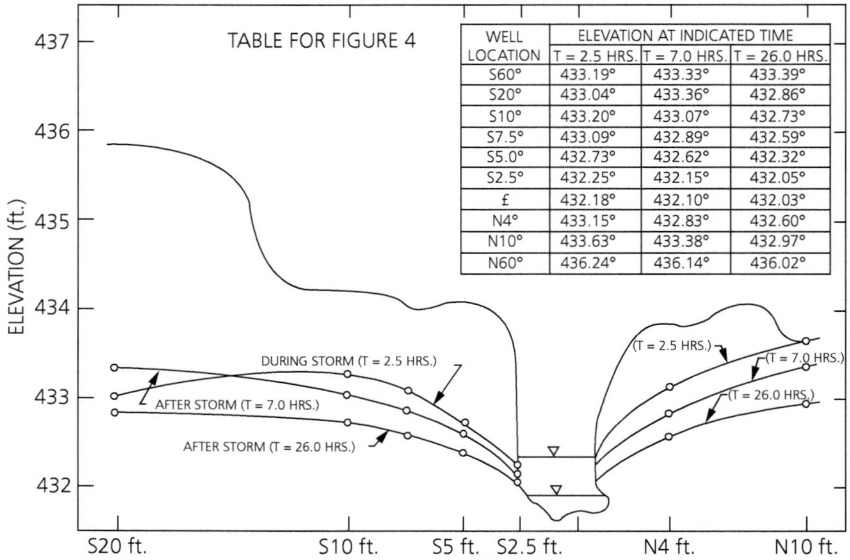

WELL LOCATION	ELEVATION AT INDICATED TIME		
	T = 2.5 HRS.	T = 7.0 HRS.	T = 26.0 HRS.
S60°	433.19°	433.33°	433.39°
S20°	433.04°	433.36°	432.86°
S10°	433.20°	433.07°	432.73°
S7.5°	433.09°	432.89°	432.59°
S5.0°	432.73°	432.62°	432.32°
S2.5°	432.25°	432.15°	432.05°
£	432.18°	432.10°	432.03°
N4°	433.15°	432.83°	432.60°
N10°	433.63°	433.38°	432.97°
N60°	436.24°	436.14°	436.02°

Figure 3.11 A 1.10″ rainstorm on Sept 4, 1966 showing water table fluctuations from Ragan's so-called "piezometer cross-section." Note the vertical exaggeration where for horizontal 1″ = 4′ and for vertical 1″ = 1′.

Source: Ragan (1968).

a ridge in the ground water table, similar to that observed under infiltration basins, formed along the length of the south side of the stream . . . the ground surface breaks fairly sharply approximately 15 feet from the stream. Near the stream the groundwater table is within two feet of the ground surface while at sixty feet from the stream the ground water is 8 feet below the ground surface. In the area of the high ground water table a small quantity of rainfall infiltrated the soil and converted the capillary fringe into a saturated zone and produced a very rapid, but localized, response in the ground water table . . . as the rainfall decreased this ridge drained and produced a more typical groundwater table profile. The groundwater table away from the stream did not respond as rapidly because of the larger depth of overlying soil. (Ragan, 1968, p. 247–249)

What Ragan is getting at here are the very sensitive dynamics of the water table due to the fine, uniform sandy material with a large capillary fringe. I'll have much more to say about this in Chapter 5. But for now, recall that a capillary fringe is a zone of tension saturation. The finer the material, the thicker the capillary fringe. So, in some environments, saturation overland flow can be dominated by rain falling on saturated areas and contributing to the storm hydrograph (as shown by Dunne and others) and where exfiltrating groundwater is a *tiny portion* of the total saturation excess response. While in other places with fine sand and shallow water tables (before the event commences), the ridging and exfiltration of groundwater can be the *main generator* of storm runoff and where surface runoff over saturated areas is small and restricted to a tiny percentage of the catchment area.

3.4 Reconciling the original variable source area concept with saturation overland flow

The mind, once stretched by a new idea, never returns to its original dimensions

—Ralph Waldo Emerson

One might wonder how Hewlett viewed the adjustments of his variable source area concept by people like Dunne and Ragan. I came across a letter exchange between Hewlett and the then Editor-in-Chief of the journal *Water Resources Research,* Luna Leopold. These letters (a snippet of one, shown in Figure 3.12) are fascinating. It hits on many of the themes of the day and things we will discuss in the next chapter: terminology, confusion on forms of subsurface stormflow, different views of catchment hydrology by engineers and foresters, etc. . .

In one of the other letters in this set of exchanges, Hewlett reflects on Dunne's work at Sleepers River in Vermont and notes that:

Dunne has misled himself by a study of a small piece of a basin, in well-drained sand, and under rather small rainstorms. His data is interesting because it supports rather than refutes the variable source concept. He was, in effect, dealing with a special case of a variable source area, which he and Ragan have chosen to call the partial source area. When the contradictions in his paper relating to subsurface stormflow, overland flow and interflow are accounted for, the phenomena Dunne describes fits the variable source area concept exactly. All he needs is more rainfall per hour or per day and he would see his partial source area become a variable one! (from a later 1970 letter from Hewlett to Leopold)

This is an extremely important point of clarification made by Hewlett to end upon. Because if you read the Dunne and Ragan papers, they do in fact refer to "partial areas" and not variable source areas per se. Again, the climate, geology and topography of North Carolina where Hewlett developed his concept vs. the glaciated terrain where Dunne and Ragan completed their work was quite different. In many ways, I might agree with Hewlett's assertion that had event size and time allowed, hillslope contributions of subsurface stormflow may have occurred (and indeed, others have observed this in neighboring forested headwater catchments at Sleepers (e.g., Kendall et al. (1999) and McGlynn et al. (1999)). We will explore how storm size and hillslope length conspire to activate subsurface stormflow at the hillslope scale when we get to our more modern views of hillslope scale runoff in Chapters 4 and 6.

Figure 3.13 shows a now-famous diagram from a 1970 paper by Hewlett and Nutter (1970, p. 78) where they update a little bit the nature of near stream response of the variable source area concept. They note that "infiltration is not usually a limiting process, that overland flow exists only to the

THE UNIVERSITY OF GEORGIA
SCHOOL OF FOREST RESOURCES
ATHENS, GEORGIA 30601

April 3, 1970

Dr. Luna B. Leopold
U. S. Geological Survey
Department of the Interior
Washington, D. C. 20242

Dear Dr. Leopold:

In response to your letter of March 27, 1970, here is a copy of a paper entitled, "The relation of the variable source area concept to hydrologic modeling", prepared for a Clemson University meeting by Wade L. Nutter and myself. Ray K. Linsley was there and, while he felt that the Stanford Watershed Model was satisfactory for most practical purposes, I believe that he did concede that there was something to the variable source idea and that possibly general river basin models could be altered to accommodate subroutines that account for a variable source area in the headwaters.

Dr. Linsley complained of a communication gap between foresters and engineers with reference to hydrologic terminology. I can only say amen to that. I have been much concerned with the rather loose terminology that we hydrologists permit ourselves in scientific writing. The definitions of "overland flow", "surface runoff", "interflow", and "underflow" are so variable that the words seem almost useless at times. We reserve the word "underflow", for example, to mean the underground movement of water parallel to the stream and usually in valley bottom gravels and sediments. This agrees with ASCE "Nomenclature for Hydraulics" 1962 manual or glossary of terms. "Interflow" I have never seen clearly defined in any glossary, nor do the agricultural people use it in a consistent way. The only meaning I can get from it generally is "any non-vertical movement of water below the soil surface but above the general water table."

I believe that you are correct in saying that USDA hydrologists tend to cling to the subsurface piping-system type of flow (interflow, underflow, whatever, but implying rapid delivery of water down the slopes through old root holes, animal burrows, cracks, etc.). I believe Dr. Charles Hursh first described subsurface flow that way when he was chairman of the AGU Committee on Subsurface Flow back in the 1940's. Ronald Whipkey (Forest Service) has pushed this view of matters but from sprinkler plots only, and without reference to actual watershed rainfall-streamflow records.

Figure 3.12 One of many letters between Hewlett and Leopold. This correspondence was in 1970 given to me by Wade Nutter who received it from Bob Curry at Montana when Bob was working on his Ph.D. ~1975.

extent implied in the term over-water flow or 'channel precipitation.'"[11] This is their capitulation to the findings of Dunne and others. Although they do go on to say that "in essence there are two pathways to evaluate, channel precipitation and subsurface discharge through the wetted perimeter of the channel and its storm period extensions." (Hewlett and Nutter, 1970, p. 78).

[11] And for those interested in more detail, this notion of channel precipitation was first quantified in Hursh and Brater (1941)—a paper we discussed in Chapter 2.

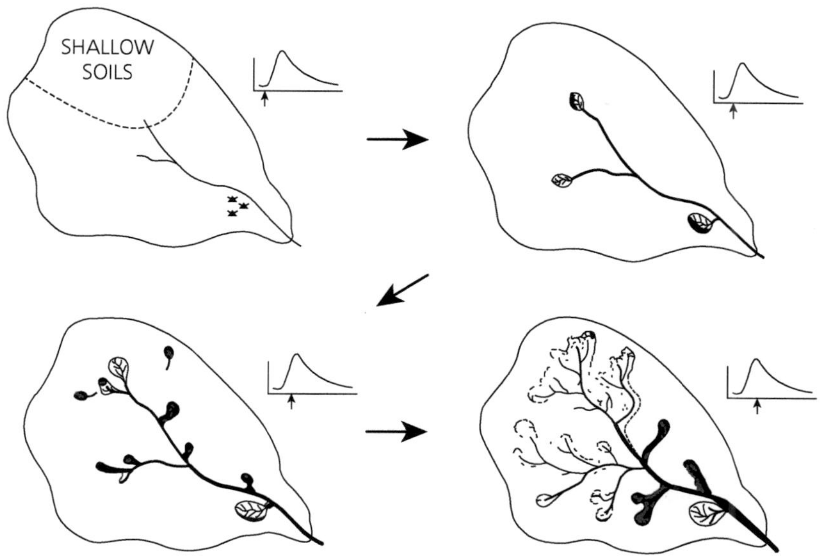

Figure 3.13 Hewlett and Nutter's "time lapse view" of a catchment showing their idea of source area expansions vis-à-vis the channel system during a storm.

Source: Hewlett and Nutter, (1970).

3.5 Summary and looking ahead

> *To acquire knowledge, one must study; but to acquire wisdom, one must observe*
>
> ***—Marilyn vos Savant***

The variable source area was a radical new concept in catchment hydrology that posited that water comes to the surface and the channel length grows when subsurface stormflow from upslope exceeds soil profile's capacity to transmit it. New concepts are uncommon in the hydrological sciences. But the variable source area concept of Hewlett and Hibbert (1967) was indeed that. Hewlett had an uncanny knack for compact organization of empirical data and observations of catchment responses that facilitated prediction of catchment behavior in different places. But even Hewlett's ideas morphed as more places were viewed through the lens of the variable source idea—expanding the concept to include more explicit recognition of near stream saturated area dynamics and the generation of saturation overland flow.

One might say today that the variable source area concept was disruptive science—the biggest contribution to catchment hydrology up until that time and a revolution in many ways, related to systems thinking. Bonell (1993) said it best when he noted that:

> by the late 1970s, acceptance of the Hewlett variable source area concept of runoff generation was universal, but the research momentum in hillslope hydrology has continued because a comprehensive understanding of process mechanisms (or delivery mechanisms) of storm runoff generation still remains elusive when referring to the headwaters of drainage basins (Bonell, 1993, p. 219)

By this, Bonell means that while the framework is solid, the details are exceptionally challenging given the multitude of conditions—topography, soil type, soil depth distribution, underlying geology, climate dynamics, etc.—that express themselves at any given site. This was true then. It is still true today. In terms of thinking about how, where and why saturation overland flow might dominate, Dunne (1983) summarized it nicely in Figure 3.14 where:

> saturation overland flow tends to dominate storm runoff in thickly vegetated landscapes with thin soils, high water tables, and long gentle concave hillslopes. Subsurface

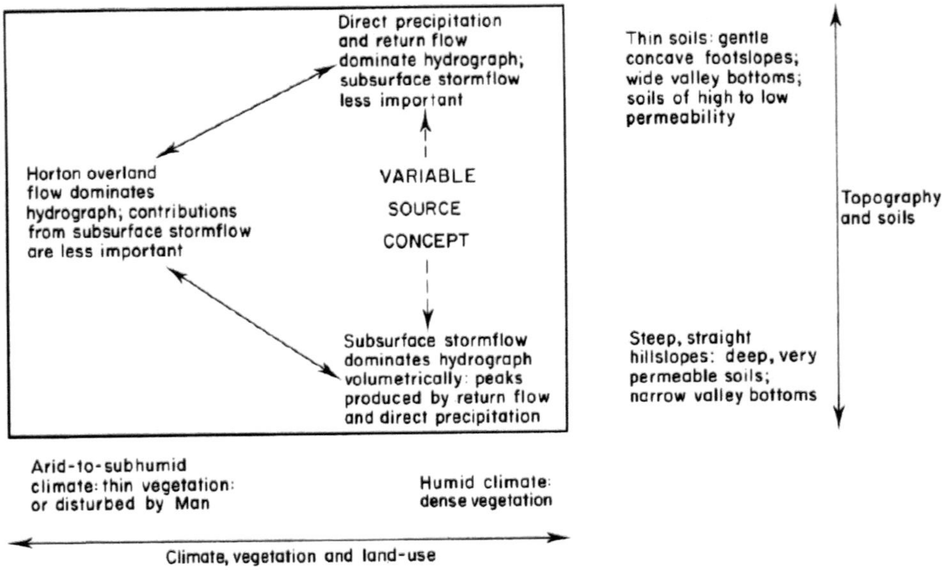

Figure 3.14 The three main runoff processes at the headwater catchment scale. based on what we have discussed so far.
Source: Original diagram by Dunne (1978).

stormflow in these regions may still be important in contributing to the recession limbs of hydrographs from small basins as well as affecting landsliding, soil drainage, and water chemistry. Thus it is necessary to combine the study of subsurface flow and saturation overland flow. (Dunne, 1983 p. 29)

Despite the huge significance of the variable source area concept, we will explore in Chapter 6 some new thinking that goes beyond the variable source area concept, linked to thresholds, connectivity and losses along flowpaths. There are still open questions that relate to the extent of the variable source area concept's applicability. We will see new ideas in Chapters 6 and 7 for integrated catchment response to precipitation inputs, and in Chapter 9, on how to diagnose how catchments store and release water.

Finally, in a recent summary of 400 global research catchments, McMillan et al. (2025) tested the Dunne and Leopold diagram shown in Figure 3.14 to see how our catchment observations match this summary. Quoting from McMillan et al. (2025, p. 500), they found that:

- arid watersheds tend to favor the shallowest (surface) or deepest (groundwater) processes, whereas humid watersheds favor mid-depth processes (soil drainage and subsurface stormflow)
- arid climates have a greater fraction of infiltration excess processes and a lesser fraction of saturation excess processes compared with humid climates
- the prevalence of subsurface stormflow varies little with aridity class, except for the most humid watersheds where a higher incidence of subsurface stormflow process is recorded.

And contrary to Figure 3.14, McMillan et al. (2025, p. 500) found that:

- saturation excess processes are more frequently described than infiltration excess processes for all aridity classes of research watersheds. These results could occur because of a myriad of controls on saturated areas—for example, saturated areas can develop because of ponding in surface depressions and topographic convergence of water, even

when the water table is well below the ground surface
• in humid watersheds, Dunne hypothesizes that thinner soils and/or gentler slopes favour surface flow, while deeper soils and/or steeper slopes favor subsurface flow. In our database, slope and soil depth show a negative correlation for individual watersheds, and we therefore analysed these variables individually. However, our results showed weak or no trend of subsurface stormflow prevalence with soil depth or slope.

We will now go beyond the Dunne and Leopold diagram process summary, and next examine the radical improvements in subsurface stormflow thinking in the period 1965–1974 in Chapter 4—improvements that continue to this day.

CHAPTER 4

Subsurface stormflow: Process advancements (1965–1974)

4.1 Subsurface stormflow revisited

What is not fully understood is not possessed
—Johan Wolfgang von Goethe

New research and new understanding of factors affecting subsurface stormflow exploded during the International Hydrological Decade (1965–1974)—a period that by some estimates resulted in the selection and gauging of >1000 headwater "representative and experimental" catchments worldwide. Before we start to look at subsurface stormflow, we must have a word on terminology. Papers in this early literature often used multiple, alternate words to describe subsurface stormflow, even in the *same article* (as noted by McMillan, 2022)! It's sad, because as we saw in Chapter 2, Hursh (1936) had already defined the term subsurface stormflow. But alternative descriptions—sometimes with a different intent—existed in this period 1965–1974, and to this day. As we noted in Chapter 2, Horton often referred to subsurface stormflow as "concealed runoff" although Horton's thought was that it is very shallow. Some papers through the 1960s spoke of "throughflow," defined by Kirkby and Chorley (1967, p. 7) as "flow which takes place physically within the soil profile." Other terms like "interflow," defined by Betson as a type of lateral flow that "often sinks in or infiltrates" (Betson et al., 1968, p. 604), date back even earlier. Subsurface stormflow though is the best aggregate term for all subsurface processes[1] that produce "stormflow"; and the latest version of the Dictionary of Geology (Kearey, 2001) now even includes the same, interchangeable definition for subsurface stormflow, throughflow and interflow.[2]

Subsurface stormflow via translatory flow was at the core of the variable source area concept. But after the saturation overland flow work by Dunne and Black (1970) and Ragan (1968), many started to ask: *is* subsurface stormflow important to the variable source area concept, given that for some areas, saturation overland flow explains all or most of the quickflow into the stream during an event? A review paper on progress in the International Hydrological Decade fueled doubt on the relevance and importance of subsurface stormflow. The paper was by Freeze (1974) and he reviewed "the ways in which water moves from hillslopes into small stream channels during and between rainfall events in upstream tributary basins" (Freeze, 1974 p. 627). In that review paper, Freeze questioned the importance and even the existence of subsurface stormflow in its connection with the storm runoff to the stream:

The prime requirement for subsurface storm flow is a shallow soil horizon of high permeability at the surface, and there is considerable evidence to support the

[1] Most senior hillslope hydrologists who I polled in the writing of this book would agree.

[2] Subsurface stormflow is the original term and one that for me captures all in a simple, definable and single term. Just like we do not add distinctions to infiltration excess overland flow or saturation excess overland flow as to whether or not it connects to the channel, I don't think we need to over-muddy the waters by adding other monikers to subsurface stormflow to define whether or not it reaches the stream in time to contribute to the storm hydrology.

Streamflow Generation. Jeffrey J. McDonnell, Oxford University Press. © Jeffrey J. McDonnell (2026). DOI: 10.1093/9780191953736.003.0004

common occurrence of such a layer. What is in doubt is whether it can deliver sufficient water to provide significant contributions to storm runoff. The quantities measured by Hewlett and Hibbert [1963] and Whipkey [1965] were small, and the authors did not relate the outflows to the channel hydrographs (Freeze, 1974, p. 632)

I can only imagine all the authors from Chapter 2 saying "what about my work?!" Freeze was heavily influenced by Dunne's work in Vermont and in particular the statement by Dunne and Black (1970, p. 478) regarding subsurface stormflow as being "too small, too late, and too insensitive to fluctuations of rainfall intensity to add significantly to stormflow in the channel at the base of the hillside"—a statement that we explored in the last chapter.

Of course, that *was* true for the small storm sizes and exceptionally short slopes monitored in the catchment at Sleepers River. But Dunne and Black were not making a universal statement about *all* places. And despite Freeze's review, there were indeed many studies completed around the end of the International Hydrological Decade that had shown clearly and conclusively the subsurface stormflow mechanics and its dominance of channel stormflow during events. One such catchment, that you will hear more about in the coming chapters, is the HJ Andrews catchment in Western Oregon. There, Harr (1977), examined 7 large rainstorms (40–183 mm) in a small headwater catchment. He found that quickflow ratios were very high, ranging from 23–51% and that subsurface stormflow as a percentage of that quickflow was more than 96% for each event! So, if one ever needed a clearer exposition of subsurface stormflow contributing directly and unmistakably to the "outflows" (i.e., the stream hydrograph), this is it!

So how could the field-defining review paper of Freeze have been so misguided?

Well, some of it gets back to themes at the end of the last chapter: place and event size. Let's look at Figure 4.1. It shows a plot of the percentage of net rainfall (i.e., the rain that reaches the soil surface as opposed to the gross rainfall above the forest canopy) plotted against quickflow[3] for

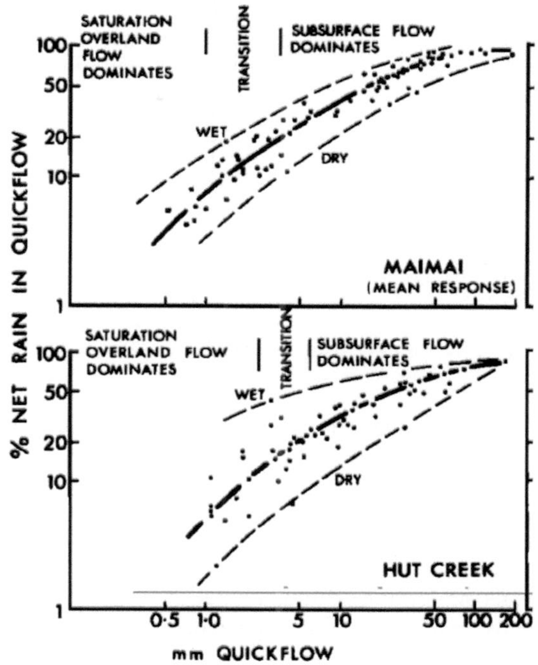

Figure 4.1 Plots of "quickflow as a proportion of storm rainfall" and "total quickflow" for two catchments, linked to the perceptual model of runoff generating processes.
Source: Pearce and McKerchar (1979).

two steep, wet catchments on the South Island of New Zealand. It shows that for small events, saturation overland flow indeed dominates total quickflow. But as the storm size increases, each of the catchments goes through a transition to subsurface stormflow dominating quickflow. In other words, subsurface stormflow is only dominant *after* sizable wetting and usually at and after the peak of the storm hydrograph.[4] This holds often for both dry and antecedent wetness conditions, but, as you see if you look carefully at Figure 4.1, wetness before the event also greatly impacts these relations.

So, we're seeing something in Figure 4.1 about how event size and antecedent wetness influence the activation of overland vs. subsurface stormflow. We will now look at refined process descriptions of subsurface stormflow that developed, following

[3] Remember, quickflow is the *stormflow portion* of the stream hydrograph.

[4] Something we will unpack in much greater detail in Chapter 6.

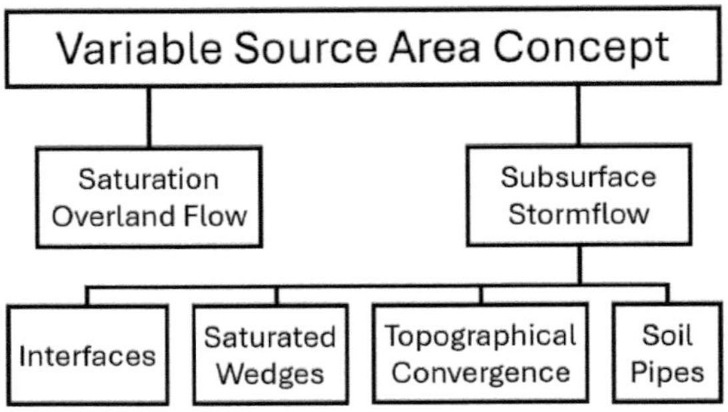

Figure 4.2 The main subsurface stormflow advancements in the few years prior and through the International Hydrological Decade.

Hewlett's ideas on translatory flow.[5] Figure 4.2 shows the four main subsurface stormflow process developments that came through the International Hydrological Decade:

- recognition of subsurface interfaces
- recognition of saturated wedges
- recognition of the importance of topographic convergence
- recognition of how high rates of subsurface stormflow could be aided and abetted by soil pipes (Figure 4.2).

4.2 The role of interfaces

Life is short, the art long, opportunity fleeting, experiment treacherous, judgment difficult

—Hippocrates

One of the first factors to be clearly identified and described mechanistically for how subsurface stormflow is generated was subsurface interfaces; meaning breaks in, or sharp reductions in, saturated hydraulic conductivity with depth. During rainfall or snowmelt events these reductions can result in temporary saturation (even if only a few centimeters deep) above that interface with subsequent downslope flow. This temporary saturation is key because of the extreme increase in the hydraulic conductivity and thus flux rate when the soil gets saturated, as we already mentioned briefly in Chapter 3.[6] In some ways, these and the subsequent studies that we will examine throughout this chapter, show that this saturation is not only important for subsurface stormflow, saturation is a *precondition* for subsurface stormflow.

Figure 4.3 shows the importance of these transient saturated zones (sometimes called perched water tables or lenses of saturation). It shows clearly the many changes with depth for permeability, soil water tension, percolation direct, etc. Note the horizontal dashed line that shows the interface. The faint dotted line above it is the capillary fringe where the soil may be very close to saturation but where there is some soil tension—i.e., the tension-saturated zone.[7] The soil profile becomes gradually less permeable with depth with a steep reduction at the interface. The soil water content (shown as percent saturation) increases downwards because of "steady percolation through the soil column" (Whipkey and Kirkby, 1978, p. 127) and there is a rapid reduction in tension (i.e., the soil suction or matric potential) toward the interface.[8]

[5] Here you might ask: But hey wait a minute, subsurface stormflow via translatory flow was at the core of the variable source area concept?! After the saturation overland flow work by Dunne and Black (1970) and Ragan (1968), Hewlett conceded that "translatory flow was merely a suggested mechanism by which the variable source concept may be working in some portions of the basin, not the concept itself" (Hewlett, pers. comm. to Luna Leopold, 1970 in another letter following the one we examined in Chapter 3).

[6] Recall too from your undergraduate hydrology class the highly non-linear relationship between water content and hydraulic conductivity: highest when saturated but often with a steep reduction as the water content is reduced.

[7] We will examine these relationships in detail in Chapter 5.

[8] Again, we will examine in Chapter 5 why this sharp shift in tension and some of the other differences between the dotted and dashed line in this figure as it relates to the capillary fringe.

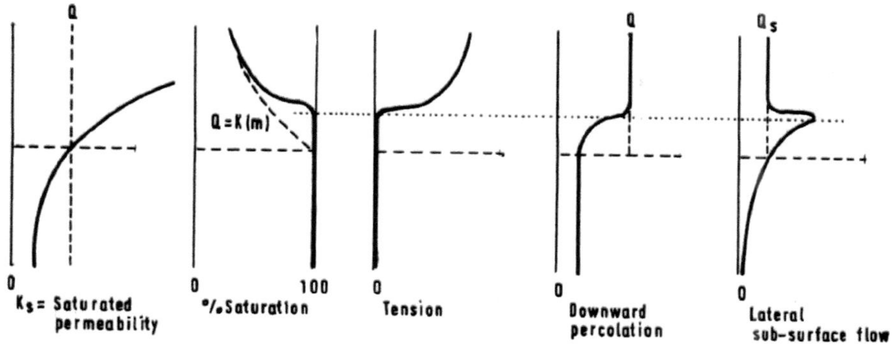

Figure 4.3 A schematic distribution of saturated permeability (saturated hydraulic conductivity), soil moisture (as percent of saturation), tension, downward percolation and subsurface stormflow for a soil which gradually becomes more impermeable with depth.

Source: Figure and quote from Whipkey and Kirkby (1978, p. 128).

The aggregate behavior of these changes at the interface are shown for downward percolation and subsurface stormflow in Figure 4.3. It also shows how water can get diverted laterally when the vertical infiltration is interrupted at the interface. Why is this important? Because even on the steepest of slopes, the direction of water movement in the unsaturated zone is vertically downward unless there is some sort of impeding layer to force it to move laterally. And the real effect of interfaces in the context of a rain or snowmelt event is to promote transient saturation, and in so doing, inducement of lateral flow.

So, interfaces are important. The paper that really revealed this was Whipkey (1965) where he first described in detail—with field measurement at an interface—how they influenced subsurface stormflow production and outflow seepage. Whipkey worked on an oak woodland hillslope in Ohio, USA. Figure 4.4 shows how he installed five troughs at depths up to 150 cm where the uppermost trough captured overland flow (over the leaf litter).

Let's now dissect these subsurface stormflow responses.

Figure 4.4b shows the outflow from Whipkey's troughs at the five different depths. The rain event is a 102 mm storm in 2 hours occurring on "dry" antecedent conditions. What do we see? Firstly, in this forested plot, there is precious little overland flow. This is to be expected—even with such an intense storm—since forest soil infiltrabilities are, unless disturbed, almost always higher than rain intensities. But look at troughs 2 and 3. What do you think is happening here? I'll wait a few minutes for you to look and think. So, what did you think of? Yes, the hydrograph for trough 2 is quite spikey with a similarly steep rise and fall. Such a steep rise and fall is usually seen in infiltration excess overland flow where everything runs off the surface. Think of flow at an interface as something akin to overland flow at the soil surface, which *itself is an interface*.[9] Now tell me about trough 3. Why the longer recession? Because, it is likely still being supplied by water from above.[10]

Figure 4.4 is a useful metaphor for the catchment response too. Flow at the catchment outlet, like flow at Whipkey's trench face, is a delicate balance of input rates, conductivity breaks and sequencing of flow from different zones in time and space. But unlike the Whipkey trenches, all we see is the integrated response at the catchment outlet: the storm hydrograph. If we had a magic way of showing the "hydrographs" for all the sub-catchment positions—near stream zones, hillslope hollows,

[9] Recall Horton calling the soil surface a "separating layer"—these subsurface interfaces can also be separating layers.

[10] This effect would come into sharper focus a few years later with the work of Betson where he noted that "the fact that interflow often sinks in or infiltrates where saturated interflow originating in shallow upslope soils is often absorbed in deeper down slope soils" (Betson et al., 1968, p. 604).

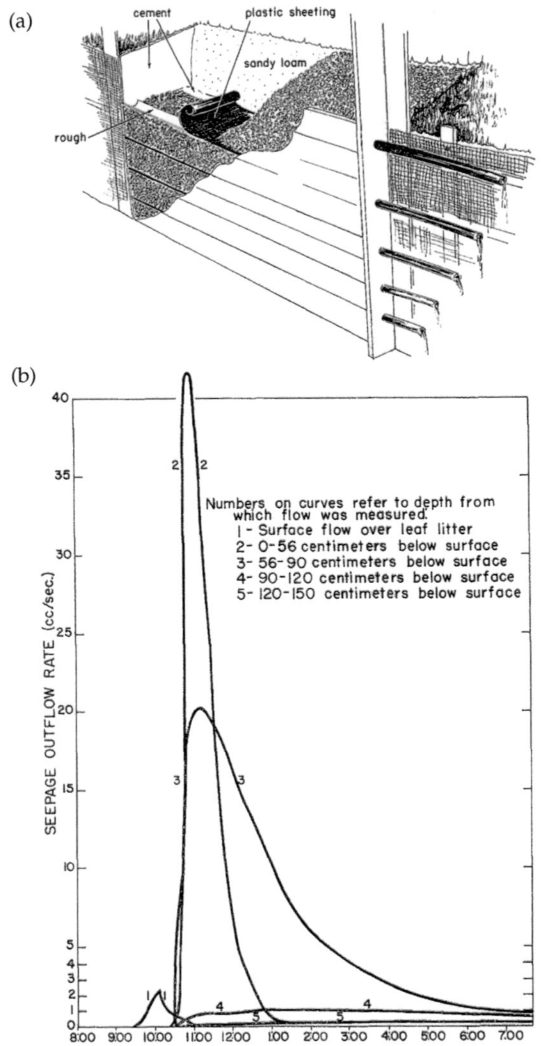

Figure 4.4 (a) Throughflow trough construction from Whipkey (1965) and re-drawn from Kirkby (1978), showing the sheet and guttering. (b) Subsurface stormflow hydrographs from a slope section that was 17 m long by 2.4 m wide for a 102 mm rain event that occurred within a 2 hr period with so-called dry antecedent conditions. Note that for a 32 mm event in 1.67 hr with so-called wet antecedent conditions for this same plot, the hydrograph responses were very different. (And not only were the trough responses different in magnitude for this other event, they behaved differently to one another compared to this event shown in the figure! I encourage to go and read the original Whipkey (1965) paper to learn more. It is one of the truly first humbling examples of complexities of subsurface stormflow—where, just when response for one set of conditions is understood, another set of conditions at the same site produces radically different responses.)

Source: Whipkey (1965).

any zones of low infiltrability, side slopes—then we'd likely see this symphony of response.[11]

I find the main results figures from Whipkey one of the most enduring and illustrative figures from early process literature. It shows the manifold subsurface stormflow problems and interactions we still face today: that different rain amounts, intensities and durations can result in very different subsurface stormflow responses and that antecedent wetness further conspires with event rainfall conditions to create very different response characteristics, even for the same site. Of course, much of this comes down to the interfaces revealed at Whipkey's trough boundaries; abrupt changes in matrix hydraulic conductivity at a soil textural or horizontal interface.

Following the Whipkey work, many would lament on the challenges of getting to grips with the controls on subsurface stormflow:

> Considering these findings, the reason that projecting runoff data from one area to another is difficult becomes apparent. No matter how similar two areas may appear on the surface, variations in the composition and depth of various soil horizons can occur that may markedly influence how any particular area within the watershed contributes to storm runoff. Clearly techniques for accounting for these soil variations must be devised before runoff data can be meaningfully projected (Betson and Marius 1969, p. 581)

Betson and Marius (1969) make a good point here: that one trenched hillslope or soil pit does not a catchment make.[12] Nevertheless, countless examples of conductivity contrasts producing event-based lateral flow would follow Whipkey's work. Some striking examples would come from the tropics[13] where the combination of high rainfall intensities and durations (associated with tropical cyclones) together with soil profiles with distinct

[11] Note that Kendall et al. (2001) have done this for the artificial, 490 m^2 Hydrohill catchment near Nanjing, China—see McDonnell (2023) for review of this site.

[12] To you non-native English speakers, this is a turn of phrase used in poetry, e.g., one well known from Richard Lovelace in the seventeenth century: "Stone walls do not a prison make / nor iron bars a cage." Who said that book writing was not fun?!

[13] Until recently, a highly under-represented geographical area for the study of hydrological processes, save for the early important contributions of people like Mike Bonell and Sampurno Bruijnzeel.

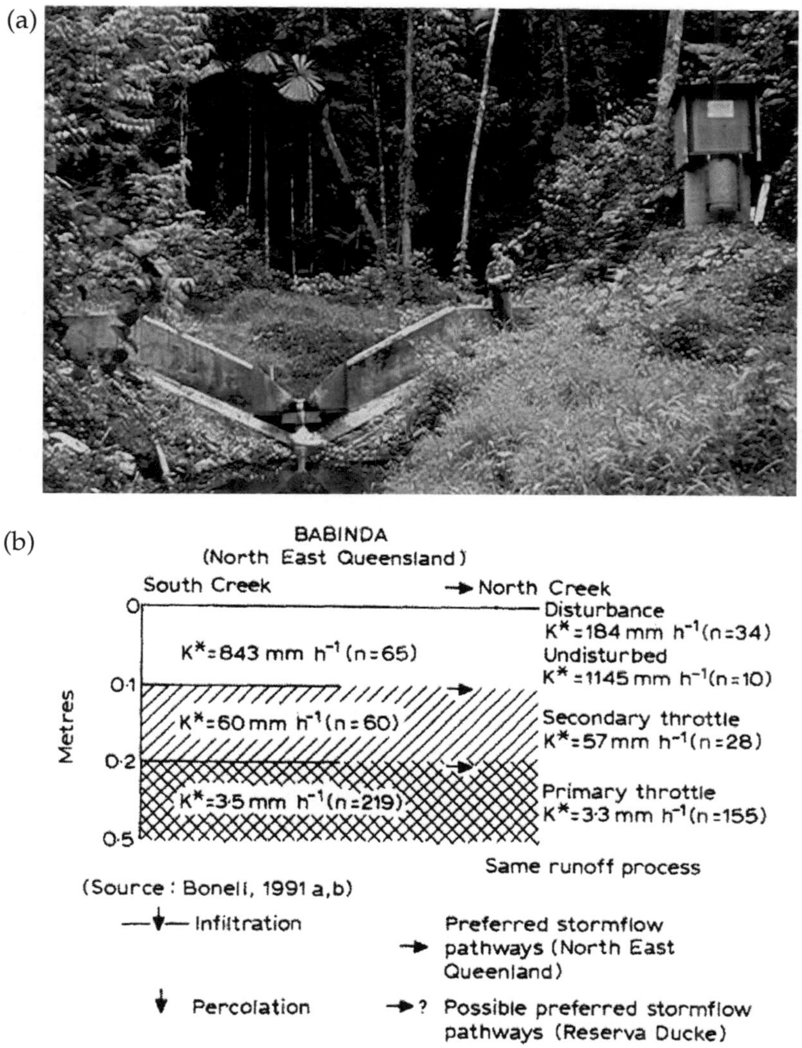

Figure 4.5 The South Creek experimental catchment near Babinda, Australia. This site and the neighboring North Creek experimental catchment were the first research catchments established in a "cyclone-prone part of the humid tropics." They affect areas of Australia, the Caribbean, the western Indian Ocean and The Philippines. Three soil layers with different hydraulic conductivities form throttles for lateral flow at the interfaces. *Source*: Photo from Chappell et al. (2012). Figure from Bonell (1993).

hydraulic conductivity layers is the norm. Mike Bonell—a hero of tropical forest hydrology through the 1970s—worked at a site in Australia called Babinda (Figure 4.5). There, conductivity breaks at three main soil horizons led to subsurface stormflow production[14] (Figure 4.5b).

[14] A very good review of all this tropical subsurface stormflow literature can be found in Elsenbeer (2001)

Again, in the Babinda example and nearly all cases of flow at an interface, the words of Betson and Marius are very pertinent vis-à-vis interfaces:

> As these soils become saturated, runoff begins in the form of both saturated interflow and surface runoff. Whether this runoff reaches the flume, however, appears to depend upon the capability of some downslope soil to reabsorb this runoff. (Betson and Marius, 1969 p. 581)

We will return to this statement in Chapter 6 as this loss along the subsurface stormflow pathway is only again recently a topic of much discussion. But key in all this analysis of interfaces is the permeability difference above and below the interface. Arnett (1974), in his piece toward the end of the International Hydrological Decade titled "Environmental factors affecting the speed and volume of topsoil interflow," goes on to say that "from this tabulation, only permeability difference emerges as generally significant, lateral seepage increasing with contrasts in permeability" (Arnett, 1974, p. 20).[15] But in the years that followed, there were other phenomena that would emerge: the next being the recognition of saturated wedges linked to the interface itself.

4.3 Saturated wedge development

It is through science that we prove, but through intuition that we discover

—Henri Poincaré

The next big development in understanding subsurface stormflow after clear-eyed understanding of the importance of interfaces was the recognition of saturated wedge development and the importance of transient saturation for inducing lateral subsurface stormflow. This is effectively taking a more holistic view of the slope (beyond an isolated pit section from studies like Whipkey) on how an interface relates to the development of flow from ridgeline to toe slope.

Where did we first learn of saturated wedges?

In some ways, saturated wedges were already seen in the Hursh diagrams of Chapter 2 and in the original variable source area diagram of Hewlett in Chapter 3. But the place where saturated wedges were really shown clearly for the first time was at East Twin Brook in Southern England. Daryl Weyman did his PhD there (supervised by Mike Kirkby at Bristol University). Weyman (1973) instrumented the slope shown in Figure 4.6 with tensiometers and calculated flow directions and flow lines. The ephemeral stream down the small gully in Figure 4.6 was gauged and he linked this to the subsurface stormflow response at the interface with his noted "impermeable" bedrock below. If we hear from Weyman directly, he states that:

> Measurements have been made of unsaturated and saturated lateral soil water flow on a convex hill-slope with a good soil cover and impermeable bedrock during natural rainstorms. The hydraulics of flow are examined in detail with particular reference to the role of breaks in vertical permeability, the change from saturated to unsaturated flow and the velocity of flow. In this instance, after rainfall, slope flow is dominated by vertical unsaturated movement towards the profile base. Preceding upslope moisture gradients result in the growth of a zone of soil saturation upwards from the slope base. (Weyman, 1973, p. 267)

It is this "upwards from the slope base" that is the important bit here. This is effectively the saturated wedge that we are exploring in this this section. Weyman goes on to say:

> Slope discharge, through the B and B/C horizons, is related to the form of the saturation zone, within which flow is lateral, according to Darcy's law. The time required for vertical percolation and the low hydraulic conductivity of the lower soil horizons result in a hillslope hydrograph which is delayed and attenuated and cannot be regarded as stormflow. During drainage the saturation zone contracts and is replaced by a lateral unsaturated flow system at the profile base which supplies discharge from the B/C horizon for up to 42 days without further recharge. It is concluded that, in general, either distinct soil horizons or impermeable bedrock are essential for the initiation of lateral flow. (Weyman, 1973, p. 267)

So Weyman is saying that interfaces are "essential for the initiation of lateral flow" but noting that the saturation that forms at them, expands upslope, like a subsurface variable source area. Weyman makes two key points:

> • There is no evidence to suggest that lateral unsaturated flow contributes to the storm response of the slope. Analysis of the potential flow nets indicates that unsaturated flow during a storm will be entirely vertical
> • Analysis also indicates that unless the wetting front is saturated, the initiation of saturated lateral flow is dependent upon some break in the vertical permeability profile of the soil. (Weyman 1973, p. 284)

[15] And more recent model analysis of interface permeability contrast by Hopp and McDonnell (2009) suggest that a 2 order of magnitude reduction is the minimum to produce transient saturation at an interface and resulting subsurface stormflow.

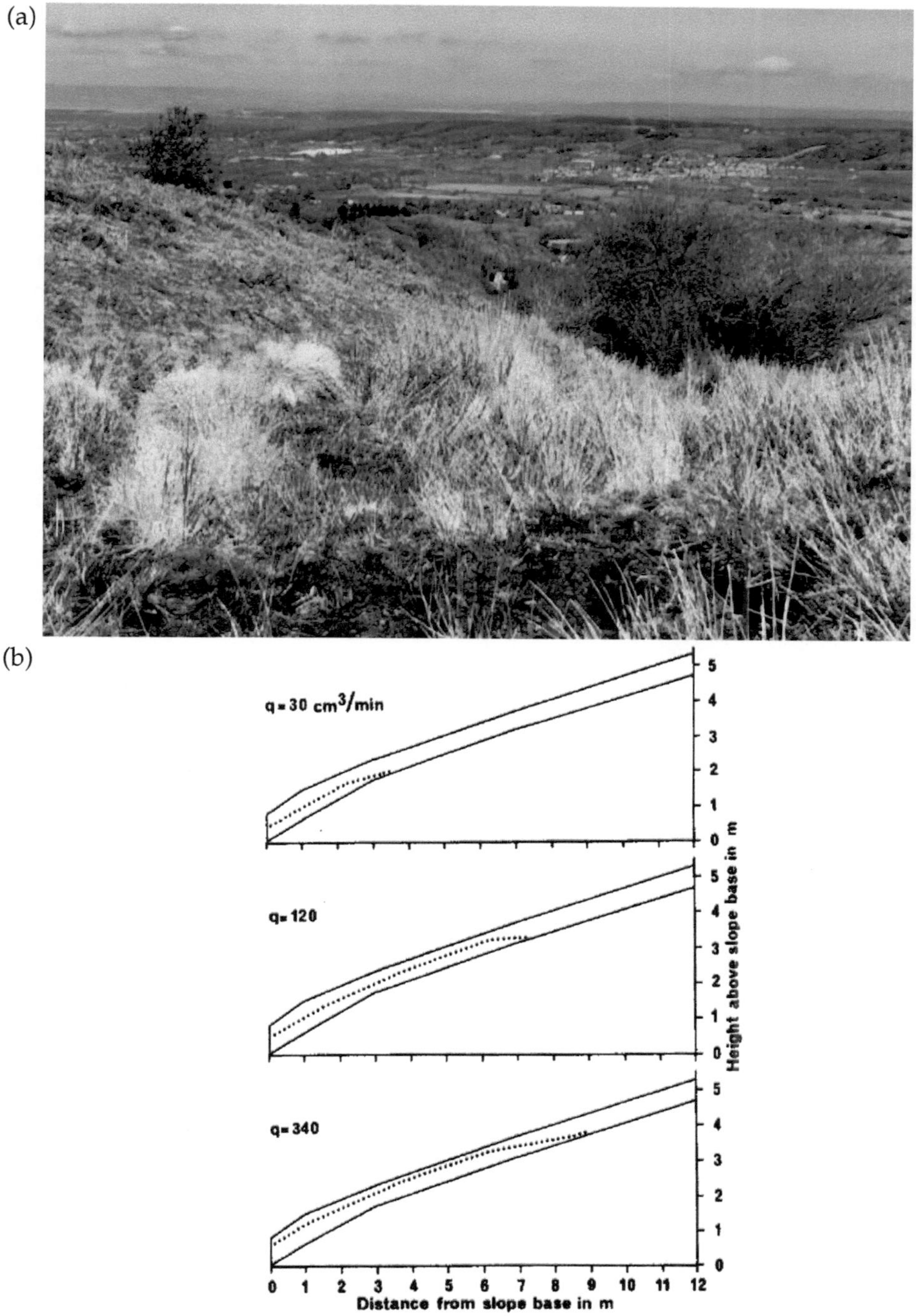

Figure 4.6 (a) The East Twin Brook hillslope. Bilberry, heather and grass occupy the site now, as they did in the early 1970s. Photograph by the author. (b) The saturated wedge development that shows the form of the transient saturated zone. The water table is shown by the dotted line for a range of discharge (q) conditions in cm^3/min.

Source: Weyman (1973).

Several key themes emerge from these two points. The most important being Weyman's observation that there was no evidence to suggest that lateral unsaturated flow contributes to the storm response of the slope. This speaks to the need for some sort of (transient) saturation as a pre-condition for subsurface stormflow. Also, when you traverse his hillslope, you get a sense of how the perceptual model gears must have been turning for Weyman. The gut of the hollow in Figure 4.6 is wet to the touch even on a dry summer's day. His work focused on the sideslopes that you can see in diagrammatic form in Figure 4.6b. This figure is strikingly similar to the Hursh diagram we saw in Chapter 2, Figure 2.8. It is notable that Weyman does not cite any of Hursh's work; but he does cite Hewlett's work. Weyman (1970) notes that the underlying sandstones, shales and quartz-conglomerates are "virtually impermeable" at the site. The side slopes are long (on the order of 300 m) and "the basin head is a very shallow topographic bowl in a summit plateau at 350 meters from which the channel drops steeply to its outfall at 230 meters" (Weyman, 1970, p. 25). Soils are thin and transmissive where a 20 cm peat layer overlies the sandy loam podsol that is about 60 cm deep.[16]

So, we have looked at interfaces and saturated wedge development from such interfaces. But everything so far has been in two dimensions. Meaning hillslope cross-sections. The next big development was to take things to three dimensions and to look at the effect of topography on the generation of subsurface stormflow.

[16] One interestingly sidenote is that the work of Weyman led directly to the development of Topmodel (Beven and Kirkby, 1979)—one of the most important and enduring conceptual rainfall–runoff models built directly upon a perceptual model. It's main architect, Keith Beven notes that a "critical observation from Darrel Weyman's work was the synchronicity of flows in a throughflow trough and in the main channel, suggesting the possibility that runoff per unit area might be approximately spatially constant; a key underlying assumption of Topmodel. This analysis of the response of the upper East Twin led to the concept of a topographic index (as $a/\tan\beta$, a as upslope contributing area per unit contour length and $\tan\beta$ as local slope)" (Beven et al., 2021, p. 527). And as we will see in the next section, contributions to subsurface stormflow are not the same on all slope sections—some matter much more than others for reasons we will now see.

4.4 The role of topographic convergence

A problem well stated is a problem half solved
—John Dewey

The natural follow-on to the discovery of saturated wedges was to explore such behavior across the catchment. Tsukamoto (1963) had already recognized the importance of so-called zero-order basins (i.e., unchannelized hollows) for runoff generation in steep, wet terrain in Japan. And Kirkby and Chorley (1967) had already outlined the main zones where the most hillslope flow might occur: at the base of slopes, in hollows, in slope profile concavities and in areas of thin or less permeable soils. Bunting (1964, p. 73) too had already referred to hollows and areas of topographic convergence as "percolines," where he noted that they were "sites of moisture accumulation and lateral downslope movement of moisture." But in terms of demonstration with data in the field, Anderson and Burt (1978) were the first to clearly show the role of topographic position on subsurface stormflow generation. In some ways Horton already alluded to this too—as did Hursh—but Tim Burt's PhD work with Malcolm Anderson at Bristol University in the late 1970s brought data to bear to define the critical importance of topographic convergence in the context of subsurface stormflow initiation and generation.

Again, the importance of saturation as a precondition for subsurface stormflow generation cannot be overstated. Burt instrumented a hillslope hollow at Bicknoller in Southern England, about another hour's drive southwest of Weyman's site that we examined earlier. If you walk both sites, Bicknoller (Figure 4.7) is noticeably steeper than Weyman's site but with less defined convergence. Two things were transformative with this new work by Burt: the first-ever spatially distributed tensiometer measurements across the slope and the first use of *recording* tensiometers (placed at 60 cm depth). These were technological advances where *both* tension and pressure could be defined and recorded as the transient water table developed on the slope (Burt, 1978).[17]

[17] For context, the 1970s was a time when most instrumentation used in streamflow generation research was home-made. Rudimentary dataloggers were available but there were no

(a)

(b)

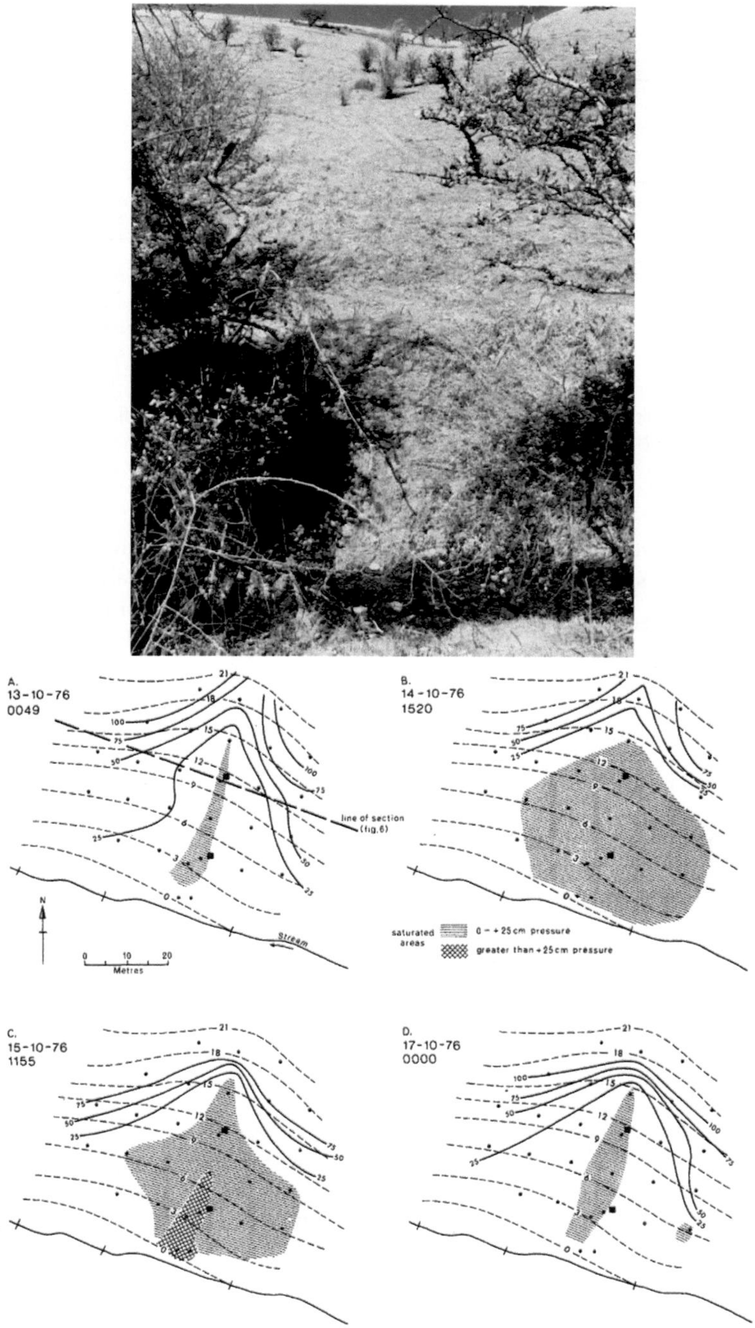

Figure 4.7 (a) The Bicknoller site of Tim Burt's PhD study of topographic convergence. Photograph by the author. (b) maps as different time snap shots with contours of matric potential (units = cmH_2O) and saturation (gray shading) at different times during a rainfall event. This mapped site is the same as shown in the photo opposite and represents the sideslopes and hollow discussed in the text above.

Source: Anderson and Burt (1978).

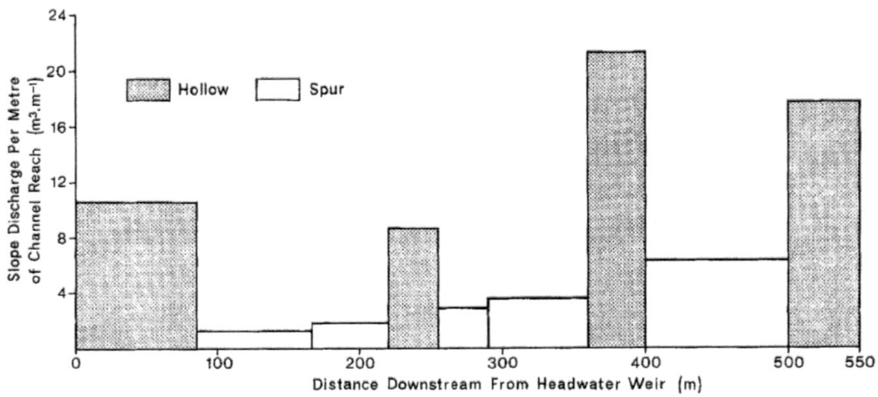

Figure 4.8 Discharge from the hillslope to the stream over a 500 m length of channel at the base of the study hillslope showing the profound importance of topographic convergence in the hillslope delivery of water to the channel.
Source: Anderson and Burt (1978).

So what did the field work at Bicknoller reveal? Anderson and Burt (1978) summarized it as follows:

Maximum saturated hillslope flow is shown to coincide with the stream discharge 'throughflow peak'. Relative to the hollow, little saturation is evident at the base of the spurs. This subcatchment model is applied to a major portion of the catchment, showing the general importance of topography in the control of stream discharge. (Anderson and Burt, 1978, p. 331)

So what they are saying, and what Figure 4.7 is showing, is that subsurface stormflow is generated in very specific spots on the hillslope: in the hollows. The so-called "spurs," meaning the nose slopes or ridges, produced little if any flow. The contour lines of the soil matric potential show a rather striking development, growth and contraction of subsurface saturation. It is much like the mapped patterns of saturation from Dunne's work from Chapter 3 at the soil surface, but now for the subsurface, based on the new recording tensiometers deployed across this hillslope. And unlike overland flow, all this is hidden from view, except for the darkest patch in Figure 4.7 where surface saturation was produced, as denoted by positive pressure potentials. Burt and Anderson are revealing here a hitherto unknown hydrological world of the subsurface stormflow pattern. We will speak much more about patterns in Chapter 6.

Anderson and Burt also looked at the impacts of these concentrated subsurface stormflow sections on streamflow. They performed dye dilution gauging in the stream into which the hollow and side slopes drained. This was done to see if their slope-based perceptual model really related to stream water additions in the topographic convergent zones. Figure 4.8 shows the pronounced effect of said topographic convergence on per unit length contributions to the stream where hollows can be seen to clearly dominate flow accumulation down valley. This trumped slope length in terms of the effects on stream delivery of hillslope water where:

convergent flow occurs (hollows) then efficiency of flow is almost independent of hillslope length. Without convergence (spurs) hillslope length exerts a highly significant control on spur discharge. (Anderson and Burt, 1978, p. 343)

These combined findings of Anderon and Burt and their related papers from the Bicknoller catchment were profoundly important for ratcheting forward understanding of subsurface stormflow processes. It finally placed interfaces and saturated wedges into a three-dimensional framework to understand

personal computers at this time. Most of the new sensors and advances in catchment hydrology were published in Technical Bulletins of the British Geomorphological Research Group (BGRG), a UK-based geomorphology organization. As a new PhD student in the 1980s, I recall this series—and this 1978 publication of Burt—being a lifeline for my own sensor development, half a world away in New Zealand.

how and when and why subsurface stormflow is produced.

Let's now look at one last perceptual model advancement through the 1960s and 70s[18] that helped to show one last big control (in some areas) for subsurface stormflow: soil pipes.

4.5 The role of soil pipes and preferential flow

Science is but an image of the truth

—Francis Bacon

Perhaps the last major component of subsurface stormflow understanding to anchor the variable source area concept (beyond the translatory flow idea alone) was the enhancement of subsurface stormflow by lateral preferential flow, especially soil pipes.[19] Hewlett's work does not really mention soil pipes or macropores, even though the earlier work in his study catchments by Hursh certainly did.[20] Why is that? One reason is that Hewlett's work was so influenced by his concrete trough models in the field that we discussed in Chapter 3—re-filled with re-packed soil. Such work destroyed soil structure. So, in effect, he erased the features of natural soils present at Coweeta. That's been an issue through the ages in soil science and agricultural soil water studies. And for perceptual model development, it is like removing a key ingredient from a recipe. It just does not taste the same. And like saturated wedges and topographic convergence, the British school of hillslope hydrology in the 1970s really helped define the soil pipe influence on subsurface stormflow.

Figure 4.9 shows the stream bank pipes on Burbage Brook, Derbyshire UK. It shows visually the outlet of pipeflow. Jones (1971) was among the first major contributions to pipeflow following the work of Bunting (1961) and others on percolines[21] in the UK that he was able to trace, noting that:

> Hydrologically, piping speeds throughflow[22] and will result in a quicker hydrograph response than might otherwise be expected (Jones, 1971, p. 608)

Jones (1971) and many others have noted that pipes are somehow intermediate between slope seepage and channel flow. And elsewhere in the UK at the end of the 1970s it was noted that soil pipes—as a key type of macropore—can greatly enhance subsurface stormflow in many environments:

> The inclusion of subsurface flow on hillslopes, or throughflow, of which pipeflow is a special type, represents the major conceptual change necessary to account for the observed pattern of runoff from many drainage basins (Gilman and Newson, 1980, p. 3)

In many cases, pipes are enlarged through eluviation (a sort of subsurface erosion). Also, in many pipeflow settings, there can be an audible gushing of water in the subsurface. Figure 4.10 shows a student of Tony Jones from the University of Aberystwyth listening for pipeflow, in a Welsh catchment in the early 1970s. Hewlett noticed this audible gurgling of macropore flow when exploring the Coweeta hillslopes with Hursh in the early days of his work at Coweeta (Wade Nutter, pers. comm., 2020). Mosley (1979) comments on this too for his work at the Maimai catchment in New Zealand. Certainly, I have heard audible gurgling on some of my pipe-riddled hillslopes when making measurements of flow at a trench face.

In steeper catchments, it is often at headscarps and other hillslope bases evacuated by landslide activity where soil pipe outlets are found. Figure 4.11 shows such a case at the Maimai M8 catchment on the South Island of New Zealand. In

[18] We will take some liberty with dates and go a few years beyond the International Hydrological Decade in one or two examples, just for completeness.

[19] For the latest review on this subject, see Beven (2018)—a paper with the provocative title "A Century of Denial: Preferential and Nonequilibrium Water Flow in Soils, 1864–1984." It goes well beyond this small chapter section here and provides the best history on soil pipes and preferential flow in the context of streamflow generation yet written.

[20] And, of course, Engler (1919) did in his early description of subsurface stormflow (recall Figure 2.6 from Chapter 2), along with many others who published early subsurface stormflow observations in the 1930s and 40s.

[21] Percolines relate to pipe networks. Described in Wales in and around the Plynlimon catchment, as (a) " first-order bedrock channels that have been overgrown by peat-forming tunnels up to 1 m in diameter" or (b) "networks of much smaller pipes in the peat, or near-surface organic horizons that result from the preferential development of desiccation cracks in the (rare) dry summers in Wales" (Beven, 2018, p. 9).

[22] Remember that throughflow here means subsurface stormflow.

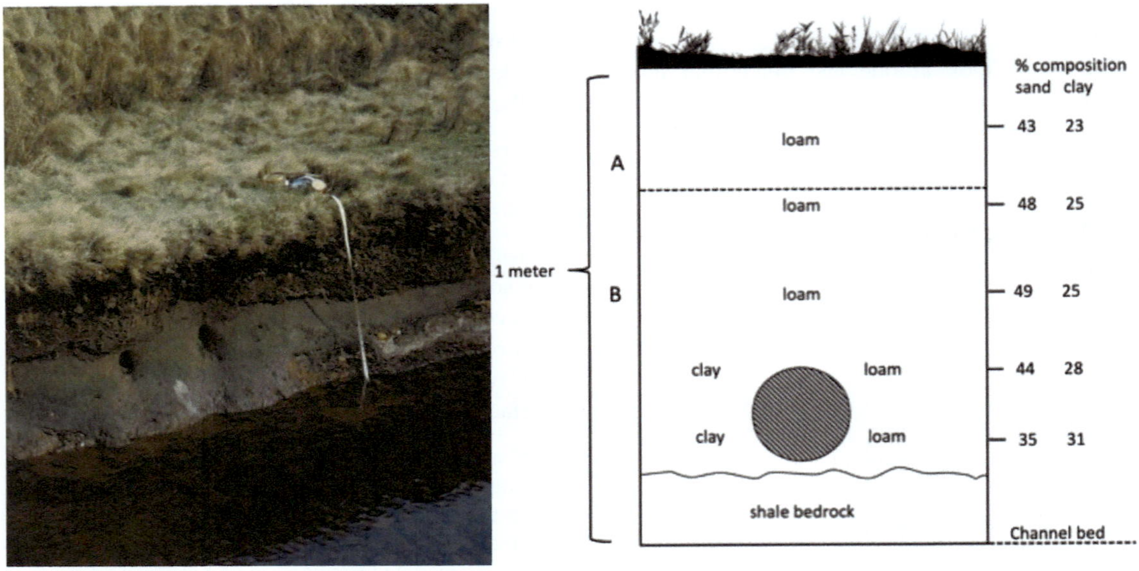

Figure 4.9 Stream bank pipes on Burbage Brook, Derbyshire.
Source: Photo from Tony Jones. Diagram from Jones (1971).

early work at the site, Mosley (1979) noted that "two pipes were discovered at the base of the B horizon through which water gushed at maximum rates of the order of 20 L/s" (Mosley, 1979, p. 800).

In perhaps the first study to really connect a perceptual model of pipeflow to integrated catchment discharge, Mosley (1979) examined the causes for the very high runoff ratios and extreme steepness of both the rising and falling limbs of the storm hydrograph for his soil pits and stream gauging sites in a small 0.3 ha sub-watershed of the M8 catchment. Figure 4.11b shows the synchronicity of the responses and the clear downslope increases in water flow volumes. His soil "pits" were ~1 m wide and he used a graduated cylinder and a stop watch to "gauge" the flow from the slope seepage (aka subsurface stormflow). Like Anderson and Burt's work, Mosley saw that hollows produced the majority of the flow; the nose slopes and spurs much less flow. Mosley would go on to say for the Maimai catchment that:

Data for a number of storms indicated that subsurface flow via 'macropores' (root channels, pipes) and seepage zones in the soil is the predominant mechanism of channel stormflow generation in storms with quickflows greater than about 1 mm. Subsurface flow from all parts of the watershed appears to contribute to stormflow even in very small storms (quickflow on the order of 3% of net precipitation). The saturated hydraulic conductivity of the soil matrix is not a limiting factor on the ability of subsurface stormflow to generate channel stormflow, because dye tracer experiments demonstrate that water may move through macropores (particularly root channels) at rates 2 orders of magnitude greater. (Mosley, 1979, p. 795)

Reflect on this quote while re-examining Figure 4.11b. What else do you see beyond the steep rise and fall of the hydrographs and the fact that flow increases from the soil pits near the ridge (lower on the plot) to nearer the steam (the upper storm hydrograph)? If you answered "that their hydrograph peaks are all synchronous!" then you would be correct. There is no lag or delay or attenuation in a downslope direction. Mosley interpretated this as part of the pipeflow influence. We will return to Mosley's interpretations—and challenges to them—in Chapter 5.

Beyond any single study catchment, perhaps the best work *mapping* soil pipe networks during the International Hydrological Decade occurred in the UK. Note that these mapped soils were often poorly drained peaty soils—different to those developed in

Figure 4.10 Listening to the audible gurgling of pipeflow water at depth.
Source: Photo by Tony Jones

the steep, forested mineral soils that we examined in New Zealand. The early mapping in the UK considered soil pipes as "a means of channel extension, by subsurface erosion and roof collapse." In other words, they add "another component to hillslope drainage that is capable of contributing stormflow to the stream" and in the way that they can "contribute to a variety of hillslope erosion processes" (Jones, 1994, p. 78). Figure 4.12 shows two forms of pipeflow mapping; one for an individual headwater catchment in Wales (Figure 4.12a) and one showing regional mapping of pipeflow likelihood for the UK.

The mapping on the left-hand side of the figure differentiates between perennial and ephemeral pipes. Jones (1979) even proposed a modification to the variable source area concept for such regions that included pipeflow and the idea of a "dynamically contributing subsurface volume." Figure 4.12b helped regionalize this thinking across much of the western UK as linked to soil drainage class.[23] This work would anticipate research in the years and decades ahead, when stable isotope tracing of water showed the importance of storage volumes in terms of mixing at the hillslope scale and streamwater composition at the catchment scale. We address all that in in the next three chapters, linked to the role of subsurface storage on modulating the age, origin and pathway of subsurface stormflow.

4.6 Summary and looking ahead

It is a capital mistake to theorize before one has data. Insensibly one begins to twist facts to suit theories, instead of theories to suit facts

—Arthur Conan Doyle

[23] A useful new review of this history is Jones (2025).

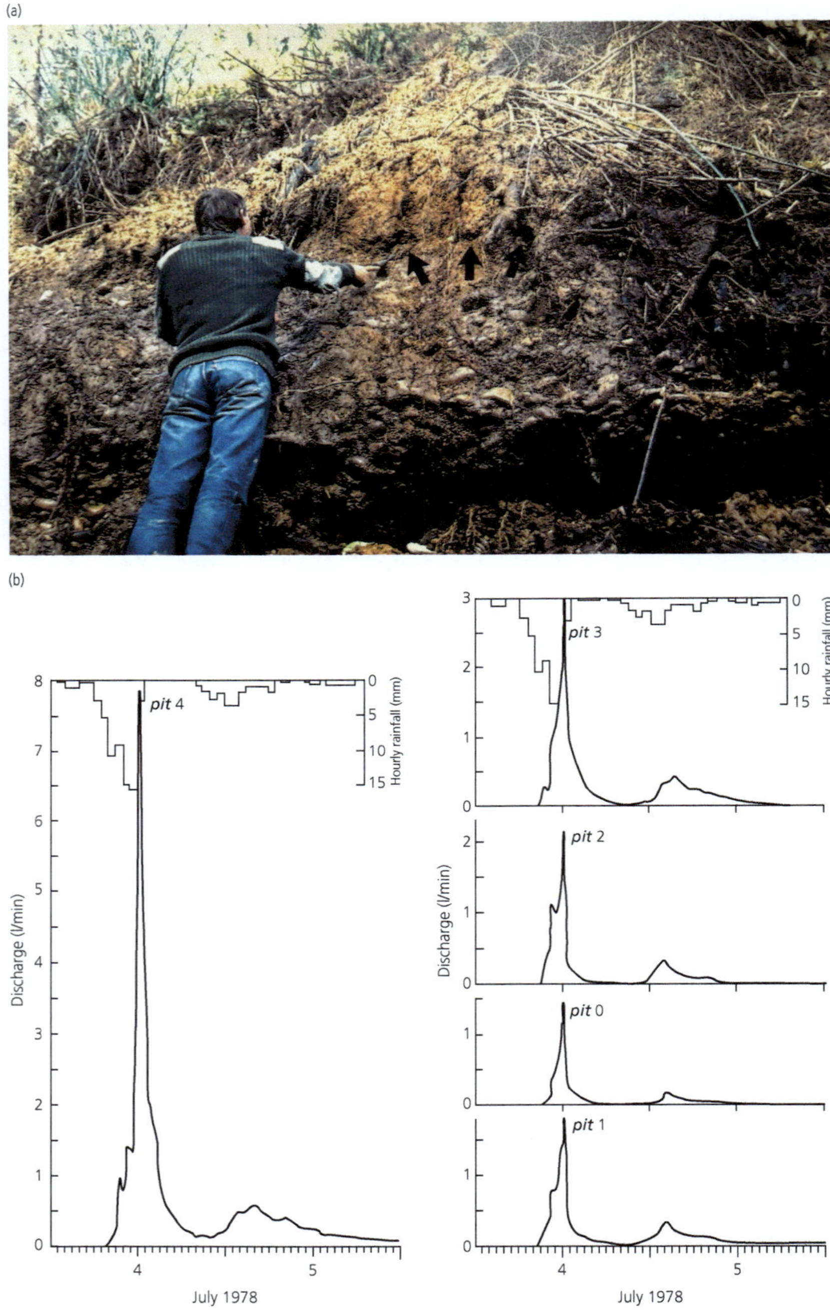

Figure 4.11 (a) Soil pipes at the interface between soil and bedrock at the Maimai catchment in New Zealand. Field assistant Steve Parker shown for scale. (b) The flow data from the Mosley (1979) hydrometric analyses.

Source: (a) Photo by the author.

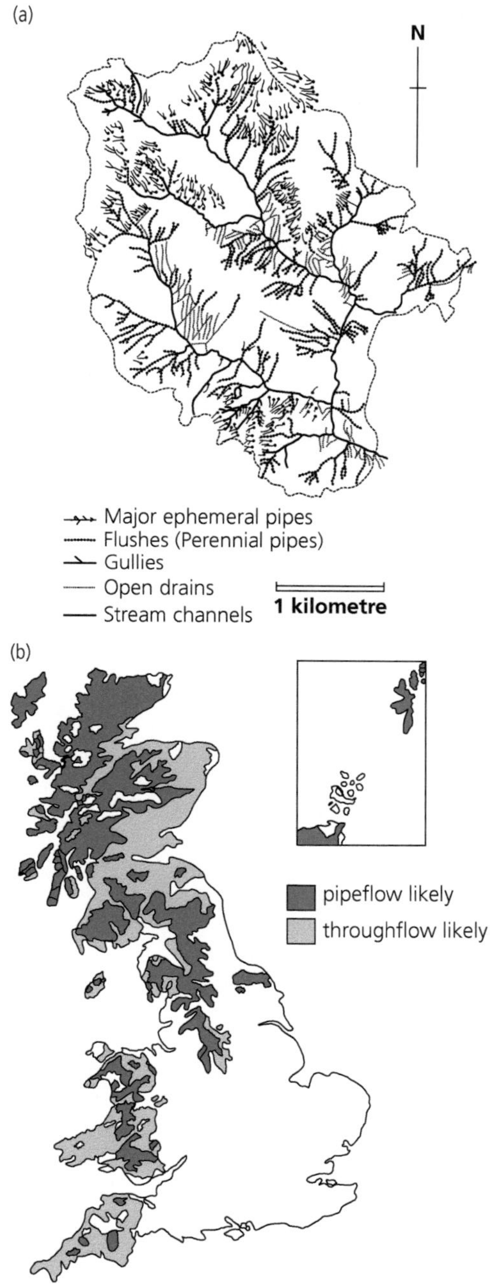

Figure 4.12 (a) A map of soil pipes and other channels in the Wye (b) A map of pipeflow likelihood and so-called "winter rain acceptance potential classes" in the uplands of Britain.

Source: Both from Gilman and Newson (1980).

We've come a long way from where we started in this chapter, questioning the importance of subsurface stormflow in light of the review of the International Hydrological Decade by Freeze (1974), where he doubted whether or not subsurface stormflow could "deliver sufficient water to provide significant contributions to storm runoff." Clearly it can! But this is a mechanism that turns on for large(r) events. The best diagram to summarize the state of subsurface stormflow research and understanding by the mid-1970s is Figure 4.13. It was the cover image for the book "Hillslope Hydrology," edited by Kirkby (1978). It shows some of the key factors (but not all) that we reviewed in this chapter: the importance of interfaces, saturated wedges, slope convergence and soil pipes.

The material that we covered in this chapter, while old is still relevant today. We continue to catalog new subsurface stormflow behaviors, including more recently, evidence of even deeper forms of subsurface contributions to the channel hydrograph on the timescale of events. Still, subsurface stormflow is a poorly understood process. There is an excellent recent review of subsurface stormflow by McGuire et al. (2024) that synthesizes both the field work and modeling of subsurface stormflow. Notwithstanding this synthesis, I would say that the lower right portion of the triangle in Figure 4.13 is still the greatest challenge for process description and perceptual model development today. And in many ways—as we will see in coming chapters—there continues to be a fundamental problem in the streamflow generation literature concerning subsurface stormflow and its occurrence. Subsurface flowpaths are the dark matter of catchment hydrology. We can see the soil. And with a shovel and some dyed water, the stained flowpaths. But what we don't see[24] —and must imagine—are the subsurface pathways water takes to the stream. Subsurface stormflow therefore is a master of deception. We'll come back to this repeatedly through the rest of the book.[25]

[24] See a great paper by Chifflard et al. (2019) with the title "How can we model subsurface stormflow at the catchment scale if we cannot measure it?"

[25] I am finishing off this book during a stint as Visiting Professor at the University of Kyoto. And the group I am sitting in

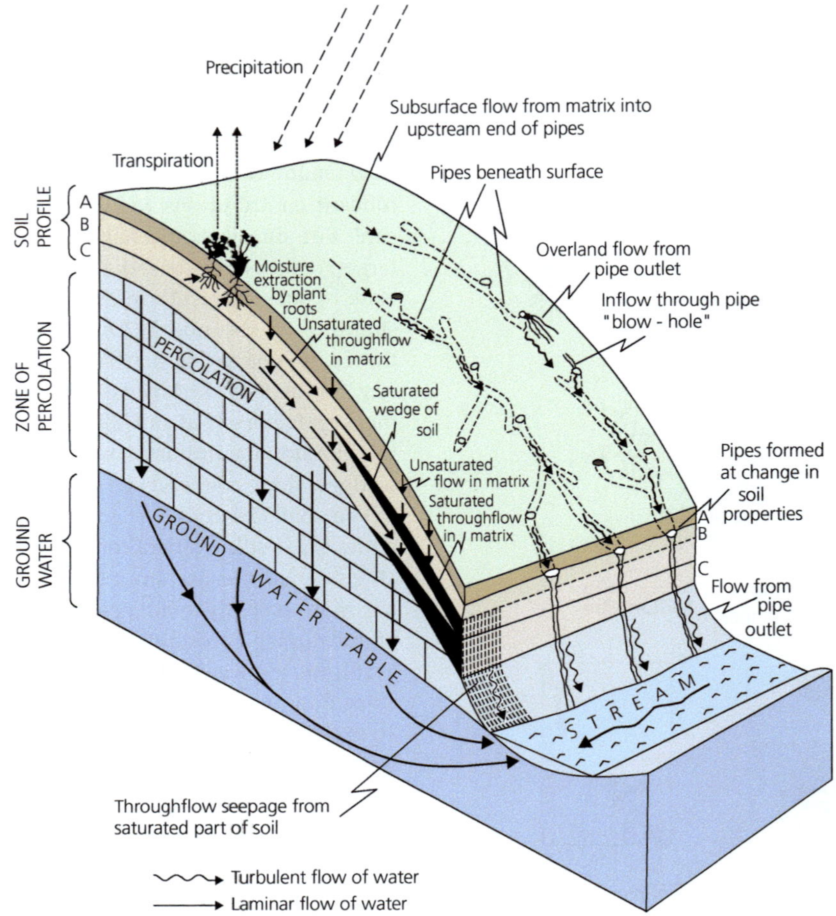

Figure 4.13 A summary of the state of the art of process descriptions for subsurface stormflow by the mid-1970s.
Source: Atkinson (1978).

at the Disaster Prevention Research Institute has a main question that they are struggling with: how can the runoff ratios for large rain events in their mountain headwater catchments approach 100% when there is no observable overland flow? In many ways, it seems as if we're still struggling with the same issues with respect to the elusive subsurface stormflow. But, the next chapter gets at some of the starting clues for an answer.

Finally, in terms of how this chapter relates to what will come in Chapter 6 and the cutting edge research of today is that now we've gone on to link subsurface stormflow processes to broader themes of connectivity and emergent behavior, and in particular, the "thresholdiness" of hillslopes. Nonetheless, the papers that we've described and discussed in this chapter are what it all stems from and understanding the arc of learning from this chapter to Chapter 6 is crucial to pushing the boundaries of knowledge forward (and to not "reinvent," unnecessarily, what these earlier benchmark papers presented). One key going forward is the needed cataloging and classification of subsurface stormflow—something the latest review of the subject strongly advocates (McGuire et al., 2024). This is also something where the early work described in this chapter will play a key role and is a theme we will return to in Chapter 9.

CHAPTER 5

The role of groundwater in storm runoff (1975–1999)

5.1 Groundwater and storm runoff

Every great advance in science has issued from a new audacity of imagination

—John Dewey

Prior to the 1970s, most process hydrologists would not have thought that groundwater could be a source of storm runoff. Recall too the discussion from early papers in Chapter 2 noting that storm runoff, even if subsurface stormflow, was not true groundwater flow. Why? Because if you back-calculated the flow rate through the porous media, you'd find, like the Mosley (1979)[1] work we examined in Chapter 4, subsurface stormflow rates were often many orders of magnitude greater than the saturated hydraulic conductivity of the porous media through which groundwater may flow. To illustrate the early skepticism about the role of groundwater in storm runoff, Figure 5.1 is a scan of a 1941 letter from Robert Horton (recall him from Chapter 2) to Howard Cook. It perhaps best-captures what the attitudes might have been toward this crazy idea that groundwater could be a source of storm runoff in the years and decades before the work that we will discuss in this chapter.

I found this letter in one the 96 boxes of Horton's materials in the US National Archives. Horton states in the paragraph immediately following this scanned section that "ephemeral ground-water flow is [a] common occurrence. . .but of relatively little consequence." Following this, the letter continues with some thoughts on wave translation of groundwater—something we will explore in this chapter—where Horton notes ". . .possibly the inventor of this theory will suggest some extra-terrestrial source of energy." He goes on to note "it is, therefore, of relatively little consequence," has "absolutely no practical value," and that we should not worry about these "unimportant matters."[2]

Following the International Hydrological Decade, studies started to accumulate at different sites around the world showing the dominance of groundwater in the storm- and snowmelt hydrograph. This was certainly a new focus. What enabled the invisible groundwater to be identified in streamflow was the tracing of the molecular movements of water using the isotopes of hydrogen (^{1}H, ^{2}H, ^{3}H) and oxygen (^{16}O, ^{18}O) that make up the water molecule. These studies were using isotopes and employing a simple two-component mass balance mixing model to quantify the amount of new water, sometimes called "event water" (rainfall or snowmelt) and old water, sometimes called "pre-event water" (e.g., groundwater) in the individual storm hydrograph, as shown in Table 5.1. Almost without exception, groundwater dominated the hydrograph peak in the early studies across a variety of headwater catchments in Canada and Europe.

What is ironic is that the "unimportant matters" that Horton referred to now define the state of the art of streamflow generation research. And as we will see in this chapter and in Chapter 7, that for streams in many headwater catchments, one could

[1] Someone whose work we examine further, later on in this chapter.

[2] This is where I would joke in class when teaching this material: "now tell us how you *really* feel, Robert"!

Streamflow Generation. Jeffrey J. McDonnell, Oxford University Press. © Jeffrey J. McDonnell (2026). DOI: 10.1093/9780191953736.003.0005

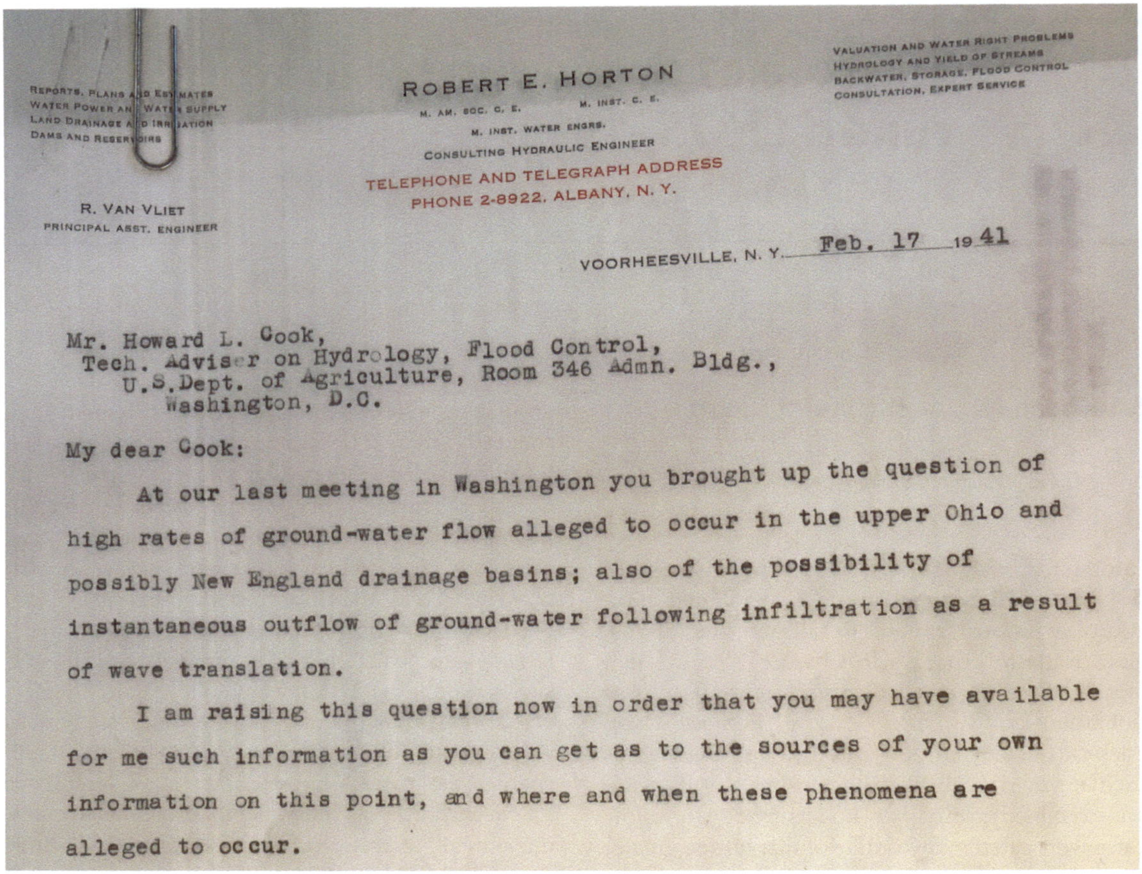
REPORTS, PLANS AND ESTIMATES
WATER POWER AND WATER SUPPLY
LAND DRAINAGE AND IRRIGATION
DAMS AND RESERVOIRS

ROBERT E. HORTON
M. AM. SOC. C. E. M. INST. C. E.
M. INST. WATER ENGRS.
CONSULTING HYDRAULIC ENGINEER
TELEPHONE AND TELEGRAPH ADDRESS
PHONE 2-8922, ALBANY, N. Y.

VALUATION AND WATER RIGHT PROBLEMS
HYDROLOGY AND YIELD OF STREAMS
BACKWATER, STORAGE, FLOOD CONTROL
CONSULTATION, EXPERT SERVICE

R. VAN VLIET
PRINCIPAL ASST. ENGINEER

VOORHEESVILLE, N. Y. Feb. 17 1941

Mr. Howard L. Cook,
Tech. Adviser on Hydrology, Flood Control,
U.S.Dept. of Agriculture, Room 346 Admn. Bldg.,
Washington, D.C.

My dear Cook:

At our last meeting in Washington you brought up the question of high rates of ground-water flow alleged to occur in the upper Ohio and possibly New England drainage basins; also of the possibility of instantaneous outflow of ground-water following infiltration as a result of wave translation.

I am raising this question now in order that you may have available for me such information as you can get as to the sources of your own information on this point, and where and when these phenomena are alleged to occur.

Figure 5.1 Scan of a 17th Feb 1941 letter from Robert Horton to Howard Cook, National Archives.
Source: McDonnell.

perhaps say that "streamflow is simply a catastrophic failure of groundwater storage!"[3]

Let's now look at four early perceptual models to explain groundwater effusion into the stream. We'll cover them in chronological order starting in 1975 and moving to ~1990:

- the substitution hypothesis
- groundwater ridging
- transmissivity feedback
- pipeflow of perched groundwater.

Each perceptual model was developed based on data from stable isotope tracing—a methodology that came into widespread usage in the 1970s and 1980s. The isotope approach, unlike most of the methods employed in streamflow generation research up until that time, was a top-down approach. Meaning that rather than making a point scale or hillslope scale measurement (e.g., of infiltration, or subsurface stormflow coming out of a trench) and then upscaling that information to the catchment scale, it was conducted *at the catchment scale*. Thus, interpretations were for streamflow generation integrated over the entire catchment, not just one piece of it.

Lastly, we should say what we mean by "groundwater." The Dictionary of Geology defines it as "the water in porous rocks beneath the water table." But in the examples in this chapter—and much

[3] To quote my friend and neighbor Adam Ward, Oregon State University.

Table 5.1 Selected hydrograph separation studies using ^{3}H (T or tritium), ^{18}O and ^{2}H (D or deuterium) up until 1979.

Authors	Snow Melt (-) or Storm Runoff (+)	Catchment size (km²)	Country	Isotope		
				T	^{18}O	D
Crouzet et al. (1970)	+	24	France	*		
Dinçer et al. (1970)	-	2.65	Czechoslovakia	*	*	
Biggin (1971)	+	1	UK	*		
Martinec et al. (1974)	-	43.4	Switzerland	*		
Mook et al. (1974)	+	650 ha	Netherlands		*	
Fritz et al. (1976)	+	22, 1.8	Canada		*	
Sklash et al. (1976)	+	73 to 700	Canada		*	
Krouse et al. (1978)	-	11.4, 111	Canada		*	
Sklash and Farvolden (1979)	+	1, 1.2, 3.9	Canada		*	

Adapted from Sklash (1990).

of what we think about in headwater catchments regarding groundwater—it is perhaps much shallower and more dynamic than only the water in porous rocks. It can be groundwater in weathered and unconsolidated material, in and around the stream (often called the alluvial aquifer) and even transient groundwater on slopes (in soil and glacial deposits) that may last for days to weeks to months. So, in this sense, groundwater is any water below the water table, even if the water table itself is highly transient.

5.2 The substitution hypothesis

A drop of water, if it could write out its own history, would explain the universe to us

—Lucy Larcom

The substitution hypothesis was the first real perceptual model for the role of groundwater in snowmelt runoff generation. It was outlined by Martinec et al. (1974) and Martinec (1975) based on their findings at the Dischma catchment in Switzerland (Figure 5.2). The substitution hypothesis was effectively a statement on piston flow displacement[4] (and dispersion) of groundwater to a stream during snowmelt whereby:

The massive infiltration of meltwater causes a corresponding increase of outflow from groundwater reserves. The infiltrated water, while slowly moving underground appears already 'in substitution' in the hydrograph represented by water stored hitherto in the subsurface reservoir (Martinec, 1975, p. 496)

What Martinec is saying is that for the steep mountain headwaters as shown in Figure 5.2, meltwater pushes out groundwater to the stream during the multi-week snowmelt period. In terms of where that meltwater went, they were further able to show that:

the greater part of the meltwater production infiltrates and contributes to the recharge of groundwater reserves (Martinec et al., 1974, p. 129)

These statements—determined from sampling at the catchment outlet—are noting that somehow, snowmelt water goes into storage, rather than running off the surface or moving as shallow subsurface stormflow and pushes out water from a previous season. In many ways, the substitution hypothesis was getting at storage release (something we will explore in detail in Chapters 6 and 7):

by this concept the quick reaction of outflow to a massive groundwater recharge is brought to agreement with the long residence time of the infiltrated water (Martinec 1975, p. 496)

Martinec was keying in on the difference between a pressure response in the stream and the velocity of the water molecule itself. One way to imagine this is with a garden hose. If you lift the hose when it is laying full of water on the ground, then water immediately comes out the other end. That is

[4] Recall how Hewlett thought of displacement in Chapter 3 linked to what he called "translatory flow."

Figure 5.2 The Dischma catchment in Switzerland where Martinec worked.
Source: https://en.wikipedia.org/wiki/Dischma#/media/File:Dischma.jpg.

like the stream hydrograph: the pressure pulse or technically, the celerity of the hydraulic potentials. But the isotope approach used by Martinec (1975) showed a different sort of speed: the velocity of the water molecules. Going back to our hose analogy, if we connect that garden hose to a tap on the wall, and turn on the tap, then water immediately comes out the end just like before. But, unlike the pressure pulse, the temperature of the water coming out of the hose does not change for several tens of seconds as the tap water slowly makes its way through the hose and to its outlet. That is like the stable isotope signal in the stream: much slower, representing the "velocity of the water molecule" or the mass flux of the water itself.

This pressure vs. velocity difference is what Martinec was referring to when he stated:

> the exact nature of the propagation is not known. Groundwater waves caused by periodic changes of pressure head (Vischer, 1970) might be involved, since they travel much quicker than the subsurface water without bringing along the infiltrated water particles. (Martinec, 1975, p. 497)

Martinec's perceptual model was in some ways, a groundwater-focused and much deeper version of Hewlett's "translatory flow" idea. Although not realized at the time (Hewlett's work was not referenced in the paper), Martinec noted that the snowmelt hydrograph reflects:

> the quickly transferred effect of infiltration on the outflow from the basin, whereas the real velocity of subsurface water particles is much smaller and results in average residence times of the order of years (Martinec, 1975, p. 497)

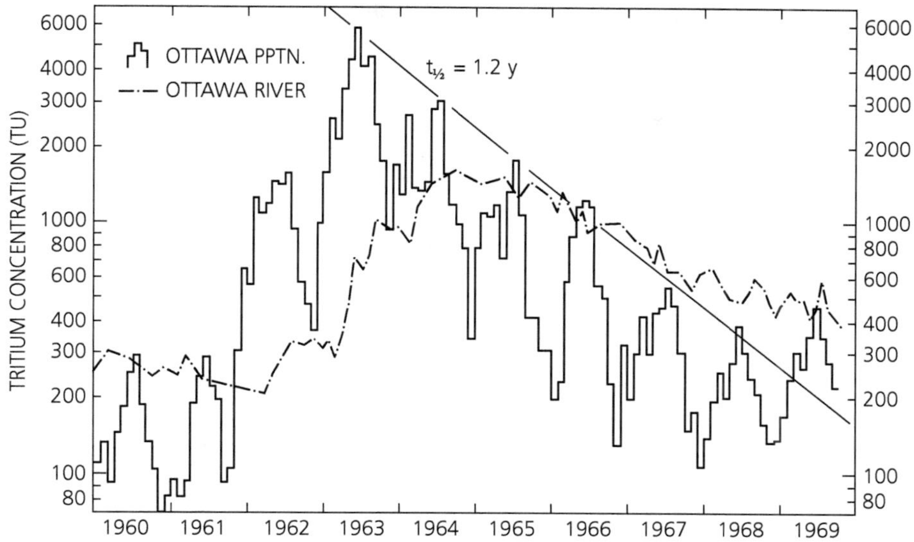

Figure 5.3 Tritium concentrations for the period 1960–1969 for Ottawa Canada
Source: Brown (1970).

Martinec was able to track these "water particles" by measuring the tritium (^3H) values in water. Unlike ^1H and ^2H, the stable isotopes of hydrogen, ^3H is a radioactive isotope of hydrogen with a half-life of 12.3 years. And while tritium is produced naturally in the atmosphere and to a lesser extent in the subsurface, these amounts are negligible compared to what was injected into the atmosphere during thermonuclear testing. That is what these early tritium studies were tracing: bomb tritium. Figure 5.3 shows the graph of tritium emplacement in the stratosphere that then rained out in precipitation events around the world—this graph showing precipitation in Ottawa, Canada and for the Ottawa river. This was an exceptionally useful tracer signal that could be used in catchment hydrology studies in the years before and after 1963[5]—the peak of tritium in precipitation around the world and the year that marked the Comprehensive Test Ban Treaty and when atmospheric nuclear tests ceased.

Martinec determined the tritium values in the stream and related them to the input values of snowmelt in collection devices known as snow lysimeters (effectively a baking sheet inserted at the base of the snowpack to intercept and collect melting water). Figure 5.4 shows their time series for streamflow (labelled "basin" in the figure) and snowmelt (labeled "lysimeter"). During the May–August period, the streamflow signal deflected toward the snowmelt signal but not fully. That amount of deflection said something meaningful about the percentage of meltwater in the stream: if the stream signal shifted to be equal to the meltwater signal, then they could say that streamwater was 100% meltwater at that point. If the streamwater signal did not deflect at all from its pre-melt baseflow signal, then they could say that streamwater had 0% meltwater. Figure 5.4 shows that in all but one of those stream sample dots, there was more groundwater than snowmelt in the stream. And if you use the sequential snowmelt signature as the "new water" frame of reference (the line marked "outflow from the lysimeter"), then all of the sampling periods contain more groundwater than snowmelt. Thus, groundwater was directly observed to be the major contributor to snowmelt season runoff.

[5] And as we will see in Chapter 8, still exceptionally useful in some places for plant water sourcing from deep soil, like the Loess Plateau in China.

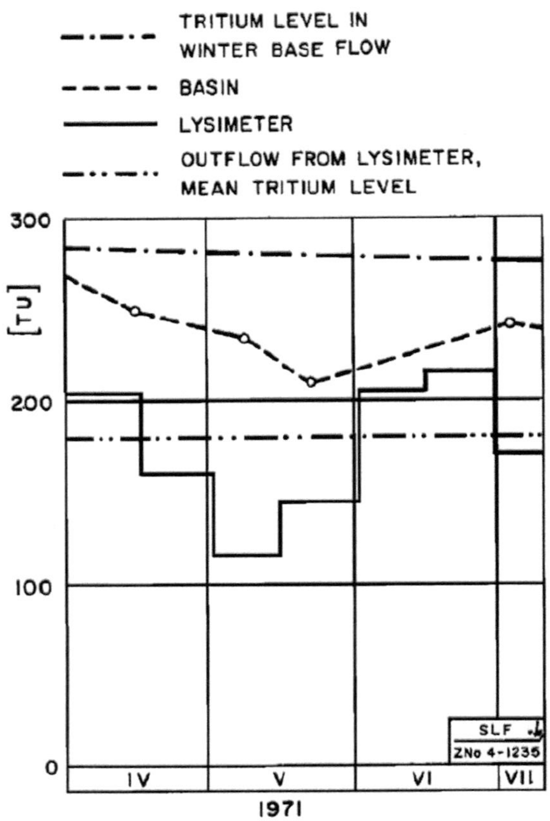

Figure 5.4 Martinec's plot of tritium concentrations in the catchment outflow (called "basin") and for meltwater from a snowmelt lysimeter. *Source*: Martinec (1975).

5.2.1 Early influences on the substitution hypothesis based on tritium concentrations

The ideas in Martinec et al. (1974) and Martinec (1975) were influenced by several earlier conference papers that were published in a companion study by members of Martinec's team (reported in Dinçer et al., 1970) for the Modry Dul catchment in Czechoslovakia and a handful of other studies using tritium as a tracer (e.g., Begemann and Libby, 1957; Eriksson, 1958; Brown, 1970). Returning to our hose analogy for the difference between celerity and velocity, Martinec's earlier work in Dinçer et al. (1970, p. 121), noted "considerable dispersion of the water molecules takes place when water is moving in porous formations." So rather than the piston flow pulse through a hose, they noted that "when consecutive mass inputs to the stream move along the flow lines they become more and more mixed." This means that when a sample of water is taken from the stream somewhere, "it contains inputs that occurred at different times."

Let's look at how this thinking might play out in terms of groundwater flow toward the stream. Figure 5.5 shows the earliest illustration of this from Eriksson (1958) who worked this out for age calculations for tritium fallout from nuclear tests where tracer inputs advect down a cross-section of a hillslope. The inputs of tracer in the most upslope position undergo the most dispersion since they have the farthest to travel to the stream. On the other hand, tracer inputs falling close to the stream get into the stream much more quickly and with less dispersion. In this way, the stream tracer signal

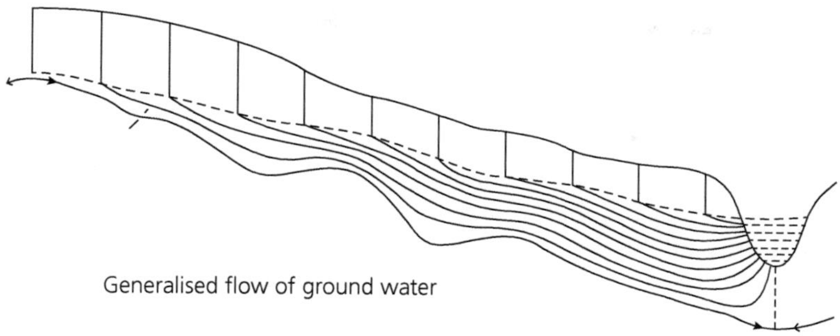

Figure 5.5 The first paper to address dispersion of the tritium signal in groundwater and showing the generalized groundwater "streamlines" toward the stream.

Source: Eriksson (1958).

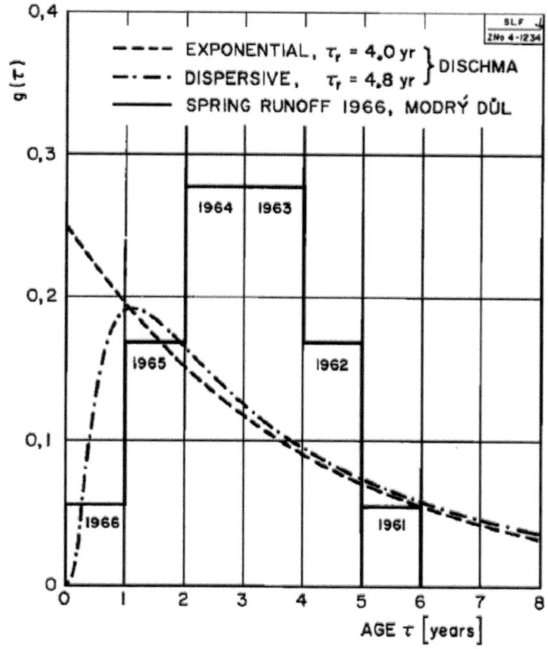

Figure 5.6 "Age distributions and residence times of groundwater in the representative basins Modry Dul (composition of base flow in spring 1966) and Dischma (period 1969–1972)."

Source: Martinec (1975).

integrates the hillslope positions with the mixture of short and long lag times.

What Martinec was able to do with the tritium signal was to quantify this dispersion and in so doing, the mean age of their groundwater flow to the stream. They used a dispersion function (Figure 5.6) to determine their "age distributions and groundwater residence times" and found that the mean age for stream baseflow[6] to be 4.0–4.8 years. Martinec rationalized this by noting that:

for a residence time of 4 yr and a travel distance of a subsurface water particle toward a surface channel, for

[6] i.e., the flow in the stream outside of the diurnal melt pulses.

example, of 1000 m, the flow velocity is 0.7 m/d, which is within the range of values assumed for groundwater flow (Martinec, 1975, p. 496)

The substitution hypothesis, while powerful for explaining the role of groundwater over the many weeks of snowmelt, was a useful, but coarse descriptor of processes that might occur on the timescale of individual rainfall or diurnal snowmelt events. Earlier work had in fact noted "pressure wave" effects where experimental additions of tritiated water were made (much like the Horton and Hawkins soil column work we discussed in Chapter 3) and that the movement of soil water is "layered," where:

new rainwater on the surface of the soil simply pushes the old water downward. Seepage water, which eventually shows up at a depth of 2 or 3 m, hours or days after a heavy rainfall, is in fact old capillary water, from soil layers immediately above, set free by a "pressure wave," the traveling velocity of which is equal to the seepage velocity of water in the wider pores. (Zimmerman et al., 1966, p. 347)

5.2.2 Subsequent sharpening of the substitution hypothesis via pressure wave dynamics

But it would be well into the 1990s[7] that sharper and more mechanistic descriptions of pressure wave dynamics would be described. The best example came from the Metman Ridge, near Coos Bay Oregon, USA. That work, led by Bill Dietrich and his many postdocs and PhD students who worked at the site, represents one of most intensive hillslope studies of all time. Montgomery et al. (1997) conducted a series of low-intensity sprinkling experiments (1.5 and 3 mm/hr) at Metman Ridge where they showed a process whereby deeper groundwater flowpaths flowing "through near-surface bedrock" can control the headwater hydrograph, illustrating "the importance of shallow bedrock flow in pore pressure development in the overlying colluvial soil." Montgomery et al. (1997) noted that:

together these observations highlight the role of interaction between flow in colluvium and near-surface bedrock in governing patterns of soil saturation, runoff production, and positive pore pressures (Montgomery et al., 1997, p. 91)

Figure 5.7 shows their site and perceptual model of groundwater movement into and out of fractured rock into the colluvium above.

In companion work at the site where irrigation rates were increased substantially, Torres et al. (1998) describe the effects of these high-intensity sprinkling experiments on the same site as:

soils with hydraulic properties that yield steep soil-water retention and characteristic curves in the near-zero pressure head range may favor the development of pressure waves. When such a soil is in the near-zero range, a rain spike may produce a slight increase in pressure head and a corresponding large increase in conductivity. The increased conductivity and the slight increase in head gradient produce a rapid release of old water as the wave propagates downward (Torres et al., 1998, p. 1876)

What Torres and colleagues are referring to here is the extremely non-linear relationship between the unsaturated hydraulic conductivity and the soil water tension or pressure head. Figure 5.8 is the relationship for their site where a 7 order of magnitude increase (!) in hydraulic conductivity occurs when the pressure head[8] goes from −1 m to zero. In this case zero represents saturation. So, even if the soils are not fully saturated, but close to saturation, the hydraulic conductivity is very high—allowing, as Torres and colleague note, *pressure propagation*.

In terms of how this all works, it is important to note that the irrigation rates in the Torres et al. work were not like in the Montgomery et al. paper; to use technical terminolgy, these simulated storm irrigations were of *biblical proportion* (linked to a companion slope stability study) with inputs intensities on the order of "370 and 5725 mm/h" over durations "0.18 to 0.50 hours" (Torres er al., 1998, p. 1867). So, the pressure wave results need to be kept in perspective given how far they were outside the normal range of multi-hour and multi-day storms for this region (rarely exceeding 10 mm/h

[7] With many papers to follow, starting with Rasmussen et al. (2000) that quantified tracer vs. pressure wave velocities through unsaturated saprolite.

[8] Other work we have examined in previous chapters called this tension or matric potential.

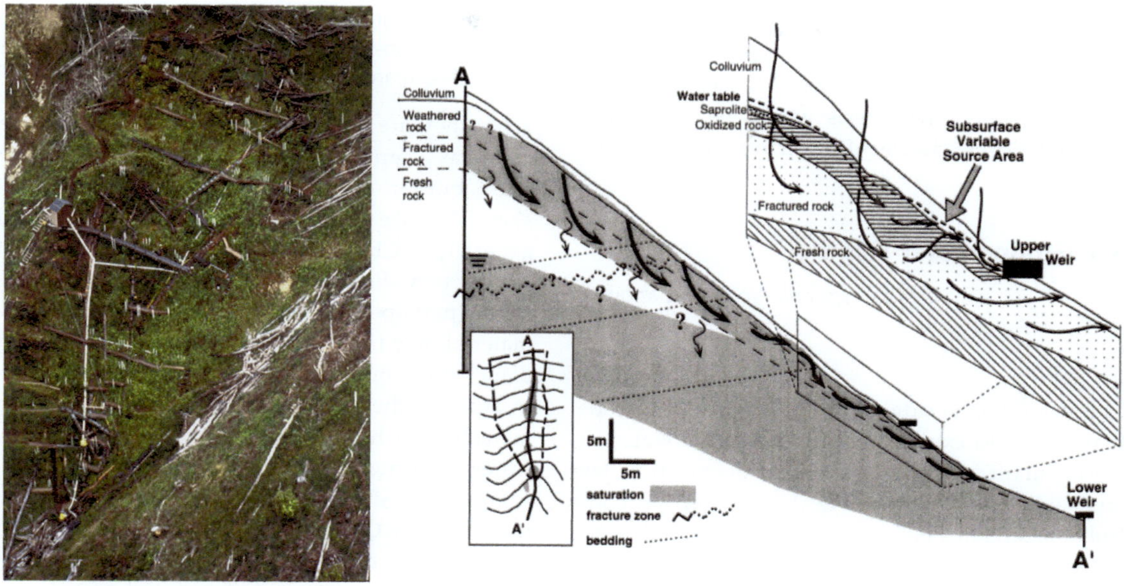

Figure 5.7 The Metman Ridge site in Western Oregon where natural events and slope irrigation work revealed key pathways deep in the hillslope and through weathered rock.

Source: Dave Montgomery, pers. comm.; figure on right from Anderson et al. (1997).

rainfall intensity[9]). Notwithstanding, the Metman Ridge (Coos Bay) study remains an important contribution to the literature on pressure wave dynamics, noting that:

once the soil attained near-zero pressure head, the unsaturated zone, saturated zone, and discharge became delicately linked such that a sudden increase in rainfall intensity generated slight but rapidly advancing pressure waves that induced slight changes in head gradient and very large changes in hydraulic conductivity. These changes produced a rapid release of stored soil-water that increased the delivery rate of water to the underlying saturated zone, leading to increased pore pressures and discharge. Moreover, pressure wave transmission may be a requisite mechanism connecting the unsaturated and saturated zones in soils that may be near saturation. Hence the hydrologic response of the landscape to precipitation events is greatly influenced by the dynamics of the unsaturated zone. (Torres et al., 1998, p. 1878)

So in summary, the substitution hypothesis is something of a general statement about groundwater being displaced into the stream during an event—what we now term pressure propagation. These findings, not unlike the variable source area concept, were disruptive and revolutionary. But unlike

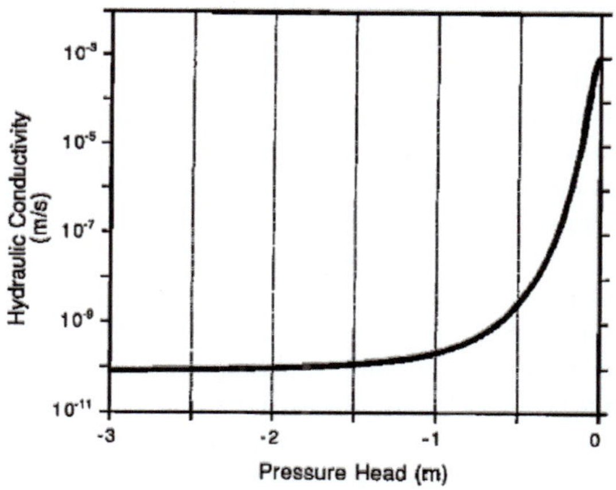

Figure 5.8 The unsaturated hydraulic conductivity function from Torres et al. (1998). Note that they computed this function and did not measure it per se. It comes from the method of van Genuchten (1980) using the sandy loam parameters of Carsel and Parrish (1988).

[9] Fun fact: the Guinness World Record for rainfall intensity is 304.8 mm in 42 minutes for a rain event in 1947 at Holt, Missouri, USA.

the variable source area concept, they were not widely taken up and fully appreciated by the community until well into the 2000s. What would follow the substitution hypothesis would be perceptual models for groundwater effusion to streams in other settings.

5.3 Groundwater ridging

If you can't explain it simply, you don't know it well enough

—Albert Einstein

The groundwater ridging hypothesis was a perceptual model to explain the role of groundwater in storm runoff generation for runoff events in flatter areas where groundwater next to the stream controls the stream's behavior. As we will see shortly, groundwater ridging is restricted in scope to headwater catchments with fine-grained material where a large capillary-fringe or tension saturated zone exists.[10] We'll get to these details shortly.

The groundwater ridging hypothesis focused on groundwater *in contact* with the stream channel margin and the reaction of that groundwater along the channel margin. The perceptual model was born out of the graduate work of Mike Sklash at the University of Waterloo in Canada. His master's thesis work was based on hydrograph separations for flood events in the 6800 km^2 Grand River catchment in Southern Ontario, Canada. For a very large flood event, Sklash et al. (1976, p. 271) used two-component hydrograph separation with ^{18}O and found that "groundwater was the major component of storm runoff, providing more than 50% of the peak discharge." Sklash's subsequent PhD work focused on a *mechanism* for groundwater effusion in a small forested catchment (in Quebec, Canada) and a small agricultural basin (in southern Ontario, Canada).

The groundwater ridging hypothesis as noted by Sklash and Farvolden (1979) was that:

> the increased groundwater discharge during runoff events is apparently related to a rapid rise along the perimeter of transient and perennial discharge areas. This groundwater ridging phenomenon probably arises from the almost instantaneous conversion of the near-surface tension-saturated capillary fringe into phreatic water. (Sklash and Farvolden, 1979, p. 45)

What they are saying is that the top of the water table has a capillary fringe that is technically saturated, but for all intents and purposes, under tension. The size of this fringe (that we discussed, generally, in Chapter 4) is very material-dependent, where very fine-grained material has a sizable fringe (perhaps some 10s of centimeters) and coarse material may have none. Adding water to the capillary fringe means that without any real change in storage, the water table rises almost instantaneously. Sklash and Farvolden termed this phreatic water—or water under pressure: groundwater.

Figure 5.9 shows one of these sites at the low-relief Hillman Creek catchment in southern Ontario, Canada. The data from the site was stage-only and not discharge or runoff per unit area. Something often forgotten in descriptions of Sklash and Farvolden (1979) is that their "observations" of groundwater ridging were defined by computer simulations—not measurements. In that model, they explored four different catchment configurations and the effects of so-called "near-stream unsaturated zone thickness" on the near-stream groundwater response to rainfall. If we listen to Sklash and Farvolden directly, they note:

> Along the perimeter of transient and perennial discharge areas, the water table and its associated capillary-fringe lie very close to the ground surface. Soon after a rain or snow-melt event begins, infiltrating water readily converts the near-surface tension saturated capillary-fringe into a pressure-saturated zone or groundwater ridge (Ragan, 1968). (Sklash and Farvolden, 1979, p. 63)

Of course, here they are referring to Ragan (1968) who we examined in Chapter 3. And like Ragan, Sklash and Farvolden saw that "increases in the hydraulic head in the near-stream groundwater occur soon after the onset of rain. These responses precede, and are apparently independent of, the upland response" (Sklash and Farvolden, 1979, p. 63). They noted too that:

> This groundwater ridge not only provides the early increased impetus for the displacement of the groundwater already in a discharge position, but it also results in an increase in the size of the groundwater discharge

[10] As we discussed briefly in Chapter 3.

Figure 5.9 The Hillman Creek catchment where Sklash and Farvolden (1979) worked.
Source: Mike Sklash

area which is essential in producing large groundwater contributions to the stream. (Sklash and Farvolden, 1979, p. 64)

So there you have it. A perceptual model to explain how groundwater could indeed result in large inputs to the stream on the time scale of an *event*. All leveraged against the earlier work we examined by Ragan and his near-stream well behavior in Vermont. Like that site where fine, rather homogeneous sand promoted large capillary fringes, the sites where Sklash and Farvolden worked also had large capillary fringes.

5.4.1 Laboratory and field-based evidence for the capillary fringe

Mechanistically, the groundwater ridging hypothesis was still lacking true field evidence, other than the measurements by Ragan (1968). It was Gillham (1984), Abdul and Gillham (1984) and Abdul and Gillham (1989) who first explained *how* rapid groundwater response could drive groundwater ridging. This work started out in the lab and later moved to the field. Figure 5.10 shows the 8×120×140 cm plexiglass box model used by Abdul and Gillham (1984). Filled with fine sand and fitted with tensiometers and an artificial stream bank, they rained on this small model and noted that:

if the zone of tension saturation extends to, or near, ground surface, the application of a small amount of water can cause an immediate rise in the water table, and the magnitude of the rise is much greater than would be expected on the basis of normal specific yield values for sandy materials (Abdul and Gillham, 1984, p. 691)

Figure 5.11 shows a moisture release curve that retains water under suction (the tension saturated zone).[11] In this schematic, the capillary fringe is

[11] This is a different sort of curve from the "Torres curve" earlier. That one also plotted pressure head (also sometimes

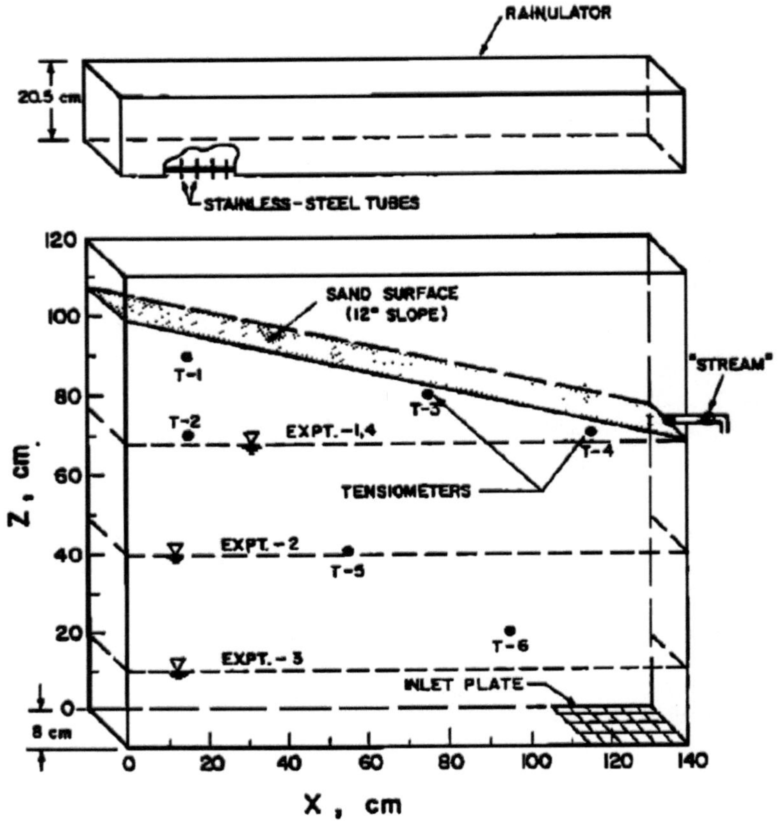

Figure 5.10 The 8×120×140 cm plexiglass box used by Abdul and Gillham (1984).

where soils remain near saturation even under applied suction or tension. There is a rapid decline in the soil moisture at an inflection point—often called the air entry value—and this is what they are referring to as the "height" of the capillary fringe: the distance in cm to the tension reading (also in units of cm or m of water) where that drop-off happens.

Abdul and Gillham (1984) further described their sand box findings (shown in Figure 5.12) where:

> because of the sloping surface of the laboratory model the rising water table caused the rapid generation of hydraulic gradients directed towards the toe of the slope (the stream) resulting in the immediate generation of groundwater discharge to the stream (Abdul and Gillham, 1984, p. 691)

This statement really helps to explain how it works. If the capillary fringe extends to the ground surface, then only a very small amount of rainfall or snowmelt is needed to result in a leap of the water table to the surface. But that leap occurs only around the channel margin. The water table back into the hillslope is largely unaffected by the precipitation event since percolating water takes longer than the event time scale to get to the deeper water table away from the stream channel. And with this rapid increase in water table, they found—just as Ragan

called suction or tension or matric potential) against hydraulic conductivity. But if we looked at similar moisture release curves for Torres and Gillham, they would be very different. Unlike the Gillam curve, the Torres curve would release water with any applied suction or tension. In other words, there would be no capillary fringe. That sort of behavior is characteristic of many coarse-grained microporous forest soils.

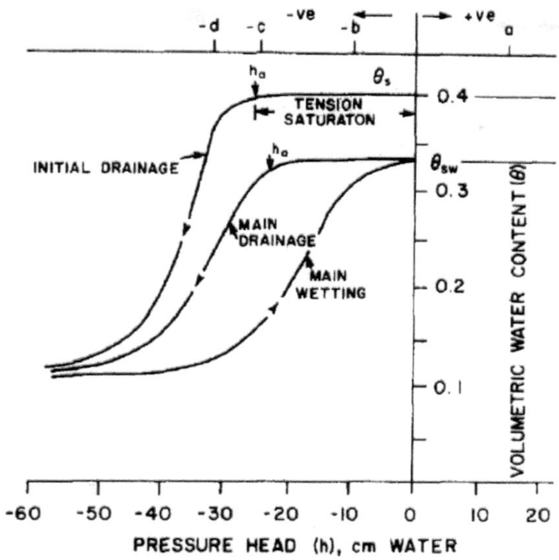

Figure 5.11 A moisture release curve showing the capillary fringe (termed here, tension saturation).
Source: Gillham (1984).

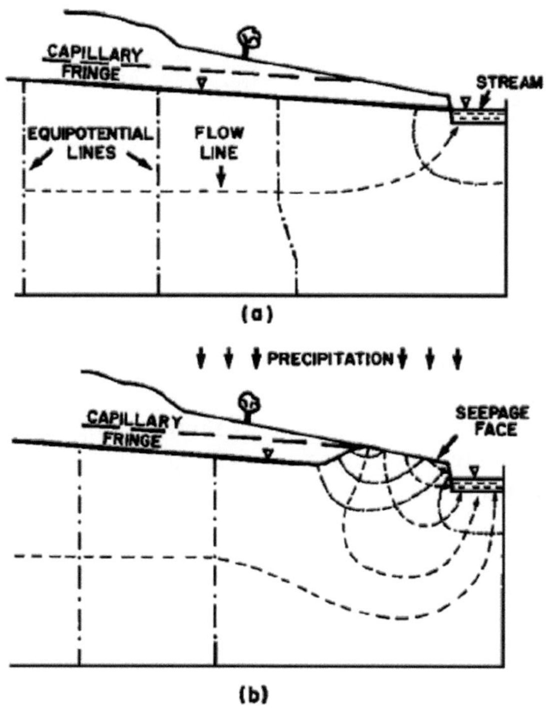

Figure 5.12 A flow net and perceptual model showing the capillary fringe and groundwater ridging (a) before a rain event and (b) immediately after the onset of storm rainfall.
Source: Gillham (1984).

(1968) found in the field at his site in Vermont that we explored in Chapter 3—a groundwater ridge where hydraulic gradients were steepened. That steepening of hydraulic gradients is what promoted the "increased gravity drainage of groundwater to the stream."

Abdul and Gillham (1984) noted that old or pre-event water dominated the hydrographs of their experimental findings, where:

> the proportions of pre-event and event water in the streamflow after steady conditions were reached were determined largely by the extent of the seepage face. This in turn would be determined by the hydraulic conductivity of the medium, the surface slope, and the rainfall intensity. (Abdul and Gillham, 1984, p. 691)

Lastly,[12] in field examination of capillary fringe impacts, Abdul and Gillham (1989), in two field experiments that examined the links between streamflow generation in a shallow water-table aquifer at Chalk River Ontario, Canada (dominated by glacial outwash sand and a sizeable capillary fringe on the order of 0.5 m), noted:

> the water-table adjacent to the stream, where the capillary fringe extended to or close to ground surface, responded rapidly to the precipitation events due to the initially low storage capacity of the medium. This rapid and large response led to the development of a water-table mound on both sides of the stream and flow nets showed that the mound resulted in the discharge of pre-event water through the seepage faces that developed on both sides of the stream. Furthermore, the mound contributed to the discharge of event water in that precipitation falling on the seepage faces was transported directly to the stream as overland flow. (Abdul and Gillham, 1989, p. 1)

So the capillary-fringe groundwater ridging behavior is very much a near-stream phenomenon that it is restricted to fine-grained sediments. Nevertheless, work continues on its manifold expressions and the energy considerations associated with its activation (e.g., Waswa and Lorenz (2015). Now let's examine another mechanism for the rapid release of groundwater to the stream in low angled terrain—transmissivity feedback.

5.5 Transmissivity feedback

Nature provides exceptions to every rule
—Margaret Fuller

The term transmissivity feedback[13] describes how groundwater can rise up into overlying and highly conductive materials to generate fast lateral flow. In a way, Hursh (1944) had already alluded to this behavior in the steeper Coweeta catchments in North Carolina where he noted:

> On sloping land and for natural soil-profiles with strongly differentiated horizons, the writer believes that another factor limiting the rise of the water-table is the tremendous increase in lateral transmission-rate as the water-table approaches the soil-surface. (Hursh, 1944, p. 744)

Hursh is noting here that, though water is able to rise into more macroporous soils, preferential flow allows rapid lateral movement of water rather than the simple 1D rise of the water table as observed at a single well. This rapid lateral flow limits the further rise of the water table—something that would later be referred to as a "hydraulic limitation" on the groundwater response (Anderson et al., 2010).

It would be Allan Rodhe and colleagues in Sweden some 40 years after Hursh who really linked all this to the rapid effusion of isotopically old groundwater in Swedish till soils during spring melt in what we know today as transmissivity feedback. The work was done at the Svartberget catchment as shown in Figure 5.13. It is a low-angled, formerly glaciated environment with extensive glacial till coverage.[14]

Compared to the substitution hypothesis and groundwater ridging, transmissivity feedback is very simple (Figure 5.14) and unrelated to any

[12] We'll return to this groundwater ridging idea and the role of capillary fringes a little bit later in this chapter. And while it has been criticized as a viable mechanism (e.g., Zaltsberg, 1986; McDonnell and Buttle, 1998; Cloke et al., 2006) for coarser forest soils, it is still an early and useful perceptual model for how groundwater can get into a stream channel on the timescale of an event in low-relief areas with fine-grained materials.

[13] A term first used in two PhD theses; ". . .for want a generally-accepted term, shall be termed 'transmissivity feedback'" (Lundin, 1982, p. 33); and then again in Bishop (1991) in his PhD thesis, now available online at https://doi.org/10.17863/CAM.120776

[14] Note Svartberget is now part of the well-known Krycklan catchment study, the largest catchment monitoring program in Europe (see Laudon et al., 2013 for details).

Figure 5.13 The gauging station in one of the study catchments in the Rodhe PhD thesis, Svartberget with catchment area 0.5 km^2.
Source: Photo by Ben Abbott.

Figure 5.14 The perceptual model of transmissivity feedback.
Source: Grip and Rodhe (1985).

capillary fringes linked to groundwater ridging or displacement ideas coming from the substitution hypothesis. Simply put:

> the slope of the groundwater surface remains almost constant, except closest to the stream. The water flow in the hillslope increases partly because the cross-sectional area (height) increases, but mainly because the added saturated layers have higher conductivity than those below. Grip and Rodhe (1985, p. 93)

So what Grip and Rodhe are saying is that for a low-relief, low-energy environment with glacial till, water would move slowly in that low conductivity till until rain or snowmelt filled the storage in the till and the water table rises into the overlying and highly conductive mineral soil.

Figure 5.15 shows data from one of Rodhe's measured sites and the runoff-groundwater relation. It shows a very non-linear increase in runoff once the water table rises up into the mineral soil. Beyond the hydrograph separations and explanation via transmissivity feedback, Rodhe (1987) also estimated the storage volumes of both soil water and groundwater using the damping ratio of the isotope signal in the stream. He concluded that these volumes

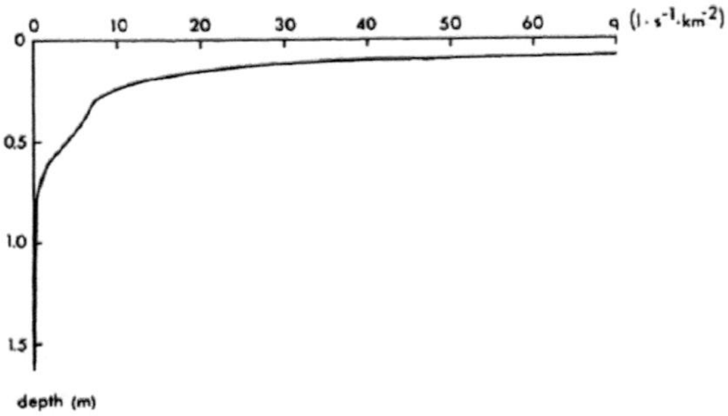

Figure 5.15 The highly nonlinear relation between the depth of the groundwater table and the amount of streamflow, as one of the earliest descriptions of what was later called transmissivity feedback, namely that streamflow increases dramatically once the water level rises into a more conductive layer.

Source: Rodhe (1987).

suggested that catchment transit times were on the order of 1.5–9.5 years, supporting the notion that groundwater dominated streamflow (and not rainfall or snowmelt).

The shortness of this small section is in no way linked to any lack of importance of transmissivity feedback as a robust perceptual model. Indeed, this section has been short since transmissivity feedback was so clearly defined early on. It is a simple and powerful idea with transmissivity feedback being certainly much more widespread than groundwater ridging, with many recent studies noting its behavior in northern peatland environments (e.g., Waddington et al., 2015), in upland glaciated terrain in the Northeast USA (e.g., Detty and McGuire, 2010) and even in follow-up studies at the Sleepers River catchment in Vermont USA—the home of saturation excess overland flow (e.g., Kendall et al., 1999).

Now let's look at one last early perceptual model for rapid groundwater release to the stream—this time in very steep and wet environments and linked to the pipeflow and perched groundwater.

5.6 Pipeflow of perched groundwater

I don't study to know more, but to ignore less
—Juana Ines de la Cruz

Our three perceptual models for groundwater effusion that we have examined thus far have all been about groundwater that was present in the catchment before the runoff event—groundwater connected to the stream and in sufficient volume to account for the volume of old (pre-event) water in the storm hydrograph. Isotope-based hydrograph separation studies continued after the Sklash and Farvolden (1979) paper (that outlined groundwater ridging) (Table 5.2). And by the mid 1980s, more and more results came from much steeper and wetter headwater catchments with incised valley bottoms. Meaning, places where there may not be enough groundwater in a discharge position to account for the volume of old or pre-event water in the stream hydrograph. By the mid-1980s, studies were already hinting that hillslope water (soil water or perched groundwater tables set up during events) may be needed to explain the isotope signal (e.g., Kennedy et al., 1986).

5.6.1 Pipeflow was assumed to be new or event water associated with storm rainfall

Two of the studies listed in Table 5.2 (Pearce et al., 1986 and Sklash et al., 1986) were from the Maimai catchment in New Zealand that we examined in Chapter 4 (as shown in Figure 5.16). Maimai is indeed a steep, wet place with very incised valley bottoms. It is also the place that Mosley (1979) had described (recall from Chapter 4) as steep slopes riddled with macropores; where soil pipes created exceptionally responsive pipeflow-enhanced runoff. In addition to his natural rainfall events, Mosley (1979) also conducted simple water-dye

Table 5.2 Selected hydrograph separation studies using ^3H (T, tritium), ^{18}O and ^2H (D, deuterium) for the period 1979–1986.

Authors	Snow Melt (-) or Storm Runoff (+)	Catchment size (km²)	Country	Isotope		
				T	^{18}O	D
Herrmann and Stichler (1980)	-	18.7	West Germany			*
Schwartz (1980)	-	368	Canada	*	*	
Sklash and Farvolden (1980)	-	2.8	Canada	*	*	
Rodhe (1981)	-	4.0, 6.8	Sweden		*	
Bottomley *et al.* (1984)	-	10.5	Canada		*	
Bottomley *et al.* (1984)	+	10.5, 1.24, 1.76	Canada		*	
Rodhe (1984)	-	several	Sweden		*	
Christophersen *et al.* (1985)	-	41 ha	Norway		*	
Bottomley *et al.* (1986)	-	10.5	Canada		*	
Hooper and Shoemaker (1986)	-	42.4	USA		*	
Jacks *et al.* (1986)	+	33.5	Sweden		*	
Kennedy *et al.* (1986)	+	620	USA	*		*
Obradovic and Sklash (1986)	-	60	Canada		*	
Pearce *et al.* (1986)	+	3.8 ha	New Zealand		*	
Sklash *et al.* (1986)	+	3.8 ha, 1.6 ha, 0.3 ha, 2.8	New Zealand			*

Adapted from Sklash (1990).

additions to his short hillslope sections. There he saw rapid breakthrough of the applied tracer through preferential flow paths. He interpreted this as the rapid subsurface stormflow of storm rainfall (i.e., event water). These dye responses were, as he noted, "300 times greater" than the saturated hydraulic conductivity of the soil. So, his perceptual model—and indeed all the pipeflow studies from steep, wet places that we examined in Chapter 4—was one of pipeflow of rain water through the hillslopes. Or as Jones (2025, p. 1269) put it in his review of this early pipeflow era: "all speculated that the pipes were either collecting surface runoff or else draining an overflowing storage that had been fed directly by rainwater infiltration upslope."

Pearce et al. (1986, pp. 1263) followed the Mosley (1979) work at Maimai and used a stable isotope tracing approach to refute the idea of macropore flow of new water, noting that "the processes which deliver 'old' water are largely responsible for hydrograph generation." They go on to say that:

> currently favored runoff mechanisms, which involve the rapid flow of 'new' water over the ground surface or through the soil matrix, soil macropores, or other rapid transit pathways, cannot explain the streamflow response. (Pearce et al., 1986, pp. 1263)

So in other words, this isotope-based work fundamentally contradicted the earlier work of Mosley. If we look at the data from Sklash et al. (1986)—the companion study to Pearce et al. (1986) from Table 5.2—isotope tracing shows that the majority of storm runoff in the stream on the timescale of an event was water that existed in the catchment before the rain stated falling: i.e., old or pre-event water. Figure 5.17 shows the rainfall (plotted in the second panel from the top) and the hydrogen isotope values (deuterium) in the range −5 to −15 per ml (on the δD or δ^2H scale in the second panel from the top). The stream starts at a value near −40 per ml and deflects upwards on the plot, toward the rainfall. But it only deflects a little bit toward the rainfall. This was quantified using the simple two-component mixing model (shown in the third panel from the top) that shows that rainfall makes up only a small percentage of the total streamflow, even at the peak of this very flashy hydrograph. Of course, this was consistent with most of the studies shown

Figure 5.16 Aerial photograph of the Maimai catchment in New Zealand. The catchment with the dotted line around it is the M8 catchment. *Source*: McDonnell et al. (2021).

in Table 5.2 and studies that came before that (and after). But now there was a major conundrum: how could soil pipes deliver pre-event water when they were thought up until that point to be deliverers of event water?!

5.6.2 Groundwater ridging at Maimai?

The question then became: what mechanism *could* explain the isotope findings at Maimai? Remember, Pearce et al. (1986) were highly skeptical of pipeflow of new water:

> Our study refutes these earlier conclusions and casts serious doubt on the relevance of macropore flow and other forms of rapid throughflow of new water in steep, forested catchments that have shallow, highly permeable soils, and are highly responsive to storm rainfall. The appropriateness of field methods which have demonstrated rapid throughflow velocities is also placed in serious doubt. (Pearce et al., 1986, p. 1270)

Sklash et al. (1986) proposed the groundwater ridging mechanism (recall from Section 5.4) to explain how the Maimai catchment could release old water so quickly to the stream, noting:

> The physical response of the Maimai streams is entirely consistent with the concept of groundwater ridging. According to this concept, a disproportionately large rise in the water table is caused by the conversion of the tension-saturated capillary fringe into phreatic water by infiltrating rain [Sklash and Farvolden, 1979; Gillham, 1984; Abdul and Gillham, 1984; Dracos, 1984, and others]. (Sklash et al., 1986, p. 1279)

Sklash et al. (1986) rationalized this for the steep, wet, Maimai catchment by noting that the response is:

> one in which saturated wedges on the lower slopes and groundwater ridges in the valley bottoms develop quickly as infiltrating rain converts the tension-saturated zone into phreatic water. This conversion rapidly increases the hydraulic gradient to the stream and promotes increased gravity drainage of groundwater. When the rainfall rate cannot sustain the groundwater discharge rate, the phreatic zone thins, the hydraulic gradient to the stream diminishes, and the stream hydrograph begins to recede. (Sklash et al., 1986, p. 1280)

It all sounded so plausible. Except it wasn't.

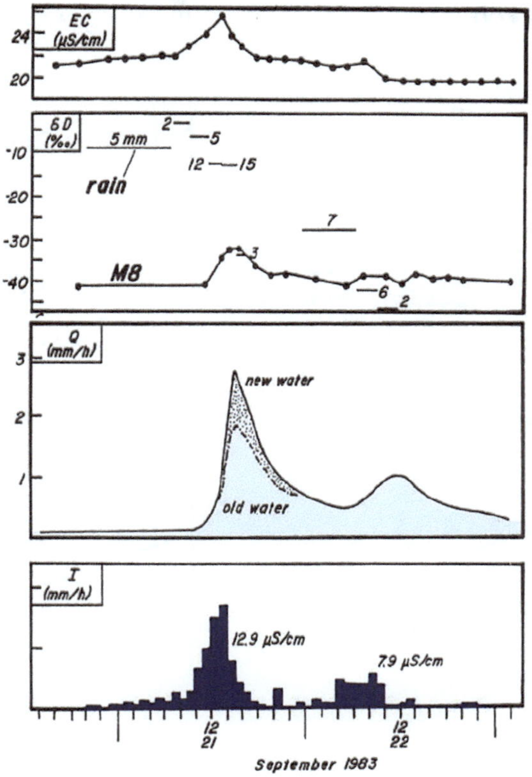

Figure 5.17 Isotope hydrograph separation at the Maimai catchment in New Zealand showing the dominance of old water.
Source: Sklash et al. (1986), re-drawn by Sklash (1990).

5.6.3 Pipeflow of old (pre-event) water

McDonnell (1990) examined the moisture release curves for the Maimai soils and found no evidence of a capillary fringe—rather that the soils released water with any applied tension. And then there was the issue that Maimai did not have enough old water in the valley bottom in a discharge position to account for the old water volume in the storm hydrographs that Sklash et al. (1986) had observed. That old water was sourced from the hillslopes—at least for larger rainfall events.

McDonnell (1990) had one additional piece of information that neither Mosley (1979) nor the Pearce et al. (1986) and Sklash et al. (1986) groups had: a network of recording tensiometers to go after and explain mechanistically how the new water input was converted to old water output. What these tensiometers indicated[15] was that preferential flow moved vertically into the soil and triggered the development of transient saturation at the soil–bedrock interface. Once transient saturation formed on the hillslopes, lateral pipeflow activated and delivered the pre-event soil water quickly downslope and into the stream. In other words, a perceptual model of pipeflow of old water (as shown in Figure 5.18 in terms of how this counters two previous perceptual models at Maimai).

Figure 5.19 shows in more detail how water bypasses to depth while the soil matrix may still have unrequited storage capacity. McDonnell (1990) noted that:

[15] Of course, tensiometers can't measure preferential flow—but bypassing of water *can* be detected with disequilibrium in matric potentials—meaning, deeper tensiometers responding ahead of shallower tensiometers.

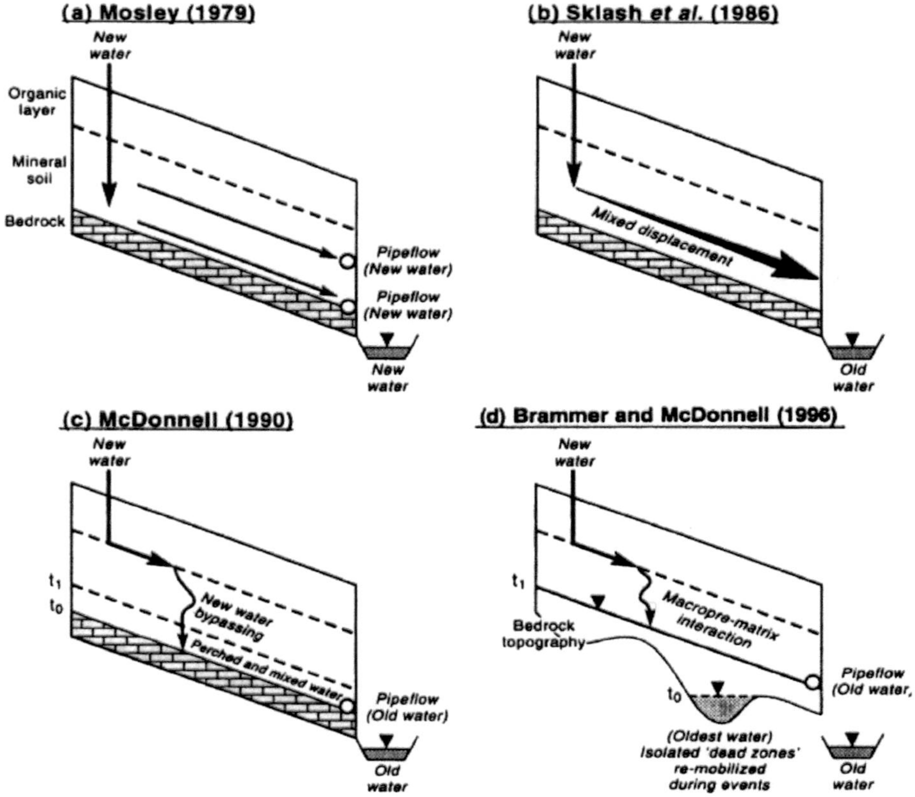

Figure 5.18 The evolving perceptual model at Maimai showing the progression of ideas as new information came in—ultimately arriving at a perceptual model of pipeflow of old (pre-event) water.

Source: Modified from McGlynn et al. (2002).

soil water diffusion to unsaturated locales is too slow, relative to its preferential drainage via a crack," and "as invading new water moves to depth, free water perches at the soil-bedrock interface, and 'backs-up' into the matrix where it mixes with a much larger volume of stored old matrix water", then "once free water exists, large pipes in the lower soil zones quickly dissipate transient water tables laterally downslope, producing a rapid throughflow response of well-mixed old water. (McDonnell, 1990, p. 2830)

Six years after these results were published from the Maimai catchment, they would be supported by studies at another steep, wet and pipe-riddled set of catchments at Plynlimon in Wales where Sklash et al. (1996) showed that:

The isotope and electrical conductivity data suggest that storm runoff in both the main pipe and the stream is overwhelmingly old water. A sensitivity analysis suggests that the old water is supplied both from near-stream groundwater and upslope groundwater delivered by the ephemeral soil pipes. (Sklash et al., 1996, p. 921)

There is a full-circle element to this follow-on work in the sense that the Plynlimon catchment is essentially[16] where Tony Jones (from Chapter 4) observed pipeflow dynamics. We now know that soil pipes, and preferential flow paths in general, have the potential to transport isotopically old water to the stream on the timescale of events, including tile-drained agricultural hillslopes (e.g., Klaus et al., 2013). We will return to the interplay between soil pipes and flow responses in Chapter 6.

[16] Technically his work was just outside of the Plynlimon catchment boundary.

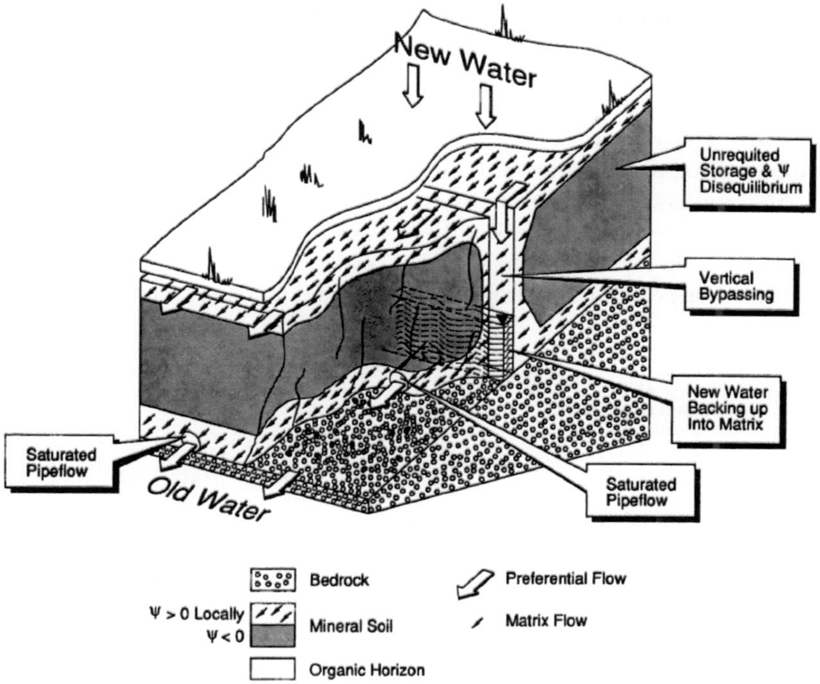

Figure 5.19 The perceptual model of McDonnell (1990) for rapid effusion of old water through soil pipes. The idea combined elements of unrequited storage and matric potential disequilibrium for generating preferential flow of "fully mixed (old) water."

5.5 Summary and looking ahead

Example is always more efficacious than precept
—Samuel Johnson

The period ~1975–1990 is when isotopes came to the fore in catchment hydrology and when perceptual model development and testing would never be the same again. In the span of less than 20 years, a field was revolutionized: a new technique that tracked raindrops through the catchment allowed for the description of the velocity of the water molecules rather than pressure propagation, and hydrograph separation using mixing models rather than some arbitrary graphical slope-line hydrograph separation.[17]

In the many decades since the first hydrograph separation studies, evidence has continued to show the dominance of stored water in the storm hydrograph. While the event water getting to the stream happens the same way everywhere (i.e., direct channel precipitation, overland flow), the old water release happens in a multitude of ways (e.g., the four early ones reviewed in this chapter)—but each being some form of *storage-release* as we will explore in Chapters 6 and 7. The most recent review by Jasechko (2019) shows that:

> sixty percent of the n=101 studies suggest that more than half of streamflow is comprised of groundwater (i.e., minimum value shown in the figure exceeds 50%). Eighty-five percent of compiled studies suggest that groundwater may comprise more than half of streamflow (i.e., maximum value shown in the figure exceeds 50%). (Jasechko, 2019, p. 907)

Again, we need to be careful about the term groundwater—as in the Jasechko reviewed studies, and indeed the work we reviewed in this chapter, groundwater could also be transient groundwater and perched water tables within the soil prolife. Nevertheless, Figure 5.19 shows this graphically

[17] Recall the quickflow separation technique from Chapter 3.

and shows the considerable geographical expansion from the tabular summaries of early studies in Tables 5.1 and 5.2. These studies are heavily weighted to forested, mid-latitude catchments and combine event(s), season(s), or year(s), as noted in a review of isotope hydrograph separation (Klaus and McDonnell, 2013) and as shown by the blue bars in Figure 5.20. Nevertheless, even at the peak of the hydrograph, the gray bars in Figure 5.20 show that stored water dominates the peak of the hydrograph.

It is worth noting that the field has now largely moved on from simple two-component hydrograph separations. There was a whole other mixing model approach with stream chemistry called end-member mixing, developed by Christophersen and Hooper (1992) that we have not discussed. Two-component mixing largely gave way to three-component models in the 1980s to include the soil water component (DeWalle et al., 1988); and then to continuous open system mixing (Harris et al., 1995)

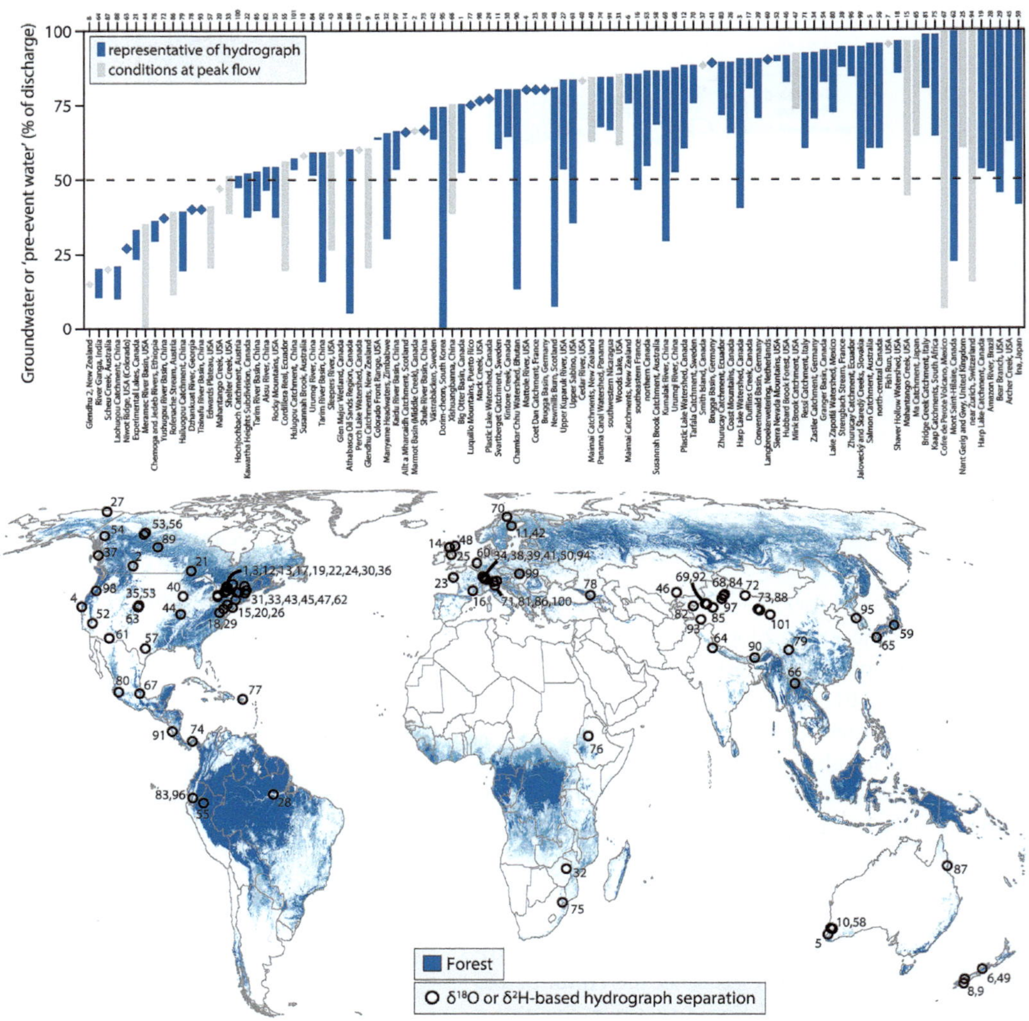

Figure 5.20 Stable-isotope-based hydrograph separations. (a) Compiled estimates of the percentage of streamflow comprised of groundwater or pre-event water based on $\delta^{18}O$ or $\delta^{2}H$ data (n=101 studies; each column represents a study).

Source: Jasechko (2019)

and a combined tracer and runoff transfer function approach called TRANSEP (Weiler et al., 2003); and most recently, the "young water fraction" approach of Kirchner (2016) that we will return to in Chapter 7.

Many open questions exist though. One relates to the fact that in some wet catchments, there may be no measurable rainfall in the storm hydrograph! This can be the ultimate test of our ability to understand the difference between celerity and velocity. Consider Figure 5.21 before we end this chapter. These are the first high-frequency isotope data that I am aware of, collected in a headwater catchment for a rainfall–runoff event. The upper panel is the

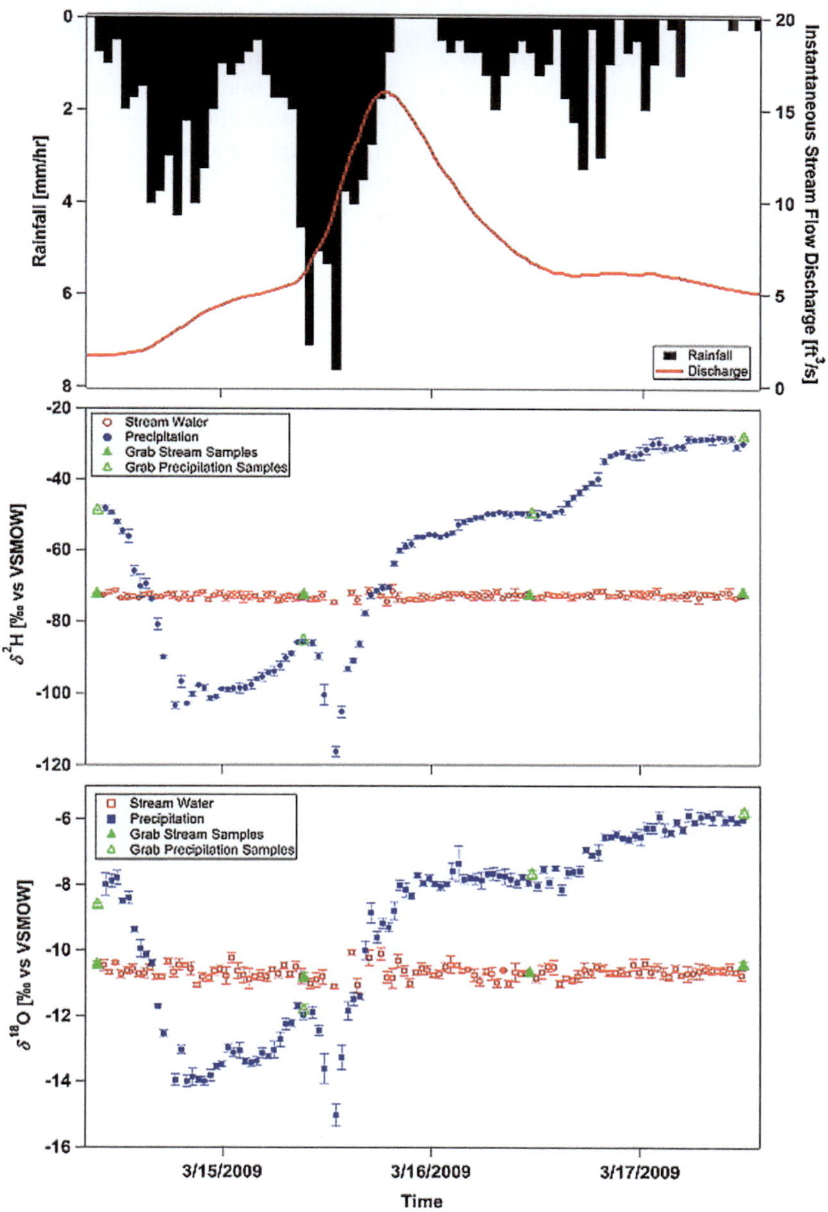

Figure 5.21 Stable isotope response for a rainfall event at the 1 km^2 WS1 catchment at the HJ Andrews catchment in Western Oregon, USA. *Source*: Berman et al. (2009).

rainfall hyetograph in black and the stream discharge in red for a 1 km^2 headwater catchment at the HJ Andrews catchment in Western Oregon, USA. The two lower panels are ~20-min interval stable isotope data for hydrogen and oxygen. What do you see? If you answered "there is no rainwater in the stream, even at the peak of the hydrograph," then you would be correct. The rainfall (blue dots) shows considerable variations in time for the event, linked to air mass trajectory, rain intensity, etc. But the streamwater isotope signal (in red) is essentially a flat line. If you were a doctor, you might say that "the patient is dead." Here, as a hydrologist, we would say that the stream responds quickly and substantially to the rainfall event (i.e., the hydrograph; the celerity) but the water in the stream during the event is water that was in the catchment prior to the rain event—pushed out in some way, shape or form. In other words, with very low velocities such that no discernible difference in stream signal is registered within the two-day period. Now this is perhaps an extreme case—it is steep and wet. Other sites, as we saw at the beginning of this chapter, show a mixture of event and pre-event water; albeit one usually more weighted to stored water than to event precipitation. But I hope this example shocks you into realizing the difference between celerity and velocity; the latter being the source of endless open questions regarding the flow pathways needed to explain this behavior—some of which we will address in Chapters 6 and 7.

Our open questions in terms of the role of groundwater in streamflow generation now go well beyond hydrograph separation. The big question now is *how old* is the old water (the pre-event water) in the storm hydrograph and throughout the full streamflow response over days, months and years in a headwater stream. Chapter 7 will get at that and some of the other modern developments vis-à-vis catchment transit time and catchment storage.

PART 2

What Do We Think We Know?

CHAPTER 6

Flowpath connectivity (~2000–present)

6.1 Thresholds and connectivity

If your science experiment needs a statistician, you need to design a better experiment
—Ernest Rutherford

Two new important themes emerged in streamflow generation research in the late 1990s: thresholds and connectivity. The studies listed in Figure 6.1[1] ushered in changes in thinking—moving beyond the linear growth of variable source areas in the surface and subsurface—and going after the sources of non-linearity and complexity in catchment systems. A good dictionary definition of non-linearity in this sense is "when the outputs are not proportional to the inputs across the whole range of the inputs." Let that sink in a tick. It means that not all precipitation events are created equal. Some may do nothing for runoff production; others may have a huge impact. And as we will see for the many other catchment systems that we will now examine, the old way of thinking of linear wet-up and dry-down (e.g., saturated wedges) would give way to a new way of thinking: catchments as systems with pieces that connect and disconnect in highly threshold-like and scale-dependent ways.

This new way of thinking of thresholds and connectivity was born out of analyses of a much larger range of storms and storm sizes and places. Here, the old adage "that to know what is enough, you need to know what is more than enough" is perhaps a good shorthand for this development. And importantly in relation to more storms and more places came examination of patterns. A key early study from Figure 6.1 (top left-hand corner) highlighting the power of looking at spatial patterns was the PhD work of Andrew Western at the University of Melbourne in Australia. His work provided one of the first clear examples of the link between spatial patterns in soil moisture and the threshold response in stormflow in the stream. We'll come back to other studies and other scales and climates shortly. But first let's look at Western and Grayson (1998) to see what they did, what they found and why it was so transformative for streamflow generation understanding.

Western and Grayson (1998) mapped soil moisture across the 10-ha Tarrawarra catchment in Southeast Australia (Figure 6.2). Their results showed striking features: during dry states, mapped patterns of soil moisture were somewhat random. But during wet periods, the soil volumetric water content was not only higher but also showed topographic organization. Think back to Anderson and Burt (1978) from Chapter 4—this is like their hillslope hollow work,[2] but now writ large at the catchment scale. And what did these patterns at Tarrawarra suggest? They suggested that during dry periods, flow was dominated by vertical transfer of water in and out of the catchment. But during wet periods, lateral flow (subsurface stormflow and saturation excess overland flow) caused the topographic organization and re-organization of soil moisture.

[1] This new connectivity thinking also affected the geomorphology community who have focused on the evolving nature of structural controls with many studies noting the importance of the interaction between topographic controls and catchment processes as the key to understanding the dynamics of hydrological connectivity (e.g., see Bracken and Croke, 2007; Wainwright et al., 2011, Bracken et al., 2013; Wohl et al., 2019).

[2] Of course, they measured matric potential and not volumetric water content.

Streamflow Generation. Jeffrey J. McDonnell, Oxford University Press. © Jeffrey J. McDonnell (2026). DOI: 10.1093/9780191953736.003.0006

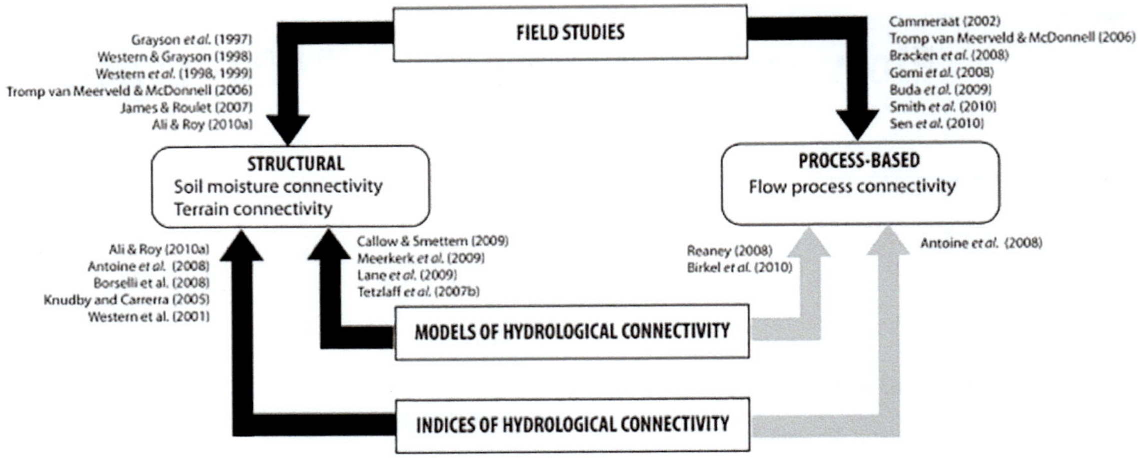

Figure 6.1 Some of the key early studies and systems thinking on connectivity. The term structural connectivity means the spatial patterns in the catchment that affect the flowpaths to the stream. The so-called functional aspects of connectivity (termed "process connectivity") gets at spatial patterns (for example, surface microtopography, or soil depth distribution) and how they interact with connected flow to affect water transfer in catchments.

Source: Bracken et al. (2013).

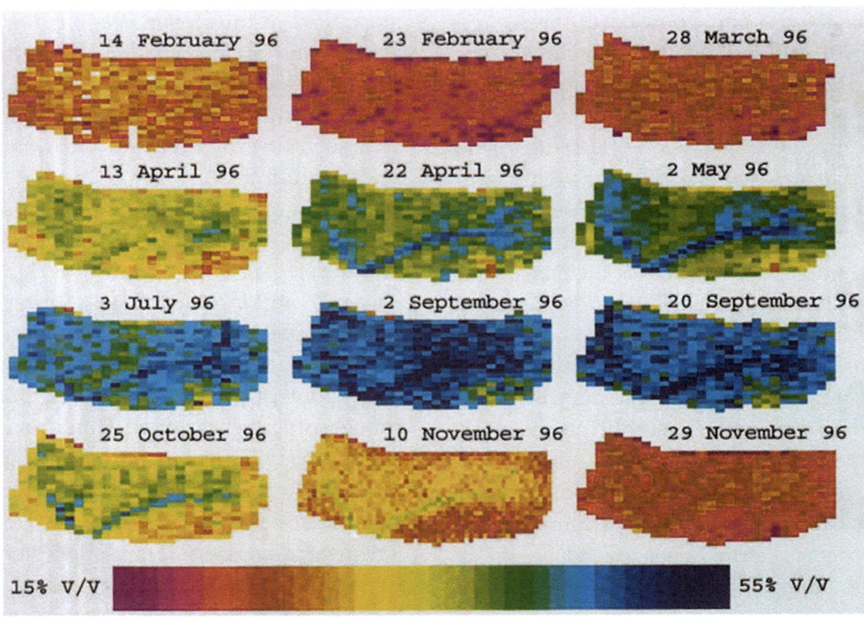

Figure 6.2 Maps showing the spatial patterns of soil moisture change across the 10-ha Tarrawarra catchment in Australia from the PhD thesis of Andrew Western. Each rectangle represents a measurement of the average volumetric water content over the upper 30 cm of the soil profile.

Source: Western and Grayson (1998), re-drawn in Western and Grayson (2000).

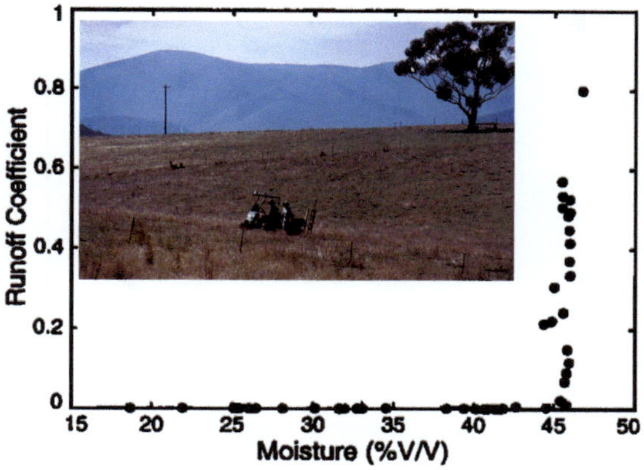

Figure 6.3 The runoff coefficient for Tarrawarra against soil moisture for days with >5 mm rainfall. See the vehicle with which they made the soil moisture measurements using TDR, all tracked by GPS in the inset.

Source: Western and Grayson (1998). Inset photo by Andrew Western.

But beyond the beautiful patterns, Western and Grayson (1998) and a number of other papers by them (e.g., Grayson et al., 1997; 1998 and others that followed into the early 2000s) showed the consequence of the spatial patterns for catchment scale runoff: a clear and sharp threshold response. Figure 6.3 shows the runoff coefficient (i.e., the runoff ratio; essentially event rainfall divided by event runoff) as a function of average soil moisture for the Tarrawarra catchment. Below a volumetric soil water content of about 45%, the runoff coefficients were effectively zero. Once this moisture content was exceeded, the runoff coefficients rose sharply with exceptionally high runoff coefficients (over 80%) for the wettest state. So, their catchment runoff response could be explained by alternating soil water patterns switching between the two preferred states; that it was not only the *average* moisture but also the *patterns* of soil moisture that mattered. And it mattered greatly given how dramatic the difference in streamflow responses were for events on either side of the moisture threshold in Figure 6.3.[3]

There would be much follow-up work on connectivity to the early Western and Grayson (1998) as shown in Figure 6.1—notably in Italy by Penna et al. (2009; 2011). These, and other examples would showcase the relation between connectivity and threshold responses, something we now think of as "emergent behavior."[4] Emergent behavior is something more than the individual runoff processes we have seen thus far (i.e., subsurface stormflow, saturation excess overland flow, infiltration excess overland flow). Think of emergence like the old saying "the whole is greater than the sum of its parts."[5] More technically, connectivity as emergent behavior is a "property or phenomenon" that is observed that "somehow transcends the level of the objects that nevertheless produce it" (de Haan, 2006, p. 293). So, connectivity of any point in the system depends on where that point is in a connected network; that when pieces of the system become linked, the response is often threshold-like in nature.[6] So, connectivity is an emergent behavior that leads to a threshold.

Let's now go beyond soil moisture patterns[7] and their associated thresholds and get to the root-cause, mechanistic controls that might drive connectivity and threshold behavior. We will start with examination of the mechanisms underlying the link between connectivity and threshold response at the hillslope scale.

6.2 Connectivity and thresholds for subsurface stormflow

Science is but an image of the truth

—Francis Bacon

The first study to reveal the clear mechanistic linkage between spatial patterns of perched water tables, connectivity and the threshold response for subsurface stormflow generation (as listed in

[3] In Canada we would call this a "hockey stick" graph, named after our beloved icon and sport.

[4] Note that the idea of emergence came much earlier to other fields like ecology and is now found throughout the scientific literature from physics to social sciences.

[5] This was first said by Aristotle who recognized "connectivity" in the context of relationships and interactions that between parts, something new is created that is "distinct from the parts themselves."

[6] With these thresholds being like tipping points for runoff.

[7] There would be many other studies elsewhere on mapped patterns of soil moisture in relation to streamflow (e.g., in Italy by Penna et al., 2011; 2015).

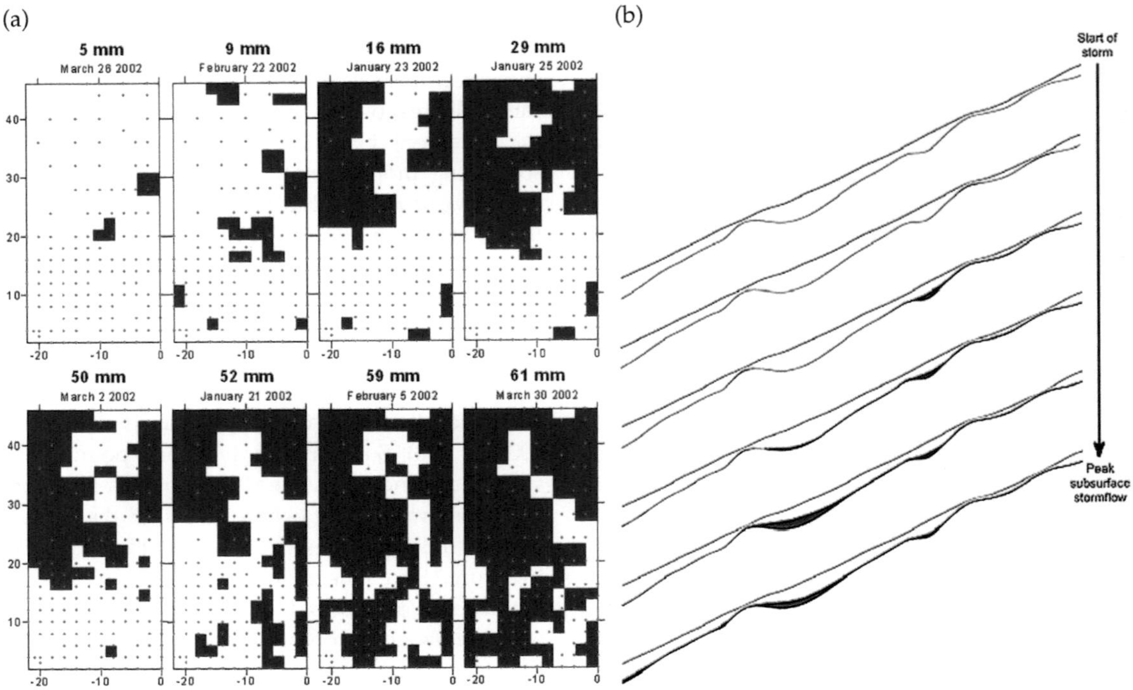

Figure 6.4 (a) Fill-and-spill behavior from a forested hillslope in Georgia, USA. Ponding produced at the soil–bedrock interface with growth and later connectivity of subsurface saturated patches resulting in a threshold runoff behavior at the slope base (see Figure 6.7). (b) The right-hand side illustrates the same behavior for a 2D cross-section of the slope, showing that such runoff may cascade from the upper slope sections to lower slope sections. (Note that Ilja first saw this behavior in a class project in her PhD using a 2D simulation of flow across the hillslope with the Hydrus 2D model. These simulations inspired the subsequent field measurements. We will discuss the dialog between experimentalist and modeler in Chapter 9.)

Source: Modified from Tromp-van Meerveld and McDonnell (2006b).

Figure 6.1) was the PhD work of Ilja van Meerveld at Oregon State University, working at the Panola catchment in Georgia, USA. Figure 6.4 shows how connectivity presented itself at that hillslope—in plan-view (Figure 6.4a) and in a cross-section (Figure 6.4b). Tromp-van Meerveld[8] and McDonnell (2006a, b) showed for this forested hillslope[9] how increasing amounts of rainfall for different events created increasingly connected patches of saturation at the soil–bedrock interface. For small events (e.g., the 5 and 9 mm events in the upper left of Figure 6.5a), only small patches of saturation (i.e., the transient perched water table) developed at the soil–bedrock interface. But as storm size increased, so did the total area of the saturated patches and their connectedness. At a rainfall event threshold of ~60 mm, there was at last connectivity between these subsurface patches of saturation and the lower slope boundary. As a result, flow at the base of the slope "turned on." With all this driven by subsurface topography and not the surface topography. We will look at the flow response from this slope in just a tick—but for now, let's understand what might have driven these patterns.

Of course, the bedrock surface map is a map of the subsurface interface; something we already learned about in Chapter 4 that could be a key factor for generating subsurface stormflow. Figure 6.5 shows a survey of the surface and subsurface topography

[8] Note that Tromp-van Meerveld is Ilja van Meerveld—a leader in our field whose surname is now simply "van Meerveld".

[9] Soils were loamy sand over a well-defined soil–"bedrock" interface where the bedrock was granite and sometimes weathered saprolite.

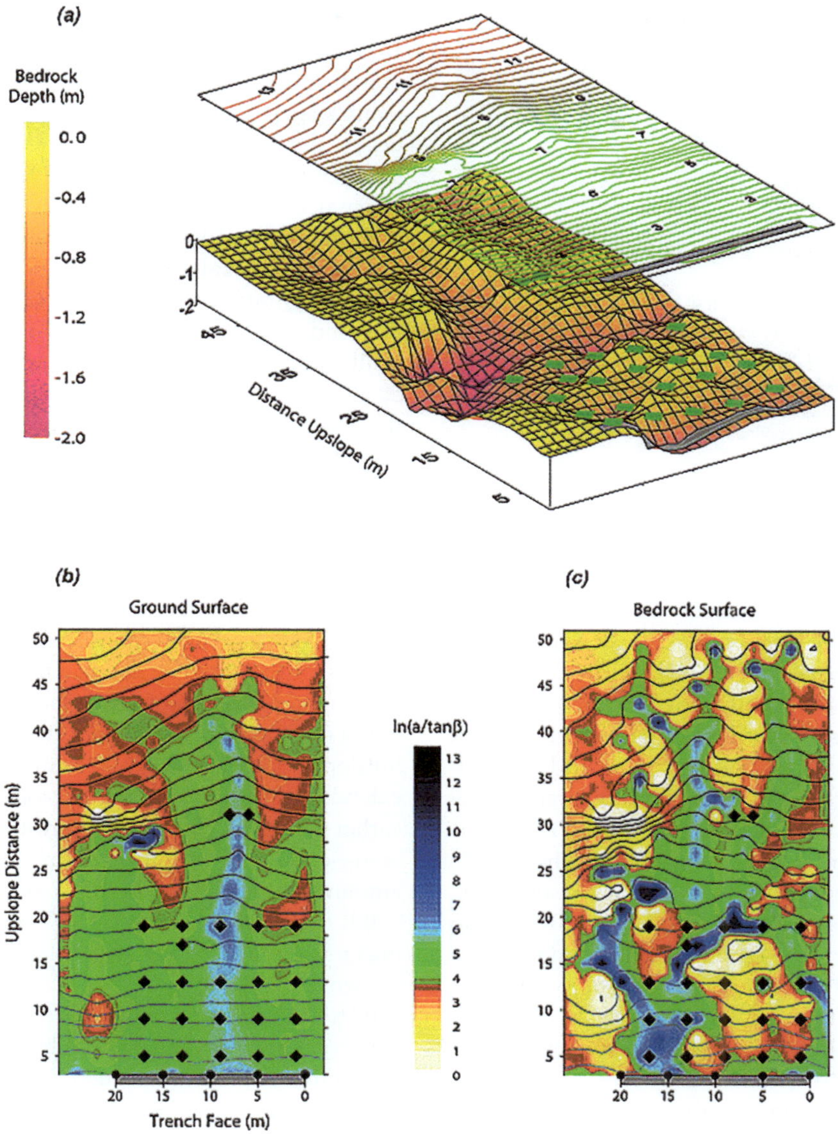

Figure 6.5 Hillslope terrain analysis for the Panola trenched hillslope showing (a) depth to bedrock, (b) topographic indices (technically, a calculation of upslope contributing area "a" to any given point, divided by the local slope angle "tan β") for both the ground surface, and (c) bedrock topography. The green (Figure 2a) and black (Figures 2b and 2c) diamonds show the co-located tensiometer positions.

Source: Freer et al. (2002).

from the trenched hillslope at Panola. It also shows a calculated topographic index[10] in Figure 6.5b and 6.5c. The patterns of saturation that we saw in Figure 6.4 show that flowpaths of water at the hillslope scale followed the topographic index patterns shown in Figure 6.5c—meaning, the water followed the topography of the subsurface interface and not

[10] Recall from a footnote in Chapter 4 that the topographic index was developed originally by Beven and Kirkby (1979) and is a way to determine the downslope flow of water.

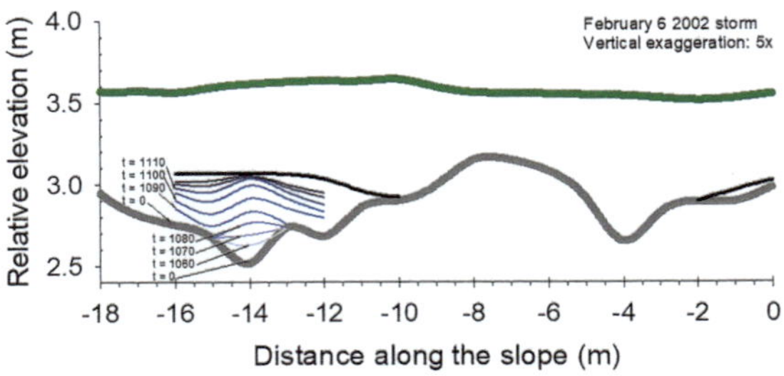

Figure 6.6 Saturation development at an interface from time (t) equals zero to 1130 minutes into a rainfall event based on continuous groundwater level measurements. This filling of water then spills laterally downslope once it rises high enough to overcome the downslope impedence to flow. The thick lines represent the surface topography and bedrock topography.

Source: Tromp-van Meerveld and McDonnell (2006b).

the surface topography.[11] The identification of subsurface topography and its role in producing patterns of subsurface stormflow had first been seen at this site (by McDonnell et al., 1996) who showed how mapped patterns of transient saturation (that produce the lateral subsurface stormflow) were controlled by this subsurface topography. Let that sink in. The key message here is simple and so obvious in hindsight: that *subsurface* topography was the key "surface" for subsurface stormflow. This is consistent for all we have looked at in Chapters 2–4—the difference being that until McDonnell et al. (1996) and then follow-on work by Tromp-van Meerveld, no one had thought to map and explore the *patterns* of this subsurface topography and how it might link to emergent behavior at the hillslope scale and flow networks thresholds for runoff generation.

We have now seen how these topographic irregularities in the subsurface topography affect the build-up and patterns of transient saturation at the interface. But how does it look if we zoom in on an irregularity? Figure 6.6 shows data for a point scale cross-section from the Panola slope—roughly 18 m wide and 1 m deep. It shows how saturation at an interface develops. Like we saw in many examples in Chapter 4, water infiltrates vertically. But the interface between the loamy sand soil and the underlying saprolite and granite on this slope at Panola meant that water "backed up" at the interface and localized saturation builds with time during a rainfall event. Of course, not *all* of the infiltrating water "backs up" at the interface; indeed at this site, ~90% of it continues vertically *past* the interface[12] (Tromp-van Meerveld et al., 2007). But some of it does back up. And the time series shown in Figure 6.6 illustrates how microtopographic depressions fill at the soil–saprolite interface[13] before water starts to spill, or flow, laterally downslope; and with the filling of the depressions connecting spatially as the storm size increases.

Finally, the upshot of these patterns of saturation at the interface and the connectivity at the hillslope scale is the threshold response at the slope base. For Panola, Figure 6.7 shows how the ~60 mm threshold of event precipitation that is needed to activate hillslope-scale fill and spill[14] forms saturation that extends to the bottom of the hillslope (that we saw earlier in Figure 6.4)—where a threshold response is seen—effectively the "turning on" of considerable amounts of connected subsurface flow.

But Panola is just one site. How might these findings relate to very different hillslopes elsewhere in the world?

It turns out that many follow-on studies in very different environments showed this same connectivity–threshold relation. Examination of the wetter and steeper Maimai hillslope, on the South Island of New Zealand—a place we examined in Chapters 4 and 5—shows a threshold response

11 Other work on a different hillslope in the same catchment showed that the control of the bedrock topography on the groundwater flow directions depends on the level of the groundwater. When the groundwater table does not rise very high, flow is controlled by the bedrock topography. When the groundwater table rises much higher than the irregularities in the bedrock topography, it is much more related to the surface topography (van Meerveld et al., 2015).

12 This "loss" is something we will discuss in more detail later in this chapter for many other sites.

13 Of course, with preferential flow often aiding and abetting the rapid movement of water to the interface—knowing that our view now is that such preferential flow is ubiquitous in soils (Clothier et al., 2008; Jarvis, 2007) and the vertical movement of water is aided by such preferential flow pathways (Beven and Germann, 2013; McGuire et al., 2024).

14 At Panola, like some other slopes, it is more than just total storm rainfall. Antecedent wetness plays a role whereby for wetter pre-storm conditions more runoff is generated.

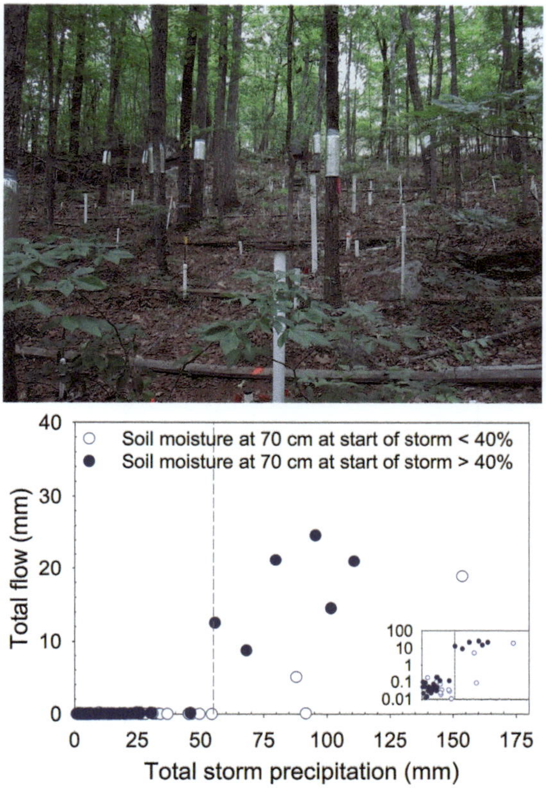

Figure 6.7 Threshold response for subsurface stormflow at the Panola hillslope. Note the almost two orders of magnitude increase in total subsurface stormflow for events larger than ~60 mm.

Source: Tromp-van Meerveld and McDonnell (2006a).

linked to connectivity of subsurface patches as per the Panola example. But as you can see in Figure 6.8a, the Maimai hillslope is considerably steeper and wetter than Panola. And this—together with differences in the permeability of the underlying bedrock—expresses itself in the amount of rainfall needed to initiate subsurface stormflow (Figure 6.8a): only ~15 mm of rainfall. Woods and Rowe (1996) excavated a 60 m wide trench along the base of a Maimai hillslope. Figure 6.8b presents their data for the whole-slope section shown in the photo. Why might that be? If you answered "wetter antecedent conditions, less permeable underlying bedrock and steeper slopes," you would be correct! Again, there is scatter in the data beyond 15 mm as a result of these other factors. But overall, the link between connectivity and threshold response is found here too (we will explore the loss along the flowpath in the next section).

But what might the pattern of flow at the subsurface interface look like? Figure 6.9a shows a photo from the Maimai hillslope. Here, blue dye was released onto the otherwise buried subsurface interface where the soil has been systematically stripped. It highlights the flowpaths at the soil–bedrock interface (Graham et al., 2010). Even on the very steep Maimai hillslope where variations in soil–bedrock topography are smaller than at Panola, the impact of the subsurface microtopography is clear: the topography of the interface drives the spatial patterns of subsurface stormflow as measured at the slope base.[15]

Now, let's travel to one last site; this time to a very low-angled slope of <5 degrees on the coastal plain of South Carolina, USA: the Savannah River Site hillslope. Here the loamy sand soil overlays a flow-impeding clay layer. Figure 6.8b shows a picture of the site, and Figure 6.9b shows a diagram of the subsurface topography and Figure 6.8b a plot of hillslope runoff totals as a function of the event size. Notice again the clear hockey-stick shaped hillslope response. But here, only a small number of events ever generate flow above the threshold—with interestingly, a similar ~60 mm rain event threshold for connectivity and whole-slope response as for the Panola hillslope. Here, the combination of a low downslope drainage efficiency and rather minimal hydraulic conductivity contrast between the coastal plain sandy soil and the clay-rich argillic horizon below results in a leaky interface above which lateral flow is only very occasionally produced. And the slope of the line after the threshold is much shallower than the slope of the line at Maimai or Panola.[16]

[15] And like the Panola hillslope, the Maimai hillslope's surface topographic index does not explain the flow distribution at the slope base (as discussed in a comment–reply by McDonnell (1997) and Woods and Rowe (1997).

[16] Of course if the hillslope produced 100% runoff then the slope of the line would follow a 1:1 line.

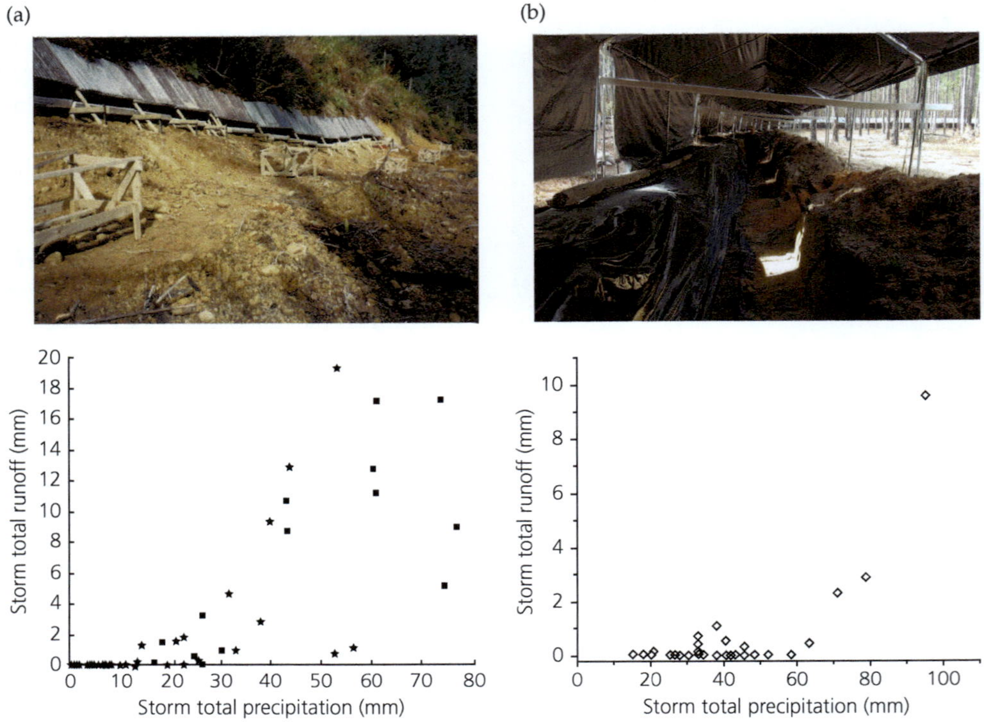

Figure 6.8 Thresholds for subsurface stormflow across 2 very different slopes: (a) Maimai, from Graham et al. (2010) and (b) the Savannah River Site in South Carolina, USA, from Du et al. (2016) (Note that Tani (1997) is one of the first clear examples of such thresholds gathered from many storms at two small headwater catchments near Okayama City, Japan). The dots represent individual storms. Note the different scale for the amount of flow.

Source: Photos by Ross Woods (top left) and Rhett Jackson (top right).

6.2.1 Pipeflow in relation to subsurface stormflow connectivity

Macropores and soil pipes can also aid and abet slope connectivity and subsurface stormflow thresholds. Uchida et al. (2005) compared the Panola hillslope that we have already looked at in this chapter and three hillslopes from Japan: the Toinotani hillslope (near Kyoto), the Jozankei hillslope (on Hokkaido Island) and the Hakyuchi hillslope (near Tokyo). Each site had soil pipes that were instrumented to record pipeflow directly. And at each site, the amounts of lateral preferential flow that were recorded were impressive. Despite the large differences between the study sites in terms of topography, climate, soil type and soil matrix hydraulic conductivity, there were several common threshold pipeflow responses that showed:

- the relationship between total rainfall amount and total pipeflow volume were highly non-linear at each site;
- the initiation of measurable pipeflow was threshold-dependent, controlled by the total rainfall and amount antecedent conditions;
- that once significant pipeflow response occurred, the maximum pipe flow rate was sensitive to the measured rainfall intensity; and
- the ratio of total pipeflow to total hillslope discharge for each site was fairly constant—regardless of total rainfall amount—once the precipitation threshold for significant pipeflow response was reached.

All in all, pipeflow can be a dominant factor in many slopes that produce sizeable amounts of subsurface stormflow. And there is likely a geomorphic

Figure 6.9 (a) Soil stripped off a ~10×10 m area on the Maimai experimental hillslope. Note the lower legs of a student at the uppermost section of the photo for scale. Photo: Chris Graham. (b) The Savannah River hillslope showing the surface topography and subsurface topography at the soil–clay (argillic layer) interface for a ~40×40 m. The elevation of the surface layer has been "lifted" to better illustrate the variability in the argillic layer. Note: blue indicates high elevation and green indicates low elevation.

Source: Du et al. (2016).

feedback at work here where chronic saturation begets pipeflow formation that is then enlarged over time by eluviation and other subsurface geomorphic processes. The main point being that soil pipes and macropores can enhance lateral flow connectivity and threshold response, just as vertical preferential flow can enhance the development of transient saturation at a subsurface interface.

Japan has perhaps been where some of the most voluminous and important work has come from on pipeflow in headwater catchments. Quite important summaries of lateral preferential flow paths and soil pipes have been made by Uchida et al. (2001). To pick one example from that benchmark review, some rather extraordinary work came in the late 1990s and early 2000s by the group of Yoshio Tsuboyama working with Roy Sidle on Honshu Island at a catchment called Hitachi Ohta. These are steep slopes (Figure 6.10a). And with many applications of chloride with concentrations measured in the field with a chloride-specific electrode, Tsuboyama et al. (1994) and Sidle et al. (2000) were able to develop a perceptual model of pipeflow connectivity (and other forms of preferential flow) with increasing wetness. Figure 6.10b shows a perceptual model of such impacts on pipeflow near an interface, where expansion of the "hydrologically active area" around the matrix–pipe interface increases with increasing wetness, thereby extending the pipeflow network upslope. The main point here is that soil pipes can greatly increase point scale drainage and connectivity.

But still the question remains: how continuous are soil pipes and preferential flow pathways downslope? Summarizing the Uchida et al. (2001) review of pipeflow in soils of the Pacific Rim, measured pipe diameters are often within a rather narrow range (with the exception of soil pipes in loess and peat soils[17]) and that the length of such root-engendered soil pipes is of the order of meters, so therefore highly discontinuous at the hillslope scale. Critically, the Uchida et al. (2001) summary suggested that the location of major pipes within the soil profile is mostly within a narrow band above the soil–bedrock interface or above a soil layer interface—something that aligns with my own experience excavating hillslopes in different places.

Hillslope excavations have revealed stunning behavior of preferential flow enhancements to connectivity at the hillslope scale. Destructive sampling at the well-studied Maimai hillslope revealed a connected, extensive preferential lateral flow network

[17] I have crawled into soil pipes in loess slopes and reached my arm into soil pipes in peat.

(a)

(b)

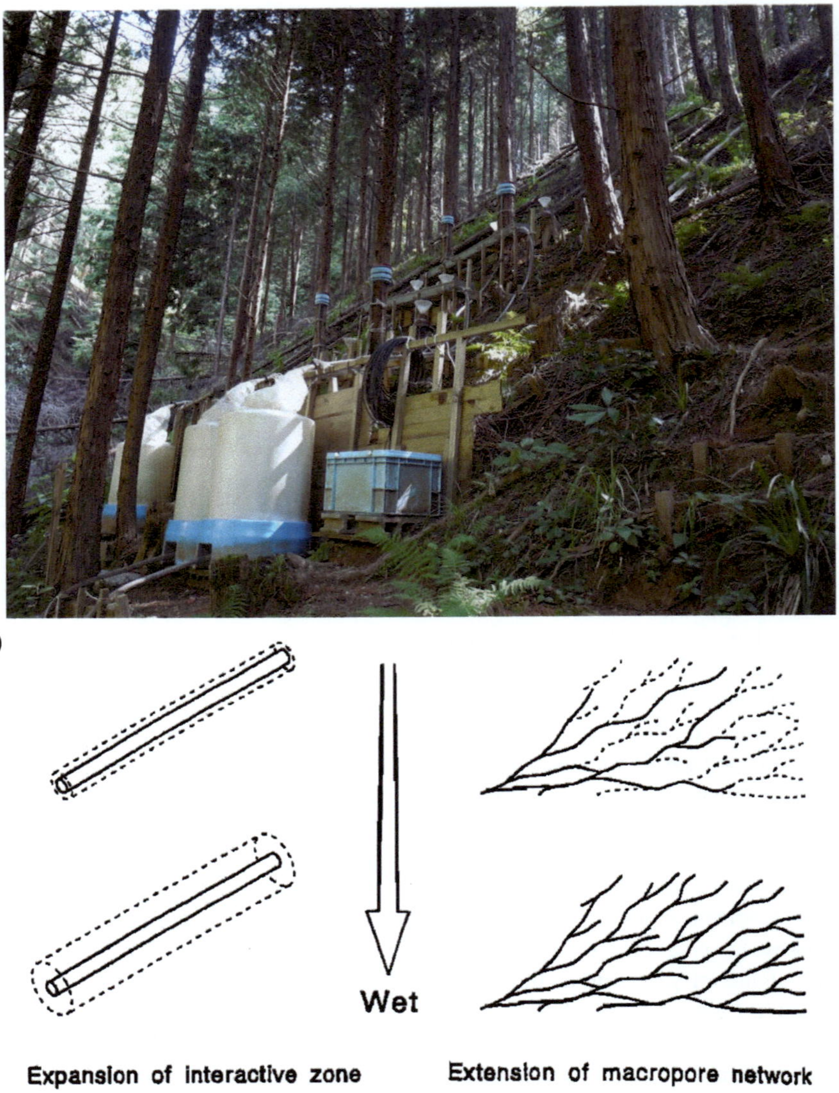

Figure 6.10 (a) The Hitachi Ohta slope site in all its glory. I've helped drag GPR antennas up and down these slopes in August and can attest that the combination of heat, humidity and steepness makes field work here difficult and the results of these studies by the Hitachi Ohta group all the more impressive. Photo from Yoshio Tsuboyama. (b) A perceptual model of how macropores and pipes interact with soil water to facilitate the network development of hydrological connectivity upslope.

Source: Tsuboyama et al. (1994) and photo courtesy of FFPRI.

at the soil–bedrock interface capable of transmitting large volumes of water quickly downslope (Graham et al., 2010; Figure 6.9a). If we travel to another Pacific Rim site on Vancouver Island in Canada, we see much more impressive forensic work where soil was removed section by section to reveal the nature of pipeflow at and near the soil–bedrock interface. This example is a much longer slope section. Anderson et al. (2009) used dye staining and excavation to examine and characterize the morphology and spatial pattern of a soil pipe and preferential flow network over a 30 m section of hillslope (Figure 6.11).

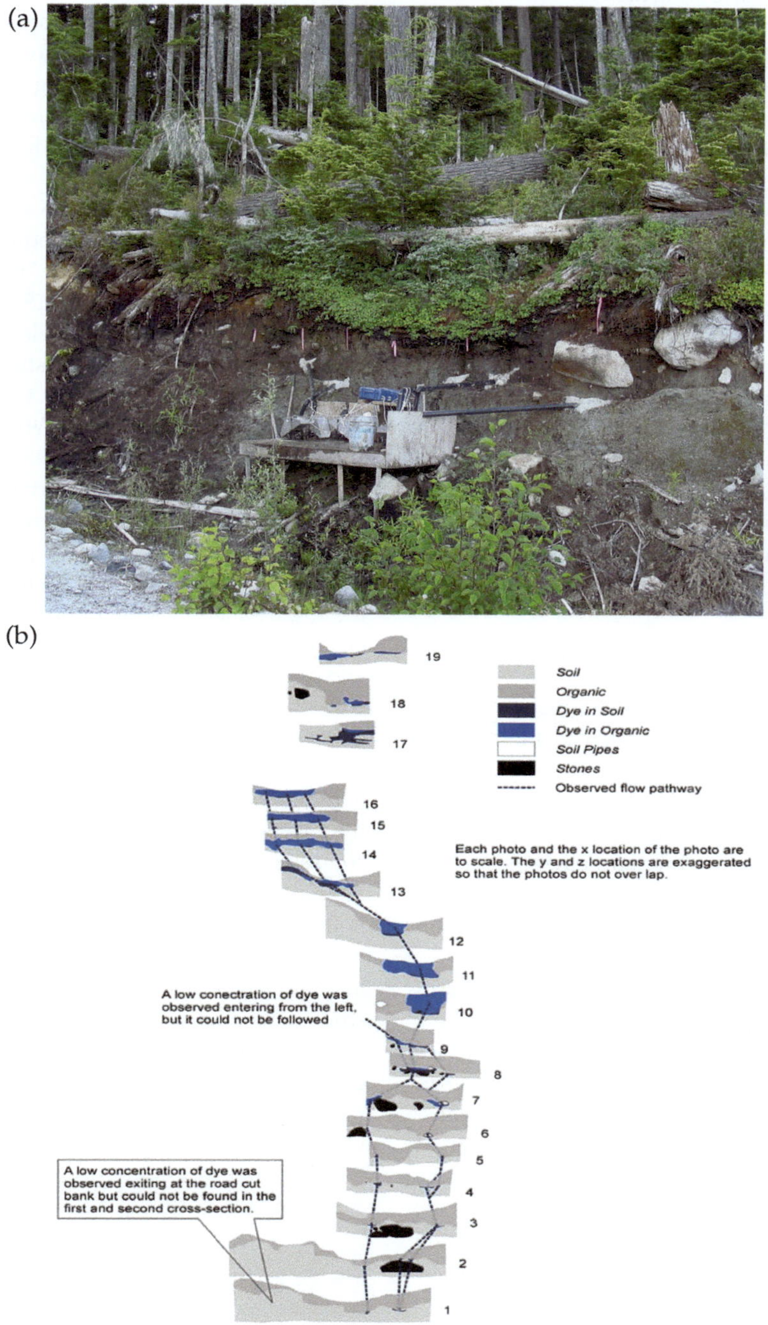

Figure 6.11 A cross-section of site on Vancouver Island, Canada showing the dye-stained soil at the hillslope scale. Note that the x-locations are "to scale" but the y and z directions are exaggerated so that, as the authors note, the photo plates "do not overlap."

Source: Anderson et al. (2009) photo courtesy of Axel Anderson.

They found that contributing areas to hillslope subsurface stormflow coincided with "highly developed and hydraulically connected preferential flow paths that had flow with little interaction with the surrounding soil matrix" (Anderson et al., 2009, p. 935). This is a rather common finding in steep, wet areas where subsurface erosion creates gaps for water movement, often at interfacial zones. Careful examination of Figure 6.11b shows how dye applied at the surface first infiltrates down to the contact between the organic layer and the soil mineral layer. It then moves further downslope, infiltrates further to then flows at the soil–bedrock interface, as revealed by the blue staining representation where "the dye staining also revealed that there is often little interaction between water in the preferential flow path and the surrounding soil matrix unless there was a constriction in the preferential flow network" (Anderson et al., 2009, p. 942).

6.2.2 Deeper subsurface connectivity

Everything we have discussed so far regarding connectivity and thresholds has been in the shallow subsurface: <1~2 m below the ground surface. But some slopes—even if they *appear* like they would produce subsurface stormflow at an interface and slope-connectivity under the right circumstances—do not. Instead, they may produce deeper forms of connectivity that drive streamflow under certain conditions.

One example of this is shown in Figure 6.12. It shows a site in the Oregon Coast range with "all the right stuff" for producing measurable subsurface stormflow: 2500 mm of annual rainfall compressed into a distinct 5-month wet season (check!); transmissive, macroporous soils (check!); an apparent interface below the soil profile (as seen with the figure with the person pointing to this visible permeability contrast; check!). This site was fitted with tipping bucket gauges to measure subsurface stormflow. We eagerly waited for the subsurface stormflow to start. We waited for the first few events of the wet season. Then a few more. Then, when nothing occurred in the first year, we waited a second year. Still nothing, despite over 2500 mm of rainfall! Why do you think this might this be? This puzzled us but in the end we realized that the upslope contributing area was too small. And, most importantly, the permeability contrast at the impeding boundary layer was not big enough to produce perched and connected transient water tables to form subsurface stormflow. Meaning, even when transient saturation occurred, water continued to infiltrate deeper into the deep saprolite below the soil, so that there was no lateral connectivity. We will come back to this issue of "loss" in the discussion below related to Figure 6.15.

This hillslope was for me, a lesson in hydrological humility. Nevertheless, the streams draining this catchment responded rapidly to rainfall events—with very high runoff ratios. Where did *that* runoff come from when there was no subsurface stormflow that we could measure and certainly no infiltration excess overland flow anywhere in the catchment to be seen? Even the area of potential saturation excess overland flow was extremely limited due to the highly incised valley bottoms. So what might be the answer? If you guessed "deep connectivity and deep subsurface stormflow" then you would be correct.[18]

We already discussed the important work at the Metman Ridge (Coos Bay) site in Oregon, USA (Chapter 5) where Montgomery et al. (1997) noted the role of the interaction between flowpaths in the colluvium and near-surface bedrock that governed the patterns of runoff generation in the headwater hollow. We suspected that something similar was happening at the site in the Oregon Coast Range in Figure 6.12. Fast forward to today and many studies have observed deeper forms of subsurface stormflow. Guo et al. (2019) has shown how preferential flow through vertical fissures in fractured rock at the Shale Hills catchment in Pennsylvania, USA link to fill and spill deeper in the subsurface, at the fractured rock–unweathered bedrock interface. And, like other studies have shown, the fill areas

[18] My first experience with deep bedrock control on runoff generation was at the 26 ha Upper Sheep Creek catchment in Idaho, USA where we had 36 piezometers installed into different layers of bedrock. That work showed that the main source of snowmelt runoff was through a deep layer of altered basalt that became connected and formed the main active contributor of runoff to the stream at peak snowmelt (Unnikrishna et al., 1995).

Figure 6.12 Not one tip in 2 years! It all seems so obvious in hindsight, but the combination of a small upslope area and not enough permeability contrast conspired to limit any measurable subsurface stormflow. The inset photo is taken just outside the main photo area and shows the transition from soil to saprolite.

Source: Photo by McDonnell.

can themselves contribute to enhanced deep percolation and groundwater recharge (e.g., Tromp-van Meerveld and Weiler, 2008; Appels et al., 2014). Figure 6.13 shows a recent example from the karst region of southern China by Zhang et al. (2024). This is a great example of where fill-and-spill connectivity may occur in shallower mineral soils, that then creates percolation and preferential flow vertically through the epikarst (the upper "skin" on the karst), that then pools in karst openings deeper in the subsurface that drive runoff. The examples of these deeper forms of runoff are exploding in recent years as drilling campaigns and geophysical campaigns discover deeper pools of water that, in some places, can contribute significantly to storm runoff.

Many studies have tried to follow the flowpath "loss" deeper into the deeper soil and weathered rock layers. And not surprisingly, a "loss" for subsurface stormflow can be a "gain" for groundwater recharge. Kosugi et al. (2006) presented perhaps the first and most detailed analysis of this, where they combined soil and bedrock tensiometers with hydrochemical measurements and water budget analyses to quantify the saturated and unsaturated infiltration from soil to bedrock. They focused on the soil–bedrock interface—where transient saturation and subsurface stormflow occur. They found that annual bedrock infiltration ranged from 35 to 55% of annual precipitation. They found also that this percentage increased as precipitation increased, suggesting what they noted was "a high level of

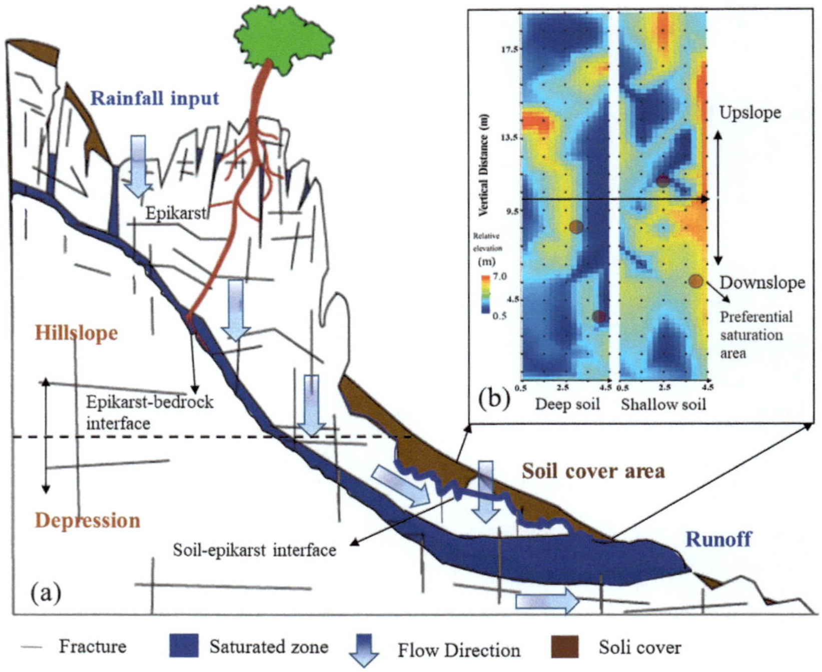

Figure 6.13 On the links between deeper subsurface runoff and fill and spill in epikarst hillslope geology in China.
Source: Zhang et al. (2024).

potential bedrock infiltration, partly explained by the high buffering capacity of the soil layer overlying the bedrock" (Kosugi et al., 2006, p. 1). This, I think, is a key observation. Think back to the exposed rock on the Maimai hillslope that we saw in Figure 6.9a. If you rained on it, little if any of the rainfall would infiltrate into the rock. But if you covered it with wet soil, the soil layer facilitates infiltration across the soil–bedrock interface and into the rock below.

The soil–bedrock interface was key for the Kosugi et al. perceptual model. They found that the "physical property of the soil layer was an important factor in controlling the generation of bedrock infiltration and saturated lateral flow over the bedrock" (Kosugi et al., 2006, p. 1). Importantly, they also found that at larger scales, the loss along the flowpath and infiltration across the soil–bedrock interface became *exfiltration* from the bedrock toward the soil layer. This accounted for >50% of their annual discharge in a small headwater catchment, mirroring some of the early findings for the Metman Ridge site at Coos Bay, Oregon, USA (that we referenced above), where, there, ~ ½ to ¾ of the applied precipitation flowed through fractured bedrock.

We already noted that ~90% of the water that gets to the soil–bedrock or soil–saprolite interface at the Panola hillslope continues vertically *past* the interface (Tromp-van Meerveld et al., 2007). Work at the Maimai catchment by Ameli et al. (2018) has quantified the fate of that loss along the flowpath at the Maimai hillslope. Figure 6.14 shows the results of integrated flow and transport analysis building on measured water table dynamics and groundwater ages for ~40 wells drilled to 10 m (by Gabrielli et al., 2018). More than 50% of groundwater recharged in the Maimai headwaters did not reappear in the same catchment. Rather, that groundwater that is recharged in the headwater catchment forms a groundwater subsidy to a larger parent watershed of which it is a part. Meaning, that out-of-catchment groundwater flow from the Maimai headwaters contributes to the larger higher-order catchment.

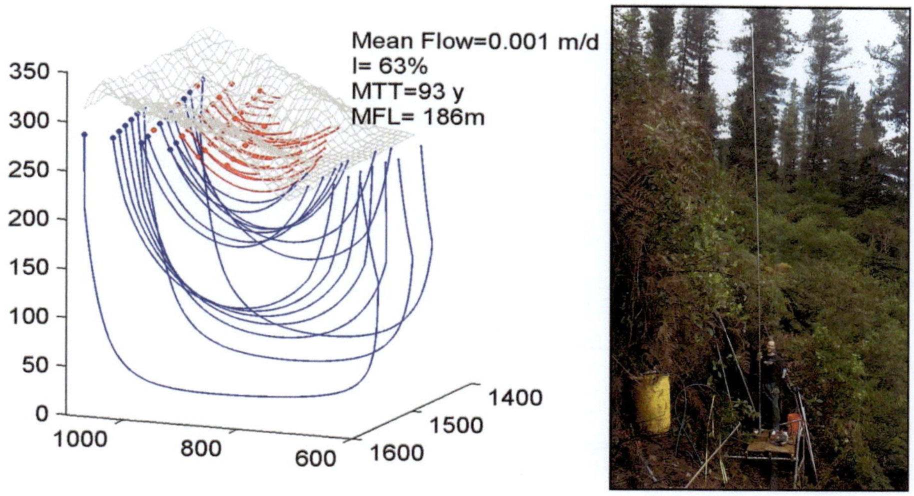

Figure 6.14 Deep groundwater flowpaths at Maimai, New Zealand where half of the groundwater recharged in the headwaters does not contribute to flow in that headwater catchment but appears rather in the parent catchment downgradient. All axis units in meters.
Source: Ameli et al. (2018). (left) Photo from Chris Gabrielli (right).

Furthermore, the relative proportion of headwater–groundwater subsidy was inversely proportional to recharge rate and directly proportional to slope angle. The result of this is that age of the headwater–groundwater subsidy (something we will explore in Chapter 7) can be many decades old, much older than the age of within-catchment groundwater flow contributions. This subsidy of water from the headwaters to their higher-order parent catchment is a research frontier and a theme we will return to in Chapter 7 when we discuss catchment storage and the residence time of stored water.[19]

6.2.3 Barriers to hillslope connectivity

We have discussed how connectivity comes about. But what limits connectivity? What are the barriers to flow? There are perhaps two obvious ones: the topographic impedance linked to the combination of slope angle and interface topographic relief; and loss along the subsurface flowpath to the formation below. Let's start with the first one by thinking about the slope angle impact on hillslope flowpath connectivity. This is something we have already seen in the three examples from Panola, Maimai and Savannah River, where all three of these hillslopes showed a balance of filling of small depressions at the soil–saprolite, soil–bedrock and soil–clay interface and then their subsequent spilling downslope once that depression storage was overcome. Hopp and McDonnell (2009) showed how this might operate for low-angled terrain like the Savannah River Site (that we saw in Figure 6.9b), where water stagnates and the fill side of the equation greatly dominates compared to spill. For steep slopes (like Maimai), there is less fill and more spill as water flows down the interface with little drainage impedance to hold back the flow (as per the photo from Figure 6.9a).

The second and more significant barrier to slope connectivity is "loss" along the subsurface flowpath. We've known about loss along the subsurface stormflow flowpath since the 1960s where Betson and Marius (1969, p. 581) noted that "whether this runoff reaches the flume, however, appears to depend upon the capability of some downslope soil

[19] Note too that inter-catchment groundwater flow is only now coming to the fore. And also, such subsidy can lead to increasing runoff ratios with scale (Asano et al., 2020) and sometimes also increases in peakflow with scale (Asano et al., 2023).

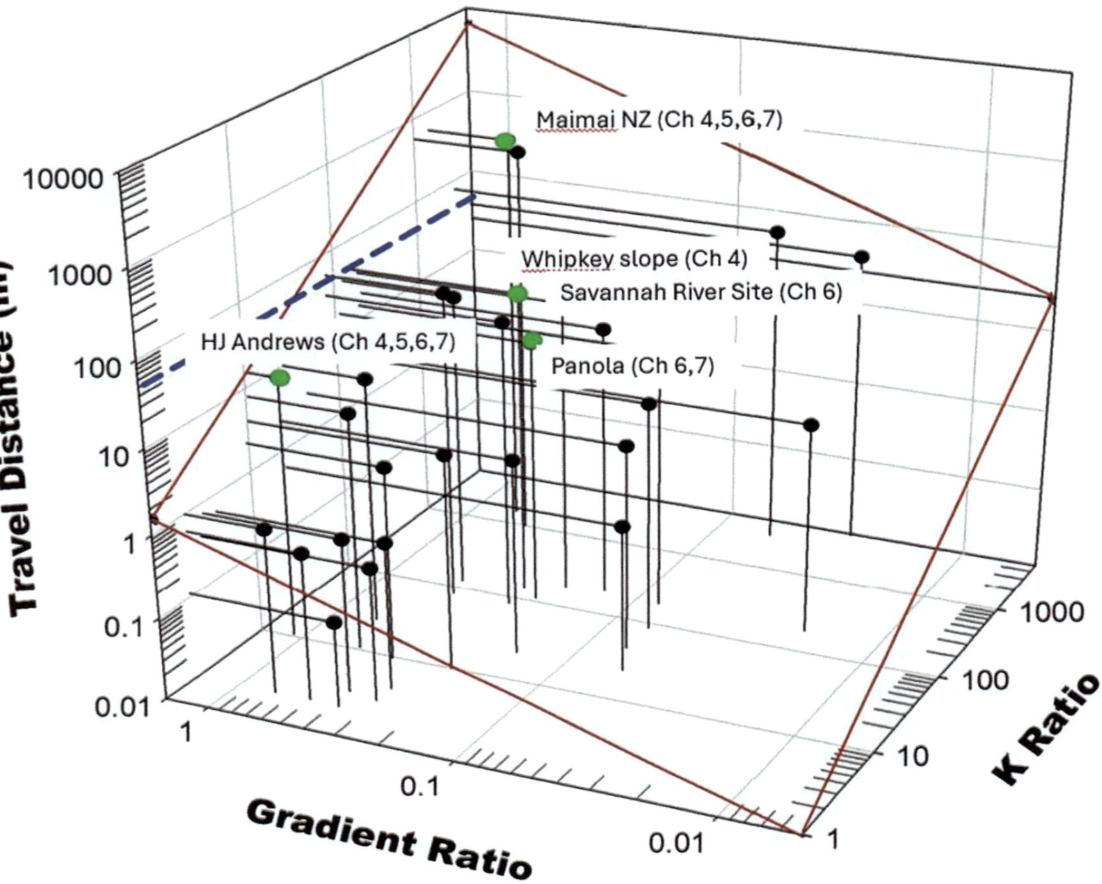

Figure 6.15 Downslope travel distances summarized for experimental hillslopes, including some of the ones we have discussed in this chapter and other chapters (in green).

Source: Modified from Klaus and Jackson (2018).

to reabsorb this runoff." Figure 6.15 shows how some of the slopes we have discussed, plus many others, plot in terms of the downslope travel distance of subsurface stormflow. This is from a synthesis by Klaus and Jackson (2018), who examined 17 hillslopes from a range of geoclimatic conditions. For each of the slopes they calculated the downslope travel distance of subsurface stormflow which they defined as "the potential distance a water parcel travels downslope above an impeding layer until it percolates into the impeding layer" (Klaus and Jackson, 2018, p. 5921). For this they used information on the slope (that they termed the "gradient ratio") and the ratio of the hydraulic conductivity of the soil above and below the main interface where subsurface stormflow is generated (their so-called "K ratio"); i.e., what is causing the restriction in infiltration, and hence local ponding and lateral flow. For their hillslope datasets from around the world (Figure 6.15), the downslope travel distance ranged from ~1 m to >100 m, with the Maimai slope as the longest downslope travel distance of any monitored site.

So, what do we see when we look at all these sites together? We know that Maimai has a very strong permeability contrast between the soil and the bedrock. Other experimental hillslopes—indeed most—have less extreme permeability contrasts.

Consequently, they have shorter downslope travel distances, with the majority with values on the order of 0.1–10 m. The key message from Figure 6.15 regarding barriers to connectivity is that a perched water table at an interface that produces subsurface stormflow does not necessarily imply that the lateral subsurface flow reaches the stream (think back to Figure 6.12 and the mysterious never-flowing hillslope trench). And as Klaus and Jackson (2018) note, such continuous connectivity is the *exception rather than the rule* for most hillslopes. This is perhaps why highly dissected catchments with short, steep slopes can sometimes produce impressive runoff amounts: there is a lower probability that the shorter slopes lose water en route to the channel.

6.3 Catchment scale connectivity

> *The weight of evidence for an extraordinary claim must be proportional to its strangeness*
>
> ***—Pierre Simon Laplace***

We have discussed connectivity at the hillslope scale. What about connectivity at the larger catchment scale? By the early 2000s,[20] we realized more and more the extreme scale dependence of streamflow generation. We have already seen examples where point scale behavior may have little impact on hillslope runoff if it is not "connected" to the large patches and network of saturation. Such scale dependence only increases as the scale of the catchment area under examination increases. One very obvious and simple sub-division of connectivity and disconnection at the catchment scale is the hillslope–stream connection. As we have seen, hillslopes can be disconnected from streams for long periods of time. But when they do activate, their contribution to the channel is often through a near-stream riparian zone of some sort. This riparian zone, unlike the hillslopes that might drain into it, is often continuously connected to the stream—at least in many humid environments. In this section we will focus mostly on humid headwater catchments where riparian groundwater is connected to streams rather than streams "losing" to groundwater (i.e., where the water table is below the stream). But we will touch on the latter in arid systems briefly in the Section 6.4.

Let's now look at some examples of catchment scale connectivity—with and without riparian zone (i.e., near-stream or valley-bottom zone) and at larger scales. The expression of the riparian zone is often related to how steep and how dissected a headwater catchment might be. Certainly, as we go to larger scales beyond the headwaters, then valleys and valley bottom zones near stream zones flatten out and increase in width, and the alluvial aquifer can be volumetrically large and itself flow in a downvalley direction, intimately linked to the stream.

6.3.1 Hillslope–stream connectivity

A headwater catchment *without* a riparian zone is an instructive—albeit unusual—place to start for understanding hillslope–stream connectivity. One example comes from the WS10 catchment at the HJ Andrews Experimental Forest in Oregon, USA. I say "without" riparian zone because in this catchment, the riparian zone was conveniently removed by a landslide—leading to a stream channel sitting on bare, unweathered rock with sideslopes issuing directly to the stream. Figure 6.16 shows the links between subsurface stormflow and streamflow in this 10-ha catchment. During the dry season, the gauged (trenched) hillslope site and the catchment produce very different amounts of flow on a per unit area basis. Then there is a ~1 month transition period where the catchment starts to wet-up and runoff is produced at both scales (hillslope and catchment) in greater amounts. Note the middle plot in Figure 6.16b shows the soil water content. There is a threshold soil moisture level there, akin to what we discussed at the beginning of this chapter with the Tarrawarra catchment. At WS10, though, during the many months that follow in the wet season, the hillslope and catchment become one; meaning that the flows on a per unit area basis produced at both scales are nearly identical. Now this could be pure luck that such a representative hillslope was

[20] Sidle et al. (2000) presented a hydrogeomorphic concept for hillslope–riparian–stream connectivity in Japan linked to the activation of different geomorphic units in the catchment: hollows, side slopes and nose slopes. If you are interested in steep, wet areas, you should consult this work.

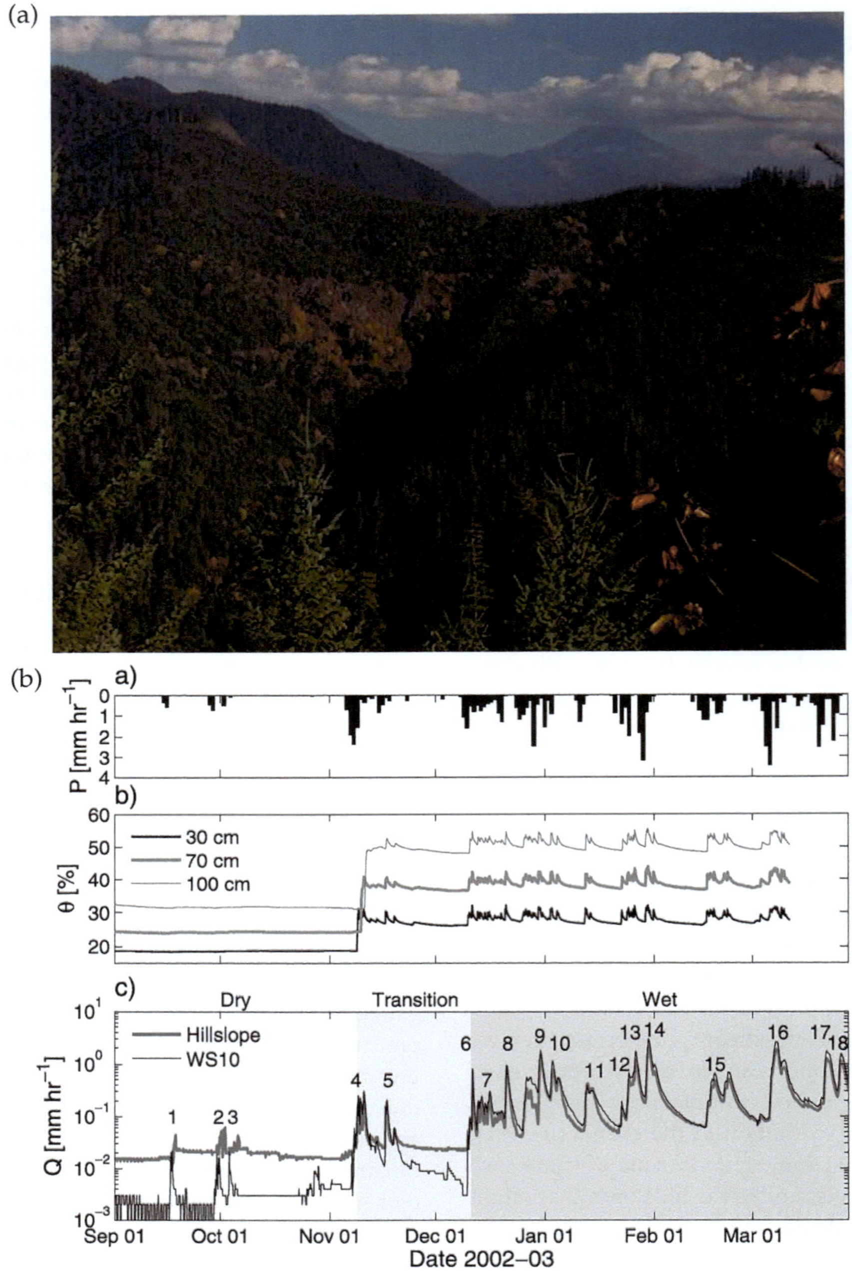

Figure 6.16 (a) The WS10 catchment (10 ha) with its trenched hillslope at the HJ Andrews Experimental Forest in Oregon, USA. (b) Time series of the hydrologic conditions of the study period: a) precipitation, b) volumetric water content from the lower nest of water content reflectometers and c) discharge from the hillslope and catchment. The shading indicates the wetting phases during this study period with full connectivity shown in the darkest patch. From McGuire and McDonnell (2010).

Source: Photo from Kevin McGuire.

chosen to be gauged; but it indeed shows that without a riparian zone, this catchment appears to be the sum of its hillslope parts during wet periods.

6.3.2 Hillslope–riparian–stream connectivity

Next let's return to the Maimai catchment that we examined in the previous section—to see an example of hillslope–stream connection mediated by a small riparian zone. Figure 6.17 shows the small Maimai headwater catchment and the gauged (trenched) hillslope section within it—the inset is the figure for the trenched hillslope that we have seen already in this chapter. Recall that each dot is a different rainfall–runoff event. The main figure for the stream shows something that is *not* threshold-like. Why might this be? It's because of the riparian zone. The riparian zone here is always[21] connected to the stream; an alluvial aquifer of sorts. Flow from the hillslope on the other hand, is driven by the processes we examined for this slope and other slopes in the previous sections of this chapter; something highly threshold-like.

If we compare the hillslope and riparian signal for a single rainfall event, rather than looking at dozens of events as dots in Figure 6.17, then Figure 6.18 shows the dance between hillslopes, riparian zone and stream for the Maimai catchment.[22] The small rainfall event on May 14 generates runoff but almost all of it comes from the riparian zone. In this storm, the rainfall amount is below the threshold of about 20 mm needed to activate the hillslope. Now look at the May 17 rainfall event and the corresponding streamflow response. The hillslope now contributes to the stream runoff: roughly a 50:50 blend of riparian zone water and hillslope-derived water. And what if we had a much larger rain event? You guessed it; the hillslope would contribute the vast majority of streamflow and its volumetric contribution would be much larger than that from the limited riparian zone volume.

Lastly, if you look carefully at Figure 6.18, it shows how hysteretic[23] these hillslope–riparian–stream signals can be. In other words, there is a timing difference for the hillslope activation; *after* the peak of the May 17 storm—where we might say that the riparian zone leads the rising limb of the event hydrograph and that the hillslope component controls the falling limb (see the timing of the red and green lines for the May 17 event in Figure 6.18).

Let's look at a still-larger riparian width and volume, this time for the 10-ha Panola catchment that we examined earlier in this chapter. This catchment contains the trenched and instrumented hillslope that we looked at in detail earlier and a 3 ha bedrock outcrop. Figure 6.19a shows a map of it with measured soil depths. Look at the riparian volume. Up to 6 m deep in places. This is a sizable alluvial aquifer by comparison to the HJ Andrews example (none!) and Maimai (some). Note also the large bedrock exposure in the catchment that produces infiltration excess overland flow. What does hillslope–riparian–stream connectivity look like here?

Figure 6.19b shows the Panola sequencing of runoff generation from a tracer perspective. It shows the streamwater sodium vs sulphate concentrations for a rainfall event against the three main end members that contribute flow to the stream: the riparian groundwater from the alluvial aquifer; the bedrock outcrop water delivered as infiltration excess overland flow; and the subsurface stormflow (aka hillslope water) as measured at the hillslope trench. Before the storms, streamwater looks like riparian groundwater. On the rising limb of the hydrograph and until the peak flow of the event (denoted by the farthest dots from the riparian groundwater end member), the stream looks like some mix of water sourced from the large bedrock outcrop and a little bit of the hillslope water signature. If the stream was a simple linear blend

[21] However, see some exciting new work by Messager et al. (2021) predicting that over ½ of the world's rivers by length cease to flow for at least one day per year suggesting that "non-perennial rivers and streams are the rule rather than the exception."

[22] Technically, this is a neighboring catchment at Maimai. But we won't complicate the story with more details.

[23] Hysteresis is another classic form of non-linearity in hydrological systems. Recall the well-known hysteresis between matric potential and volumetric water content that you likely learned about as an undergraduate. Or, if not there, your experiences changing money at an airport: one rate when converting your money into a new currency and another rate converting that new currency back to your home country's money—a more painful example of hysteresis!

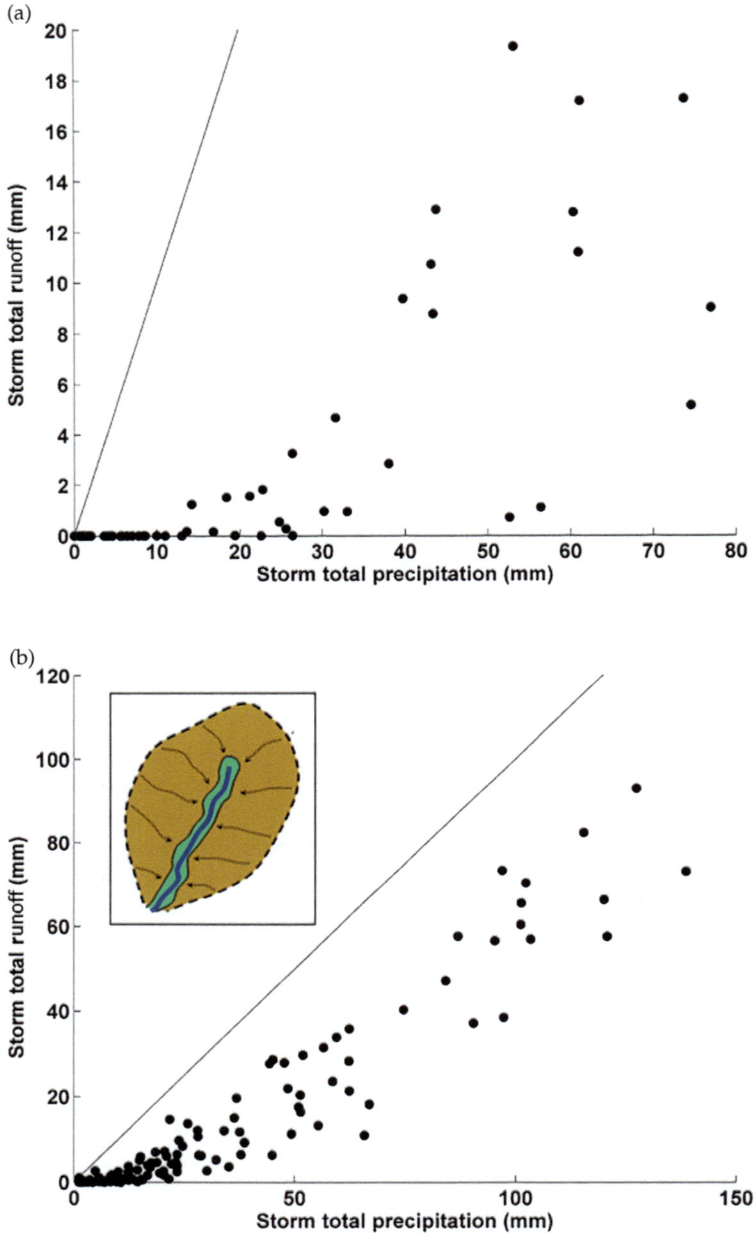

Figure 6.17 Comparison of hillslope (inset; see also in Figure 6.8a) and catchment (main figure) response to rainfall in the Maimai catchment, New Zealand. Each dot is a separate rainfall–runoff event.

Source: Graham et al. (2010).

between riparian zone groundwater and outcrop water, then the water samples (the dots) would plot along the single mixing line between these two end members. But these data suggest that the chemistry is "pulled" slightly toward the hillslope water end member. Again, a bigger storm would

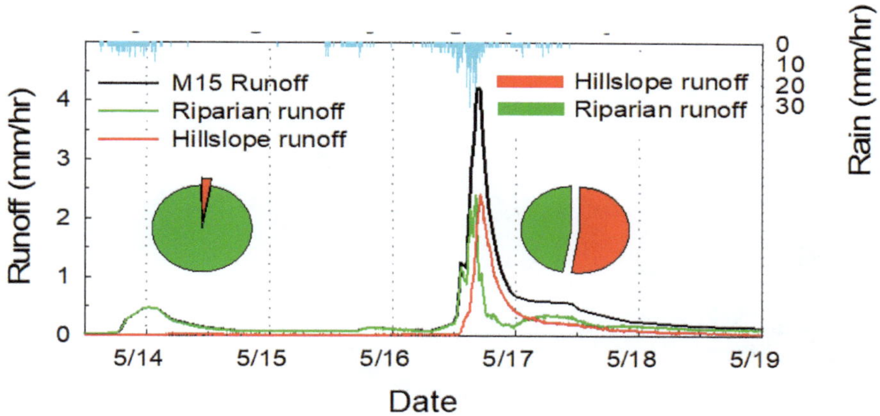

Figure 6.18 The effect of connectivity and thresholds on the contribution of riparian and hillslope water to streamflow at the catchment scale. *Source*: McGlynn et al. (2004), re-drawn by Brian McGlynn (pers. comm).

eventually shift that streamwater signature even further over to the right (i.e., to signify more hillslope water—subsurface stormflow—contribution to the stream). It's a mixing problem—literately and figuratively. How large is the volume of water in the alluvial aquifer and how much water from different topographic positions might be needed to shift that tracer signature? That too depends on how big the event input is and the antecedent wetness.

6.3.3 Groundwater dominance of connectivity

Of course, in very humid catchments, especially ones where the alluvial aquifer controls almost all of storm runoff, groundwater can have a completely dominant role on the hillslope–riparian–streamflow connection for all but the biggest events. This can happen in two ways. The first is when groundwater dominates low flows. And then when groundwater rises to the surface in the riparian zone to create saturated areas for saturation excess overland flow, that overland flow takes over in terms of the dominant source. This is similar to what we reviewed for the early work on saturation excess overland flow. But now, through the lense of more storms and more places, we can again see a transition. Figure 6.20 shows this for the chronically wet Bruntland Burn catchment in Scotland where groundwater is in close proximity to the soil surface in the near-stream zone all the time (Tetzlaff et al., 2014). Soulsby et al. (2006) showed that rain that falls onto saturated areas causes overland flow to occur, diluting the stream chemistry, so that most, if not all, of the water appearing in the stream is rainfall. In other words, riparian–streamflow connectivity without necessarily a hillslope component (except in unusually large events).

The second way groundwater can have a completely dominant role on the hillslope–riparian–streamflow connection is via transmissivity feedback, as we discussed in Chapter 5. Figure 6.21a shows work from the Hubbard Brook catchment in New Hampshire, USA. It shows the spatial patterns of the shallow water table extent and duration, and the seasonality of groundwater connectivity to the headwater stream. During drier conditions in the growing season, the groundwater table at in the catchment is "discontinuous and patchy" where "only the wettest near-stream footslope areas were consistently hydrologically connected with the stream network" (Detty and McGuire, 2010a, p. 2222). We will discuss the aspects of groundwater control on stream baseflow in Chapter 7. But for now, let's look at the thresholds that can be seen for such behavior, the first of which was noted by Detty and McGuire (2010a). Figure 6.21b shows the threshold relationship between quick flow depths and both a maximum water table height index and a sum of gross precipitation and antecedent soil moisture index. Like we saw for *hillslope* threshold

(a)

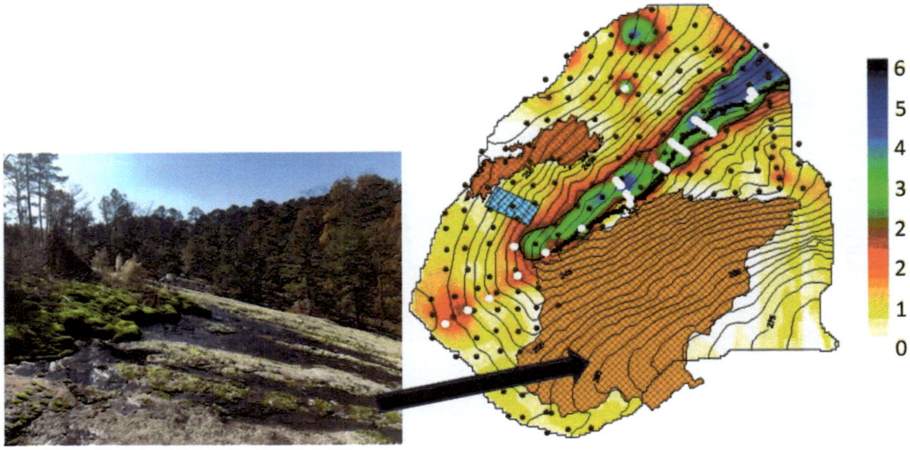

(b)

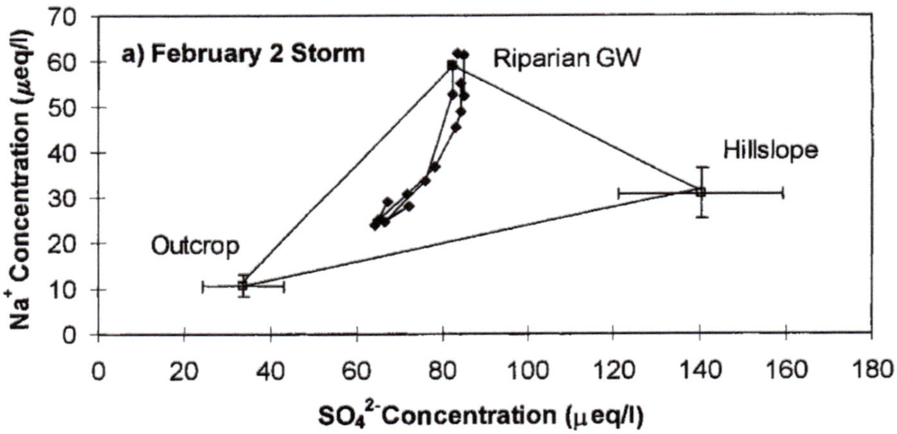

Figure 6.19 (a) Soil depth across the Panola catchment where the blue rectangle is the hillslope that we examined earlier in this chapter. Depths in meters (and based on the Herculean soil depth measurements of Al Zumbuhl—an exceptional Masters student of mine at SUNY-ESF in the mid-1990s.). The photo shows a portion of the 0.3 ha bedrock outcrop where infiltration excess is generated for storms of all sizes. The black dots are a 20 m survey grid of the measurements of depth to bedrock using a cone penetrometer. (b) The mixing plot is a top-down way of seeing connectivity. Here a bivariate mixing plot of Na and SO_4 concentration for a rainstorm. It shows connectivity from a simple geochemical mixing perspective. The end members on the mixing plot relate back to the zones noted on the map.

Source: Photo from Brent Aulenbach, map from Freer et al. (2002) and mixing data from Burns et al. (2001).

relationships between precipitation amounts and flow, these relations indicate that below the threshold, catchment scale runoff is generated mainly in the near-stream zones. Above the threshold, the groundwater contributing area expands laterally up and into its hillslopes. A key message from Figure 6.21 is the link between water table depth and the topographic wetness index (recall that definition—the upslope area at any point in the catchment divided by its local slope angle). Therefore, areas near the catchment divide have low topographic wetness index values; those areas downslope and near the stream have high values. In a series of papers, Rinderer et al. (2016; 2017; 2018) built upon this work in the 20-ha Studibach catchment in Switzerland where again the combination of rainfall amount, antecedent wetness conditions and topographic position influenced the groundwater

Figure 6.20 (a) Saturated areas. (b) These fill and spill with macro-, meso- and microtopography affecting the expansion and connection to the stream in the Bruntland Burn catchment, Scotland.

Source: Photo on left from Chris Soulsby; photo on right by the author.

response and its link to the channel during events.

6.3.4 Connectivity at larger catchment scales

The first study to show hillslope–riparian–stream connectivity at larger catchment scales was in the Tenderfoot Creek catchment in Montana, USA. Figure 6.22 shows what the catchment looks like. Jencso et al. (2009) performed some exceptional analysis linking measured hillslope–stream water table connection of wells at two dozen transects to look at what they called "the linkage between the dominant catchment landscape elements (hillslopes and riparian zones) and the channel network" (Jencso et al., 2009, p. 1). They monitored 24 wells along the stream network and found a relationship between the upslope accumulated area to a well and the percentage of time of the water year that a hillslope–riparian–stream water table connection existed. That relationship was striking with an $r^2 = 0.91$. They then applied this relationship to the entire stream network as shown in Figure 6.22 and found that "internal catchment landscape structure (topography and topology) is a first-order control on runoff source area and whole catchment response characteristics" (Jencso et al., 2009, p. 1). If you study Figure 6.22b carefully, you can see some interesting aspects of this hillslope–riparian–stream connectivity where there can be striking asymmetry along the network. In other words, one side of the stream may have a strong connectivity and the other side not. This is like the Burt and Anderson work from Chapter 4 but now shown in space and across scale.

6.4 Connectivity in arid and semi-arid catchments

> *Knowledge is the intellectual manipulation of carefully verified observations*
>
> ***—Sigmund Freud***

All of our examples of connectivity thus far have been from humid regions. What about connectivity in arid and semi-arid places? Does it play a role there? The answer is yes.

6.4.1 Connectivity and thresholds at the hillslope scale

For infiltration excess overland flow, like subsurface stormflow, most of our perceptual models prior

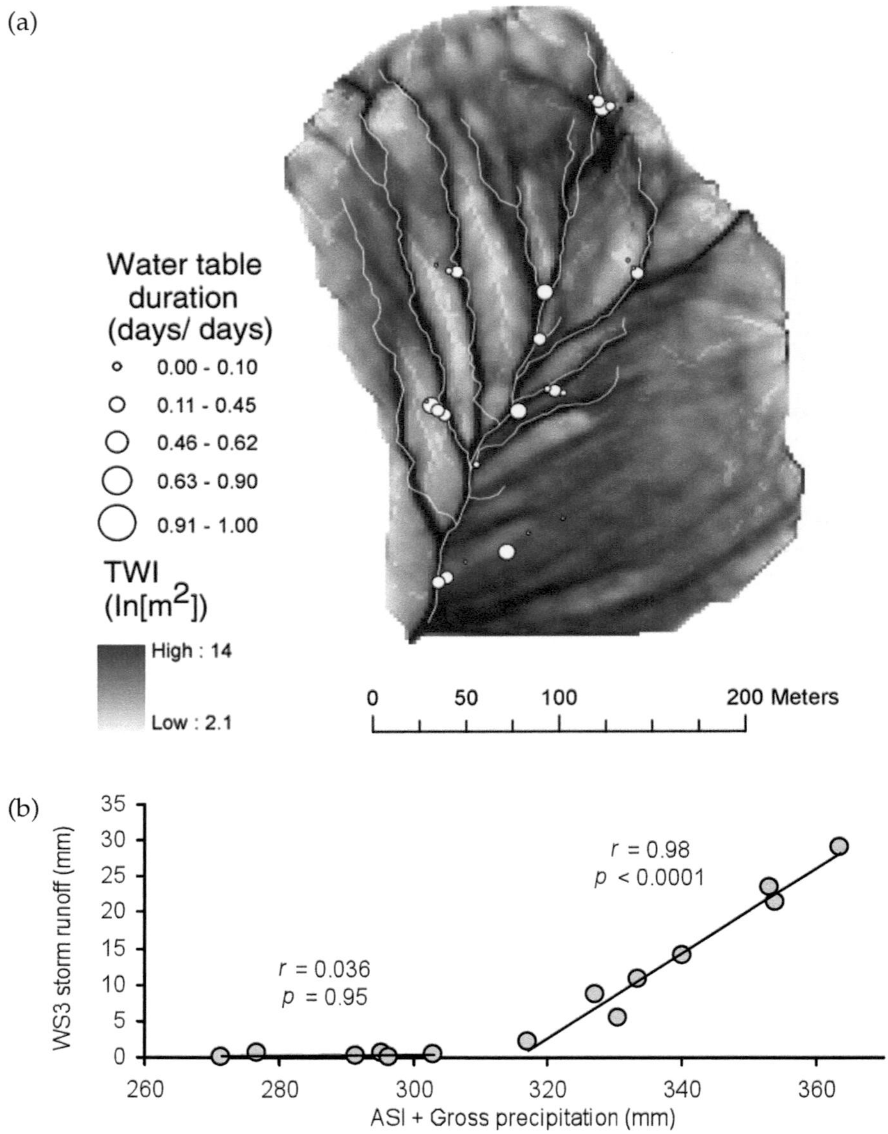

Figure 6.21 Potent perceptual findings from Detty and McGuire (2010) showing water table duration in the map, during their so-called growing season and transition period with the topographic index shown at a 5 m resolution. Below the map the relation between the total quick flow runoff depths and a soil wetness index consisting of total precipitation and the amount of water stored in the upper 20–50 cm of the soil, showing a clear threshold for flow, related to groundwater position, itself controlled by precipitation and antecedent precipitation.

to the modern era had two-dimensional, rectilinear slopes with smooth, straight soil surfaces. And as we did for subsurface stormflow, we will now explore heterogeneous slopes and their microtopography and how that matters for flow concentration, connectivity, thresholds and loss along the flowpath. Great strides have been made in recent years in understanding the microtopographic effects on slope connectivity for infiltration excess overland flow (Darboux et al., 2002b; Peñuela et al., 2013;

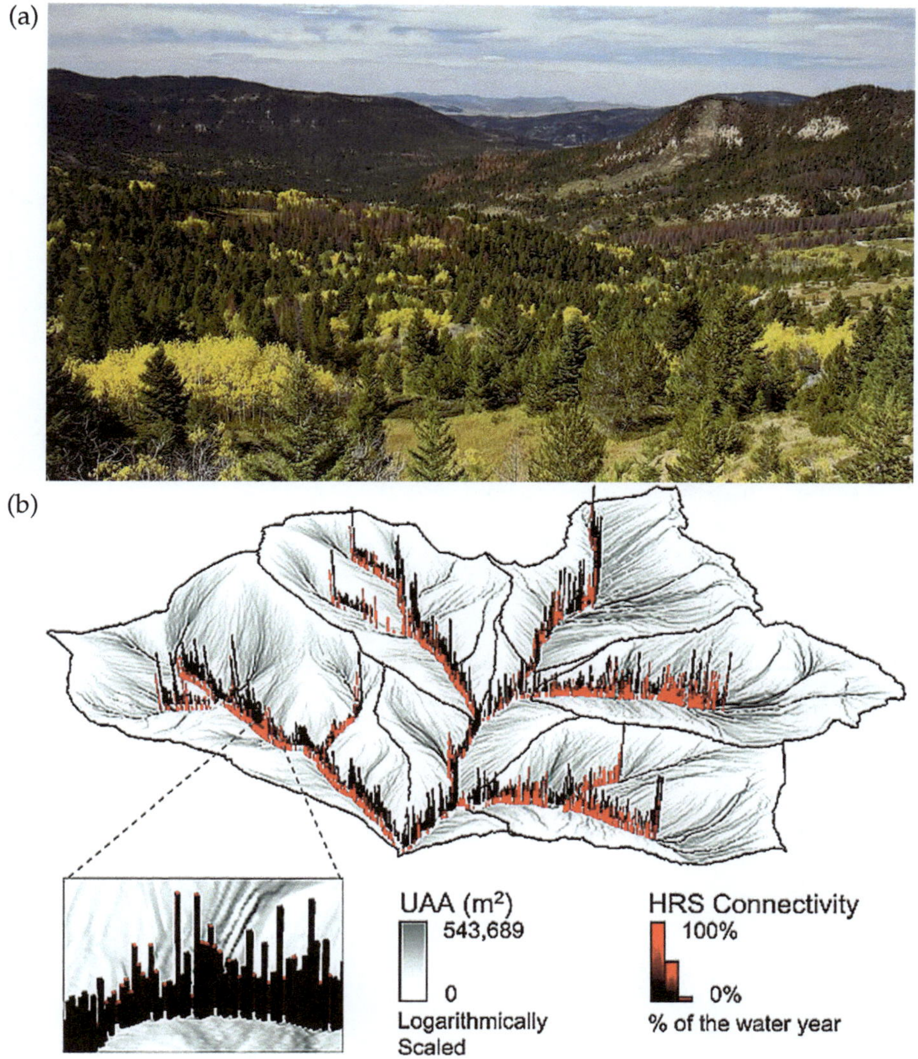

Figure 6.22 The Tenderfoot Creek catchment showing hillslope–riparian–stream connectivity across scale.

Source: Jencso et al. (2009) and photo from https://www.pewtrusts.org/en/research-and-analysis/articles/2020/06/18/conservation-gains-missed-opportunities-in-helena-lewis-and-clark-national-forest.

Kirkby and Cerdà, 2021) and threshold response. Figure 6.23a shows photos of such microtopography at the surface for a semi-arid agricultural hillslope. You can see and imagine how patches of saturation in the small surface depressions develop and connect in ways that are strikingly similar to what we examined for subsurface stormflow at a subsurface interface.

Perhaps the best modern definition of infiltration excess overland flow is the one offered by Peñuela et al. (2016, p. 1596) where overland flow is "a spatially distributed process of gradual filling and connection of depressions." This of course is linked to surface microtopographic structures and their major effect on surface storage, connectivity and runoff initiation (Onstad, 1984; Darboux et al.,

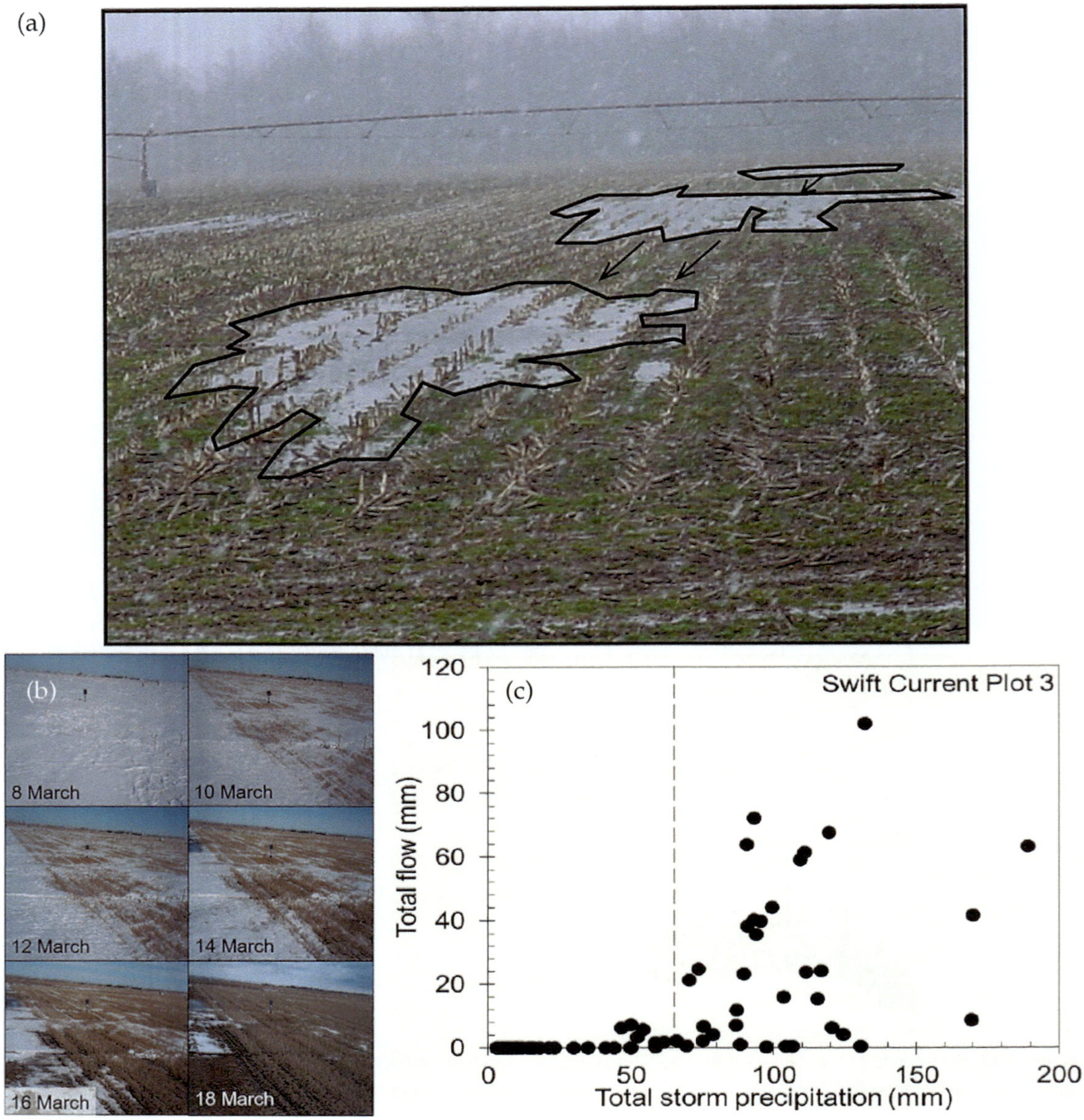

Figure 6.23 (a) A North Dakota agricultural hillslope showing filling and spilling during infiltration excess overland flow. Photo: Michael Chu, North Dakota State University. (b) Snowmelt on a Saskatchewan 5 ha hillslope plot. Note the snow melts onto frozen ground (frozen often down to 3 m!). The infiltration excess overland flow produced by snow melting on this frozen ground is dictated largely by the soil water contents in the previous fall at the time of freezing. (c) The threshold for the Saskatchewan slope.

Source: Data and lower right photo from Anna Coles.

2002a) and overland flow triggering (Darboux et al., 2002b). These rough natural surfaces exhibit runoff thresholds (Chu et al., 2013) and have been shown by many to be impacted by rainfall characteristics, soil properties and initial soil moisture (Antoine et al., 2009; Peñuela et al., 2013).

Figure 6.23b shows infiltration excess overland flow behavior for snowmelt onto frozen ground on a semi-arid hillslope in Saskatchewan in Western Canada. The 5-ha slope is gauged and Figure 6.23c shows the threshold response between inputs (total storm precipitation and melt water equivalent during the spring melt season). Again, this is very similar to what we looked at for subsurface stormflow, but perhaps with a rather messier threshold-like pattern than the thresholds we saw earlier in the chapter. Why would there be such increased scatter in this example? These events are for many years' worth of data collected for this dry place with ~350 mm of precipitation per year. For rain events on the slope in the summer period, most of the rainfall infiltrates. It takes one humdinger of a rain event to form whole-hillslope-scale runoff connectivity. But when frozen, the slope is *much* more responsive to inputs, and a lower input amount (in the form of snowmelt) is needed to generate runoff as infiltration excess overland flow. When frozen, the microtopography of the soil surface becomes all-important at channeling ponded water downslope. If it freezes in a dry state in the fall season (meaning soil moisture immediately prior to freeze-up is low) then, many months later when the snow starts to melt, most of the meltwater infiltrates and the dots in Figure 6.23c) are below the threshold. But when it freezes *wet* in the fall, this can really have an effect many months later where infiltrating meltwater encounters a very low infiltrability frozen soil surface and large amounts of infiltration excess overland flow ensues. And while there are not many studies yet that have traced with stable isotopes such flow off the surface of slopes that generate infiltration excess, Coles and McDonnell (2018) showed that for this Saskatchewan site, almost all of the snowmelt runoff sampled at the outlet of the hillslope was new water. I'll have more to say on such things in Chapter 7.

If we look at an even drier site—with similar annual rainfall but much higher rates of evapotranspiration—we see a similar threshold response; now in relation to storm rainfall only. Figure 6.24b shows such data for a 1.9-ha plot at the Walnut Gulch experimental catchment in Arizona, USA. Again, the threshold is slightly messier than the subsurface stormflow plots due to variations in rainfall intensity, duration and the specific location of intense rainfall within the catchment. In very arid sites, soil crusting can also be a factor for infiltration excess overland flow, as noted in early studies by Morin and Benyamini (1977). Early work by Hoogmoed and Stroosnijder (1984) described too how raindrop impact can break down surface aggregates. Wheater (2025) provides a summary of the role of biological soil crusts on semi-arid hillslopes in Spain (Maestre et al., 2011) and China (Chen and Duan, 2015). We will come back to the rainfall variability issue in a moment. But for now, let's next think about loss along the flowpath.

Loss along the flowpath also comes into play for infiltration excess overland flow and is much easier to see and measure than the examples we looked at for subsurface stormflow. Early work in the Negev desert by Yair et al. (1980) and Yair (1990; 1992) showed how loss along the flowpath worked for infiltration excess overland flow environments: it increases with distance along the flowpath.[24] Some particularly good datasets have emerged in recent years to help to show these effects. One useful example is from the Sierra de Enguera Soil Erosion Experimental Station in Eastern Spain. Unlike many of the other measuring sites that exist around the world,[25] the runoff at Sierra de Enguera is measured on different-sized hillslope plots that range from 1–16 m in length. Figure 6.25 shows the effect of slope length on the generation of infiltration excess overland flow. The runoff coefficient clearly decreases with slope length. Kirkby and Cerdà (2021, p. 10) note that these decreases with distance are "particularly strong for smaller storms." They noted that the runoff coefficient continues to decrease downslope, "but approaches an upper limit in which additional rainfall is almost balanced by infiltration." In other words, the opportunity for overland flow to be lost to infiltration increases with

[24] See also Dunne et al. (1991) for benchmark work for how infiltration excess overland flow is affected by the combination of rainfall, vegetation and microtopography.

[25] Many countries now have networks of overland flow plots to characterize point-scale overland flow. They are often reported in "plot years," meaning the number of plots times the number of years of record. This number is in the 1000s and tens of thousands for many developed countries.

(a)

(b)

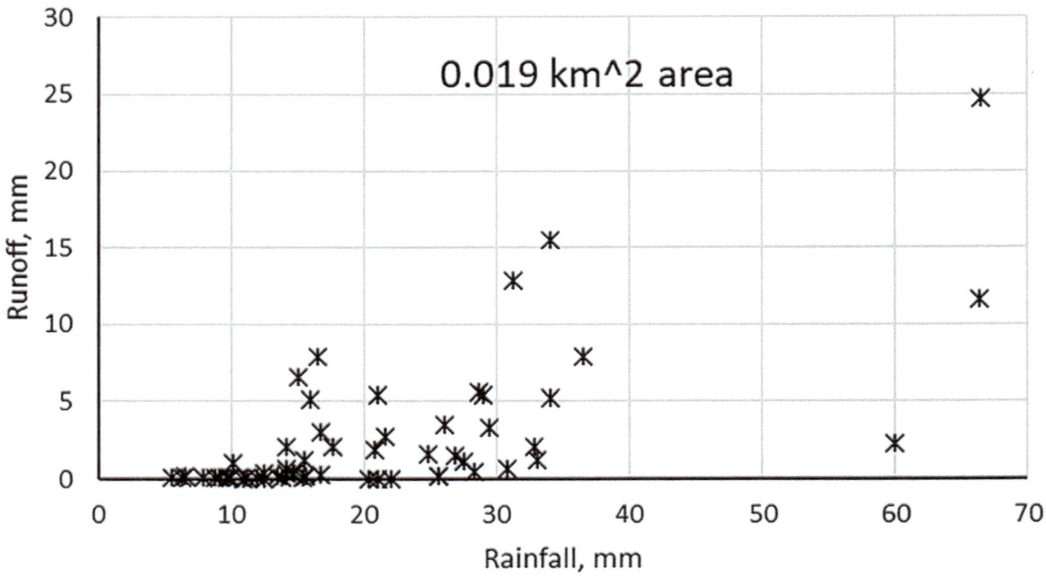

Figure 6.24 (a) Microtopography in a semi-arid hillslope (from the Anzo Borrega Foundation). (b) The threshold runoff response for the Kendall site, a 0.019 km^2 sub-catchment of the Walnut Gulch experimental catchment in Arizona, USA.

Source: Memberu Meyes, pers. comm., 2025.

(a)

(b)

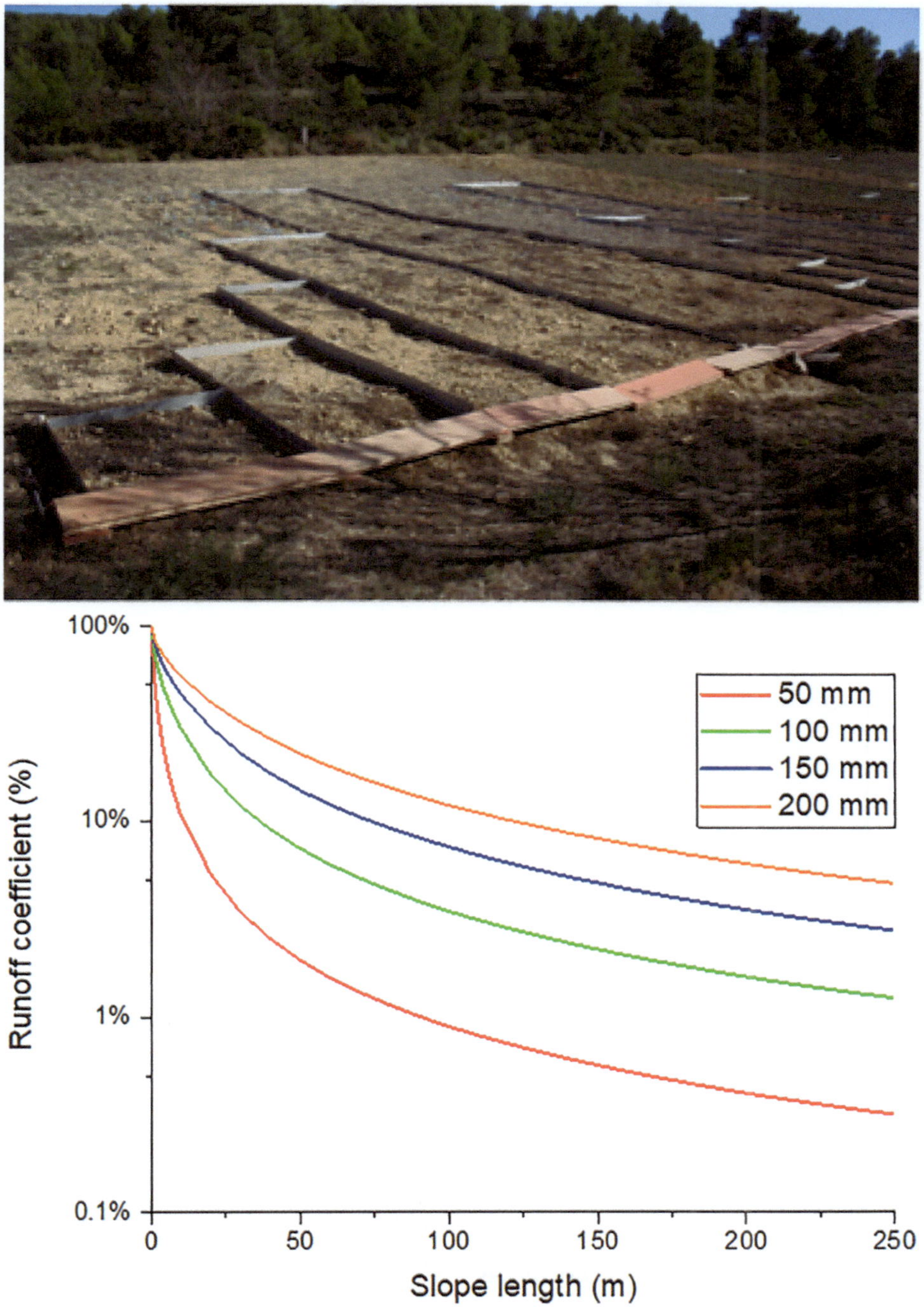

Figure 6.25 (a) The Sierra de Enguera site in Spain showing some of the long-term overland flow plots section. (b) Extrapolated response of the runoff coefficient against the length of the slope section for different-sized events.

Source: Kirkby and Cerdà (2021).

(a)

(b)

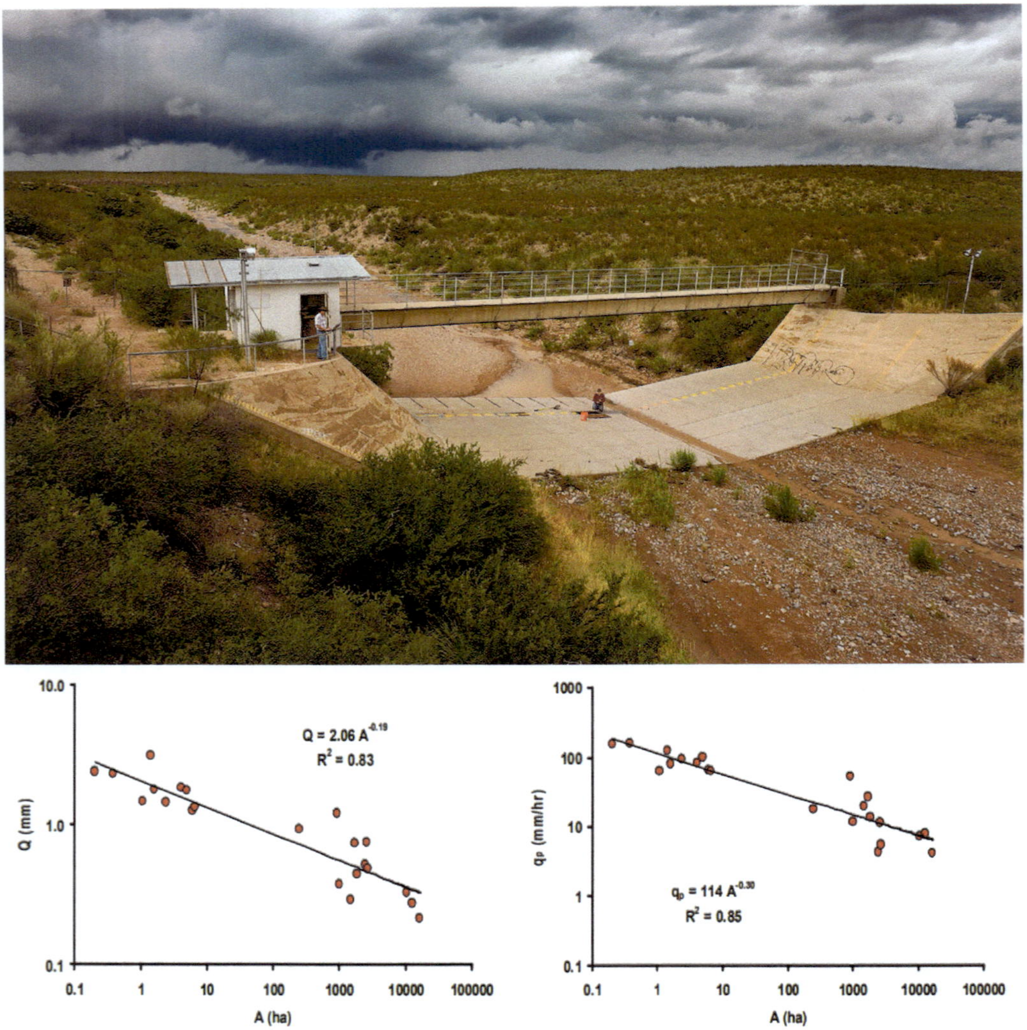

Figure 6.26 (a) The Walnut Gulch catchment in Arizona, USA. Photo from Dave Goodrich. (b) Graphs showing the relationship between the runoff (average storm runoff Q and maximum peak discharge (q_p) and catchment area (A). Data from 1963–1996.

Source: Plots from https://www.ars.usda.gov/ARSUserFiles/20221000/WGBrochure.pdf

the length of the plot. Now, how might this express itself at the catchment scale?

6.4.2 Catchment scale connectivity in dry places

If we think about hillslope–riparian–stream connectivity in arid environments, the water table can often be 10s or >100 m below the stream channel. So often hillslope–riparian–stream connectivity in headwater catchments in arid zones is simply hillslope–stream connectivity—via infiltration excess overland flow. In arid sites in particular, point-scale measurement of infiltration excess overland flow suggests that it is generated almost everywhere. But in most cases, little of this plot-scale or even hillslope runoff is recorded in the streamflow hydrograph. Why? Figure 6.26a shows a photo of the overall Walnut Gulch catchment:

annually it produces about 2 mm of runoff from the ~350 mm of annual precipitation. The rest of the precipitation infiltrates and most of that evaporates or is transpired (Renard et al., 2008). Flow at the catchment outlet—both average flow and peak flow—simply decreases with catchment area (Figure 6.26b). This is like the hillslope plots for the Sierra de Enguera site in Spain now writ large at catchment scales. Again, as infiltration excess overland flow continues downslope and downvalley, large losses occur and much of the streamflow infiltrates along the flowpath. There is little hillslope–riparian–stream connection like we saw for humid catchments earlier in the chapter because the stream channel is itself dry before the event. Therefore, the water table is not connected to the channel; it is *below* the stream bed so that even the stream loses water downvalley.

6.5 Summary and looking ahead

> *We have multitudes of facts but we require, as they accumulate, organizations of them into higher knowledge; we require generalizations and working hypotheses*
>
> ***—John Hughlings-Jackson***

This chapter begins the section of the book that I might call "the modern era" and covers what we think we know in terms of flowpath connectivity. The modern era can best be defined as starting sometime in the 1990–2000 period. The knowledge gained in this era was linked to two key factors:

- The first easy use of electronic datalogging[26] along with the related availability of commercial sensors and software.
- The availability of high-resolution digital elevation data.[27]

Together, these two factors[28] initiated a great pivot in thinking, enabled by the ability to view more events of varying magnitude at single locations and comparative analysis of multiple locations in different climate, geological and topographic settings.

Figure 6.27 summarizes the general picture of connectivity, thresholds and losses along the flowpath linked to a steadier drumbeat of deeper connections to the stream. This image is different to all the previous 2D drawings that we examined in earlier chapters. The modern era showed that soil depth is highly variable, and the soil–subsoil interface (with weathered or unweathered bedrock) is "bumpy." And this subsurface bumpiness, or heterogeneity, is a key piece of the connectivity–threshold function and to the barriers to connectivity. Similarly, in dry areas where infiltration excess overland flow is important, it is the bumpiness or microtopography of the surface that is important.

Another thing that we learned in this chapter is that most forms of runoff collapse down to connectivity, with a thresholds response once connectivity is achieved and loss along the flowpath, even during connected conditions. Figure 6.28 shows something that contrasts somewhat with traditional variable source area thinking. While not universal, this sort of fill-and-spill behavior is perhaps best described by flow that cascades down a system rather than variable source areas reaching up into a system (although both can happen). Remarkably, while infiltration excess overland flow and subsurface stormflow were seen as fundamentally separate processes, our current thinking is that they are governed by largely the same types of factors (McDonnell, 2013). Figure 6.28 illustrates the linkages that unite all shallow runoff forms—above-

[26] The first Campbell Scientific datalogger was sold in 1975. But by the late 1980s, plug and play sensors (e.g., tensimeters) to connect to Campbell loggers were still not available. The first <$100 temperature logger sold—the HOBO datalogger—was in 1991 by Onset.

[27] In May 2000, the Clinton administration of the USA turned off the so-called Selective Availability and GPS accuracy improved from ~100 m accuracy to 10–20 m. Of course, now we enjoy sub-meter to centimeter accuracy and drone-based LIDAR data. In the early 1980s, most slope surveys I was involved with used a hand level and stadia rod. By the mid-late 1980s, I recall the use of "total stations"—but these were very cumbersome units for slope surveys in steep terrain, like Maimai, where I still recall moving the reflector around the forested slope so that my assistant could see it and determine the distance and angle.

[28] Of course, this is also likened to the widespread usage of stable isotope tracing that we saw in Chapter 5 that catapulted forward again in ~2006 with the invention and commercial availability of the laser spectrometer.

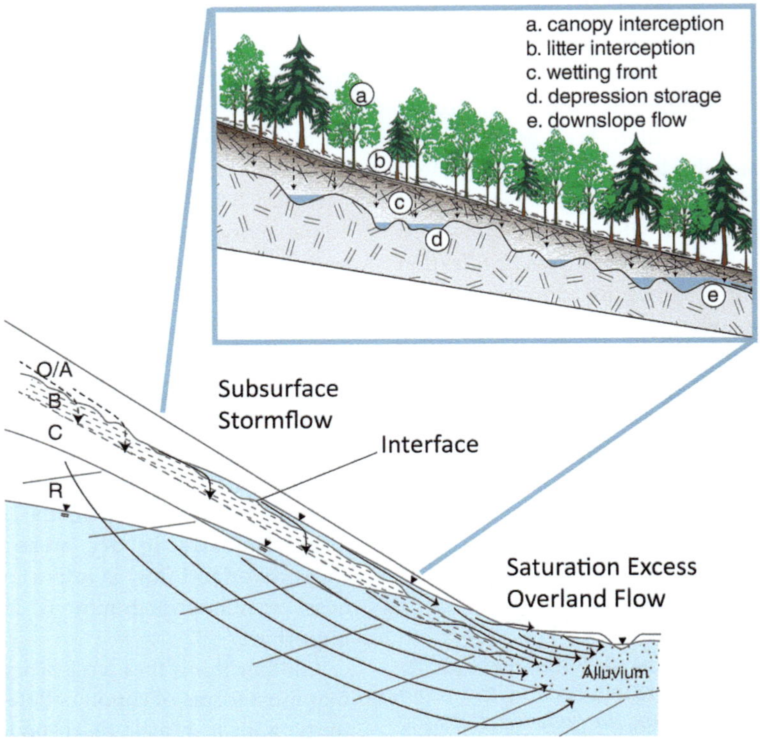

Figure 6.27 Subsurface stormflow sitting above deeper flowpaths from slope to stream. The inset shows the event-activated subsurface stormflow that we explored in this section.

Source: Adapted from McGuire et al. (2024).

and belowground—where it seems to boil down to a sequence (from the lower panel to the upper panel) that leads to emergent behavior at the hillslope scale: the filling and spilling of small depressions, continued loss to the subsurface, connectivity of expanded saturated patches, a threshold response at the base of the hillslope and ultimate filtering in a catchment comprised of multiple hillslope sections. It turns out that infiltration excess overland flow is very much like these subsurface flow patterns—except that the interface to induce lateral flow is at the soil surface.

Our goal of moving forward in streamflow generation research is to detect, understand and predict connectivity at hillslope and catchment scales. Blume and van Meerveld (2015) remains the best comprehensive overview of methods to determine slope-stream connectivity. They note the importance of future, focused efforts on connectivity assessment through site intercomparisons with an express purpose of searching for similarity measures that may allow us to "capture the wider picture of the mechanisms and factors that control hillslope–stream connectivity, and its effects on flow and transport at the catchment scale" (Blume and van Meerveld, 2015, p. 177). Figure 6.29 illustrates this vision where there is connectivity up and down the hillslope, up and down the stream reach and vertically through the subsurface profile.

Finally, you may be asking yourself after reading this chapter: were all the early studies that we looked at in Chapters 2–4 somehow wrong in their process descriptions or perceptual models? The answer to that would be no. The reason they did not see or appreciate the role of flowpath connectivity in the early studies boiled down to the fact that there were too few storms at too few

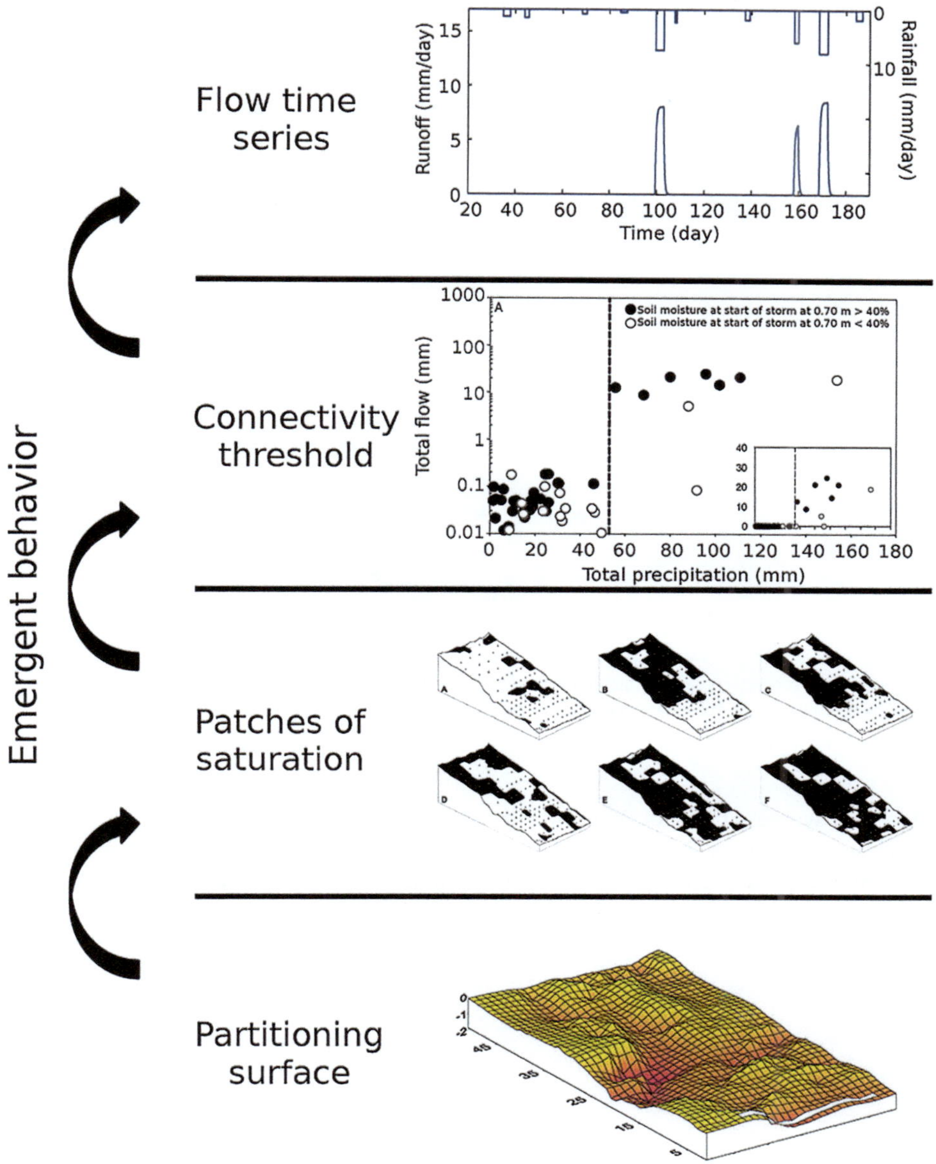

Figure 6.28 Emergent behavior from a partitioning surface. The "surface" may be the soil surface or the soil–bedrock interface or some other subsoil interface. Connectivity may be patches of saturation that are generated at the soil surface or in the subsurface at an interface driven by infiltration excess or even saturation excess.

Source: McDonnell (2013). Modified from the original diagram developed by Ciaran Harman, Johns Hopkins University. Note the figure is based on plots from Freer et al. (2002) and Tromp-van Meerveld and McDonnell (2006a,b) that were explored earlier in this chapter.

sites. Recall the old adage that "you never know what is enough unless you know what is more than enough." It took well over 100 monitored rain events to reveal the patterns seen for connectivity at Panola. What we learned in this chapter is that often below some sort of rain amount (and rain combined

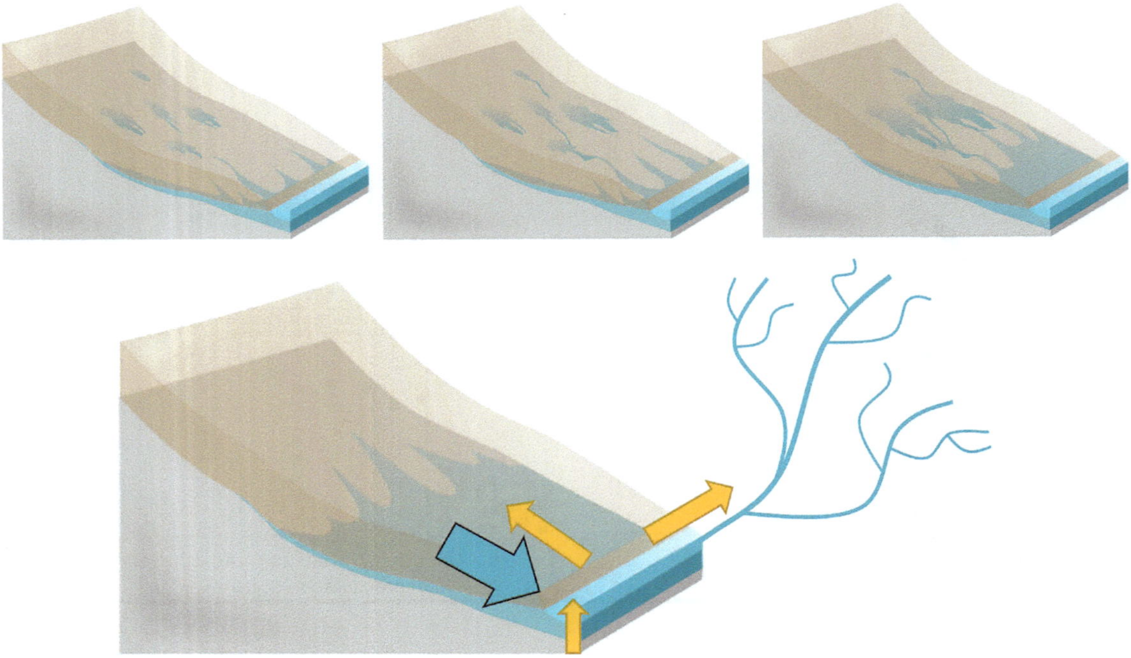

Figure 6.29 The many ways connectivity can play out through the catchment system. This can happen in a cascading way down the slope (upper panels) or through the upslope extension of surface and subsurface contributing areas (lower panel).

Source: Modified after Blume and van Meerveld (2015).

with an antecedent wetness threshold), many hillslope and catchment positions simply do not shed water. They store it. So, a modern view of runoff is now "storage excess." We'll look at this storage now in detail in Chapter 7 as it relates to the *age* of water in the stream; the catchment transit time.

CHAPTER 7

Streamwater age and catchment storage (2010–present)

7.1 Catchment transit time

Philosophical problems arise when language goes on holiday[1]

—Ludwig Wittgenstein

Water age has become a key focus of streamflow generation research—accelerating through the 1990s and 2000s and really taking off from ~2010 onward when theoretical hydrologists developing new mathematical applications got involved. And perhaps most importantly, when more and more datasets were collected to enable intercomparison of transit times from different catchments. The collection of these datasets was, in large part, made possible by the development of laser spectrometers[2] that enabled more groups to analyze water samples for stable isotopes.

Recall that we defined catchment transit time in Chapter 5 and noted the difference between water celerity and velocity—a distinction not found in most textbooks. To re-cap—and to avoid philosophical problems!—water velocity defines the mass flux of the water itself. That is what we will focus on in this chapter. From the earliest transit time analyses that we examined in Chapter 5, up until ~2010, catchment transit time was conceptualized as shown in Figure 7.1. We will try to stay true to field-based empirical data in this chapter, but we will need to stray briefly here and there into methods, as catchment transit time is not something that can be directly "measured"[3] like flow from a trench or streamflow at a gauging station. Rather, catchment transit time is calculated using a theoretical framework and some sort of mathematical function that describes the transit time distribution of the flowpaths that water takes from its input as precipitation to get to a particular point in the catchment; or how storage is released from the catchment to become stream discharge. To take a human analogy, it is like a person's age; it starts on the day we are born. So, the birth date of the water molecule entering the catchment for calculating transit time is the time it arrives at the soil surface as rain, snowmelt or throughfall. We estimate ages in the form of residence time and transit time. Transit time is the length of time it takes for water to *leave* storage. That is what we are describing for streamflow. Residence time is the time *in* storage. We'll get to that in the later sections of this chapter.

Figure 7.1 shows how the catchment transit times were calculated for many examples in this chapter. To achieve an estimate of the distribution (lower graph of Figure 7.1), the input tracer concentration of precipitation is convolved with a mathematical function (lower center in Figure 7.1)—usually a function representing transport distributions to

[1] Language relating to streamwater age had been on an exotic, all-expenses-paid vacation prior to ~2010! The literature back then was very sloppy with respect to words like residence time and transit time. And I am guilty for adding to some philosophical problems linked to terminology in the early papers!

[2] As noted in Chapter 5 where laser spectrometers became commercially available ~2006 and made stable isotope measurement available to everyone, regardless of background or training.

[3] Except in rare manipulative experiments where catchment inputs can be adequately controlled, like the covered roof experiments by Rodhe et al. (1996) or current work underway at the LEO slope at Biosphere 2 in Arizona.

Streamflow Generation. Jeffrey J. McDonnell, Oxford University Press. © Jeffrey J. McDonnell (2026). DOI: 10.1093/9780191953736.003.0007

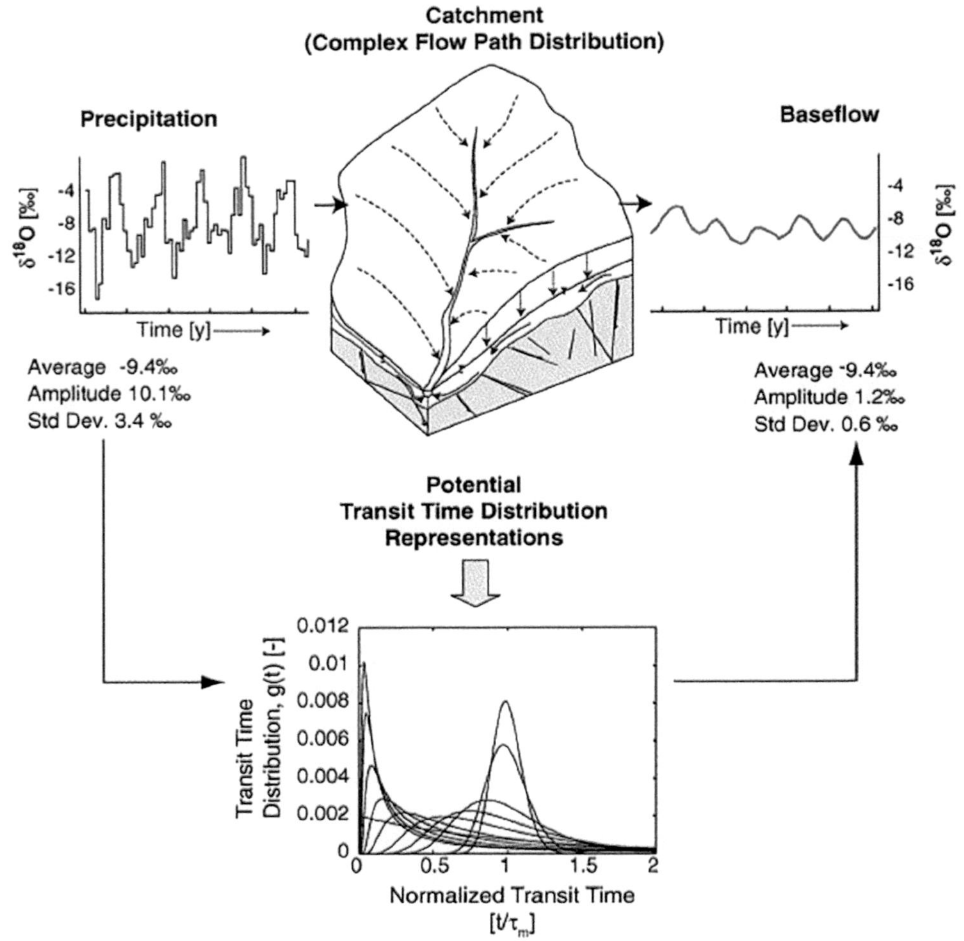

Figure 7.1 Example of how precipitation inputs to the stream are filtered through the catchment yielding a lagged and damped baseflow signal with less high-frequency variation. The transit time distribution (TTD) describes the filtering, as shown in the lowermost graph.
Source: From McGuire and McDonnell (2006).

achieve the best fit possible to the lagged and damped streamwater output time series of stable isotope data. Notice how the time series of tracers show how storage *in* and transport *through* the catchment lags and damps the tracer output relative to the tracer input. If the catchment was covered by a parking lot with zero infiltrability, then the tracer output (right-hand side of Figure 7.1) would look nearly identical to the left-hand side. But the catchment in this case—like most catchments—has subsurface storage and acts like a filter (as we examined in Chapter 6). The complex distribution of catchment flow paths is represented by the transit time distribution function, that describes the integrated behavior of tracer transport through the catchment as a whole.[4]

Figure 7.2 shows a real-life example of the Figure 7.1 approach using a best-fit for an exponential distribution function for the WS10 at the HJ Andrews catchment, Oregon.[5] The catchment

[4] For the most recent review on all this, see Benettin et al. (2022)

[5] Recall that we discussed this catchment in Chapter 6

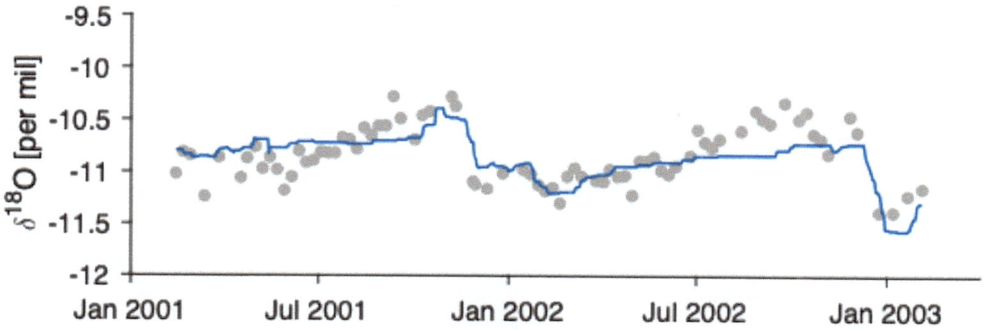

Figure 7.2 Time series for streamflow ^{18}O for the WS10 catchment at the HJ Andrews with the best fit using an exponential distribution function—i.e., the blue line running through the time series of isotope data.

Source: McGuire et al. (2005), re-drawn from McGuire, pers. comm.

transit time is then the mean of that function. In this example, catchment mean transit time is 1.2 years (McGuire et al., 2005). You will notice that this example uses ^{18}O as the tracer; others we will look at will use ^{2}H as the tracer. They are largely interchangeable for such purposes in humid regions.[6] While many of the early catchment transit time calculations we examined in Chapter 5 used ^{3}H (tritium) as a tracer for calculating catchment transit time, almost all the transit time studies from the mid-1980s onward used oxygen or hydrogen isotopes to quantify transit times (as shown in Figure 7.1).[7]

If we return to a trio of catchments that we examined in Chapter 6, we can see how different storage–discharge relations result in very different streamwater ages. Figure 7.3 shows data for the steep, always wet and mostly "connected" Maimai catchment from New Zealand; the steep, seasonally wet and sometimes connected WS10 HJ Andrews catchment in western Oregon, USA (that we just looked at); and the mostly dry but occasionally wet and rarely connected Panola catchment in Georgia, USA. Figure 7.3 shows the relation between discharge and storage and catchment mean transit time for each. The figure shows how climate and catchment geology can play out in terms of the interactions between flow, storage changes and catchment transit times.

For the Maimai catchment, there is little seasonality—the catchment is always in a wet state with rainfall on average every few days through the year. The thin transmissive soils over poorly permeable bedrock results in a flow response with a minimal storage threshold (as we saw in Chapter 6). Thus, discharge moves vertically up and down the green line but with little corresponding storage change. Consequently, the catchment mean transit time is very short, of the order of a few months (as reported by Pearce et al., 1986). For the HJ Andrews catchment (WS 10), the climate is different. It has a wet Mediterranean climate,[8] with similar total annual rainfall inputs to Maimai (>2000 mm); but with a well-defined winter wet season and summer dry season, and deeper soils compared to Maimai. As a result, the blue line in Figure 7.3 traverses a wet season *and* a dry season. Consequently, the system drains during the effectively rainfall-free 3–4-month summer season. As a result, the dynamic storage is larger than in Maimai and streamwater mean transit times for the HJ Andrews catchments are greater, of the order of 1–3 years[9] (as noted by McGuire et al., 2005). Lastly, the Panola catchment

[6] But in arid regions, they can yield different results, as shown for isotope hydrograph separations by Lyon et al. (2010). We will discuss the reasons for these differences as it relates to plant water sourcing in Chapter 8.

[7] This is because as bomb tritium washed out of global precipitation, it became an ever-more-difficult measurement challenge to detect ever-lower tritium values in precipitation and streamflow.

[8] i.e., wet winters and dry summers.

[9] We will discuss the scaling relations of these baseflow mean transit times at the HJ Andrews catchment later in the chapter.

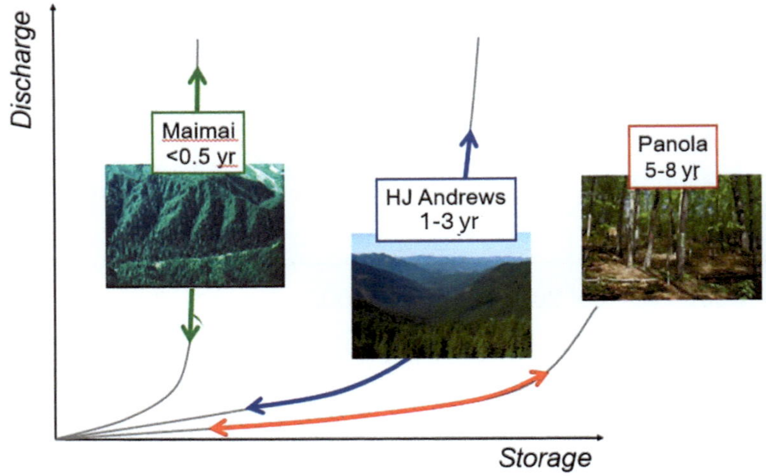

Figure 7.3 Three familiar catchments now—Maimai, HJ Andrews and Panola—with their flow dynamics linked to their dynamic storage and catchment mean transit times.

Source: Modified from an original unpublished diagram by Takahiro Sayama, University of Kyoto.

in Georgia, USA is water stressed by comparison to Maimai and HJ Andrews. It also has an enormous storage capacity in the saprolite and the riparian zone aquifer—where with additions to storage there is little corresponding increase in streamflow. Consequently, streamflow mean transit times at Panola are much longer, of the order of 5–8 years[10] (as shown in Peters et al., 2014).

Viewed through the connectivity-threshold lens of Chapter 6, you will recall that there were strong similarities in processes for these three catchments. But viewed through a transit time lens, Maimai, HJ Andrews and Panola are radically different. One other piece that came into focus after 2010 was the recognition that water in storage was often older than the streamwater transit time. This was first outlined theoretically by Berghuijs and Kirchner (2017) where they laid out a rationale for the contrasting ages of groundwater and streamflow. Reflecting on this dichotomy at Maimai, HJ Andrews and Panola, groundwater age dating (i.e., the residence time of water in storage) at each of these sites revealed groundwater ages far in excess of their streamwater transit times. Drilling at Maimai has shown groundwater ages tens of meters upslope from the stream channel at ~10 m depth on the order of 23 years (Gabrielli et al., 2018). Drilling at HJ Andrews into the fractured rock on the WS10 side slopes has shown groundwater ages from ~5–10 m bore holes on the order of 55 years (Cody Hale, pers. comm.); and at Panola, groundwater was age-dated to 26 to 27 years in a borehole ~500 m downstream of the Panola gauging station (Burns et al., 2003). As Berghuijs and Kirchner (2017, p. 8925) note "the age contrast between rivers and aquifers seems paradoxical: catchments' aquifers feed their rivers, so how can they be so much older than the rivers that drain them"? The answer to this question—and many others related to catchment transit time—is storage.

[10] We will discuss how these transit times vary with flow rate for Panola later in this chapter.

7.2 Critical zone storage

All the stream that's roaring by,
Came out of a needle's eye

—William Butler Yeats

A storage-based view of catchments developed in the early 2000s with key field contributions by Spence (2010) and field-based intercomparisons by McNamara et al. (2011). But the new storage-based view of streamflow generation also came from developments in critical zone science that emerged in the early 2000s with the creation of 10 "critical zone observatories" (CZOs)[11] in the USA and many dozens of other CZOs that followed in Europe, China and other parts of the world. The term critical zone usually means the storage between the top of the vegetation down to the bottom of the aquifer.[12] So, while the early studies we examined in Chapters 1–4 addressed soil water storage, the new critical zone lens extended the thinking and perceptual models much deeper. We saw in Chapter 5 how glacial till and glacial outwash[13] material below the mineral soil could influence groundwater connections to the stream. The critical zone ideas that emerged post-2000s would shine a spotlight on other deeper materials including flow and transport in weathered rock in the form of saprolite and saprock.[14] The first global assessment of the distribution of critical zone storage thickness (i.e., the depth to unweathered rock) by Xu and Liu (2017) found, perhaps not surprisingly, that it is controlled by factors like climate, vegetation height, water table depth, groundwater thickness, topography and bedrock lithology. They suggest that critical zone thickness ranges from 0.7 to 224 m, with a global average value of 37 m across continental areas and thickest in mid latitudes (subtropical to temperate zones). In other words, deep (!) in comparison to most of the perceptual models discussed up until this point.

But so far, we have only discussed the term storage in a general sense. In the context of streamflow generation there are two components to this storage: active storage (sometimes called dynamic storage) and passive storage (sometimes called residual storage). These elements of storage and their impact on flow and transport would only come into real focus starting ~2010 with new theory (Kirchner, 2009), new datasets (Birkel et al., 2009) and new modeling (Fenicia et al., 2010). All of our references to storage in Chapter 6 were linked to active storage. To solve the puzzle from the previous section where groundwater ages at Maimai, HJ Andrews and Panola were greater than their streamflow transit times, we need the missing piece: passive storage.

Without getting too detailed, this distinction between active and passive is perhaps less binary and really more of a continuum. Technically, passive storage is related to the tail of the transit time distribution. That is because such low streamflow contributions from "passive" storage are difficult to resolve. Interestingly, some of the early papers by Piotr Maloszewski (that we referred to in Chapter 4) discussed "stagnant water," so this idea with tracer studies has been around quite a while. Those models allowed for diffusion into the stagnant or immobile zone, but he and others also described cases where there is a dead-water component that does not circulate tracers. So how might we define passive storage? It refers to *the portion of subsurface water that does not actively contribute to streamflow or baseflow under typical hydrologic conditions*.[15] So this passive storage is the water that remains below the threshold of connectivity with surface water systems (or storage that remains after discharge rates have slowed significantly). In essence, passive storage represents subsurface water that does not directly influence streamflow but still interacts with environmental tracers. As a result, the volume of water involved in tracer circulation is often much greater than the volume required to support rainfall–runoff responses.[16] One of the first studies

[11] Essentially, well-instrumented catchments but with a focus on processes below the standard "soil–bedrock" interface.

[12] The technical definition is the "the heterogeneous, near surface environment in which complex interactions involving rock, soil, water, air and living organisms regulate the natural habitat and determine availability of life sustaining resources," National Research Council (NRC, 2001, p. 2).

[13] Till is often described as a boulder clay whereas outwash sand can be extremely well sorted sands—an example of which was our discussion of capillary fringes and groundwater ridging in Chapters 3 and 5.

[14] Saprolite is chemically altered rock. It is deeply weathered, chemically altered rock that retains the original rock's fabric, but is a soft, friable material that can often be broken apart by hand. It has much higher porosity than the parent rock. Think of it then as weathered bedrock with increased porosity. Saprock has some remaining hardness and is often dominated by fissures. So, flow in saprolite is often more akin to matrix-like flow and flow in saprock is more akin to fracture-flow.

[15] Thanks to Kevin McGuire (pers. comm.) for helping to shape this definition.

[16] We will come back to this in relation to the question of where is the bottom of a catchment? In this regard, passive storage could be different than a dead-zone where tracer never mixes or circulates. Lastly, the passive storage issue is also a timescale issue. In other words, time matters, where longer times mean opportunity for tracer connection back to surface

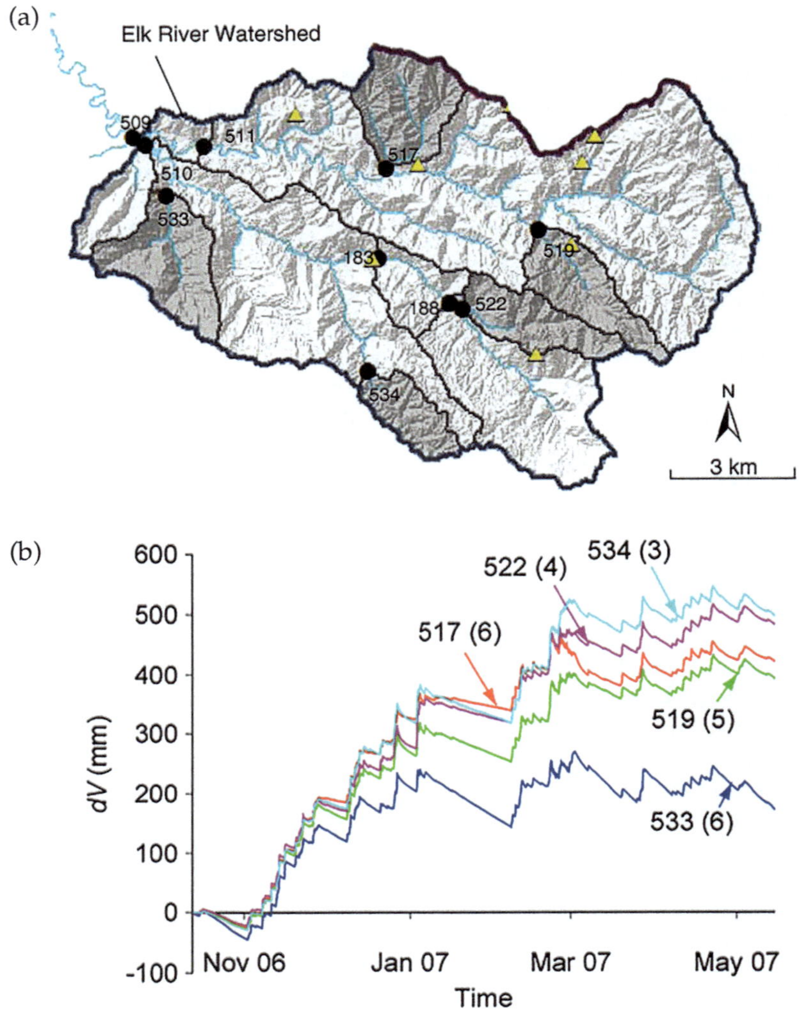

Figure 7.4 (a) The Elk River catchment in North California, USA. (b) Changes in storage (dV) with time for 5 of the 10 catchments. Catchment numbers are shown with their corresponding catchment areas in km^2 in the parentheses (ranging from 3 to 6 km^2).
Source: Sayama et al. (2011).

to show this clearly with data was Tetzlaff et al. (2014) who noted how stable isotopes could show a residual catchment storage volumes of >1000 mm; some two orders of magnitude greater than the dynamic storage changes (in their case <20 mm), responsible for generating most of their storm events.

water even in small quantities. This has been commented upon by Stark and Stieglitz (2000) as the "sting of the fractal tail."

Our discussion of storage parallels stems from what we learned in Chapter 6, where effectively all forms of runoff are linked to storage excess. Let's start now with a focus on active or dynamic storage—linking to themes in Chapter 6 but from the viewpoint of storage change rather than streamflow per se. The rest of the chapter will focus mostly on the combination of active and passive storage in the context of transit time. In a model it is easy to specify the two types of storage mathematically.

In the real world, the distinction between active and passive storage is more challenging and nuanced.

If we look at perhaps the cleanest and clearest example of dynamic storage at the catchment scale, the work of Sayama et al. (2011) for the 112 km^2 Elk River catchment in California, USA helps illustrate the dynamic storage filling and draining. Figure 7.4a shows a set of gauged catchments from 3 to 6 km^2 in the area within the Elk River catchment. Storage change is shown for the five gauging stations in the Elk River catchment for the rainy season from mid-October to mid-May. Sayama et al. (2011) used a simple water balance analysis to estimate dynamic storage changes for several headwater catchments. This was possible because this catchment was highly seasonal: months of drought in the summer, followed by months of steady rainfall in the winter. In other words, a classic Mediterranean type climate. Storage change was estimated as rainfall minus discharge minus evapotranspiration. Figure 7.4b shows this through the annual wet-up phase for the site. Prior to November, the catchment received little to no rainfall for several months. Figure 7.4b shows a pronounced dynamic storage limit and then "storage excess" pattern; i.e., the catchments store significant amounts of rainfall with little corresponding change in runoff in the beginning of the wet season and release considerably more water to the streams after their storage capacities have been filled. The amount of rainfall required to fill the storages in the individual catchments varied between ~150 and 500 mm, noted by the leveling of the dV curves in Figure 7.4b. Sayama et al. (2011) presented a "hydrologically active bedrock hypothesis" to explain these differences in dynamic storage, whereby the amount of water a catchment can store is influenced by the filling of unrequited storage in weathered bedrock—not unlike the examples from Chapter 6. In these Elk River catchments, the amount of water required to activate rapid rainfall–runoff response was larger for steeper watersheds where the restricted expansion of seepage from bedrock to the soil limited the connectivity between stored water and stream runoff.[17]

There have been many such studies since the Sayama et al. (2011) work. For instance, Jehn et al. (2021) analyzed 88 catchments in Hesse, Germany and found even more complex storage filling and spilling relations, linked also to soil permeability, hydraulic conductivity, geology and soil type. But the question now is: how old is the water in the dynamic and passive storage components? When the first studies by Birkel et al. (2009) started comparing estimates of storage by the water balance method and estimates of storage based on isotope tracer data (where storage is simply the catchment transit time divided by the discharge), the tracer-based estimates of storage were much larger—pointing to passive storage unaccounted for in hydrometric approaches. Let's now look at the residence time of water stored in the critical zone.

7.3 Residence time of stored water in the critical zone

Thoroughly conscious ignorance is the prelude to every real advance in science

—James Clerk Maxwell

Early studies of the residence time of water in storage focused on shallow forms of storage: soil water. Most of the early work is from forested headwater catchment settings. One of the first early examples using this convolution integral approach for a time series of soil water stable isotope *patterns* was done at the Maimai catchment in New Zealand. Figure 7.5a shows that soil water mean residence time ranged from about 10 to 100 days for positions across the headwater catchment with clear directional increases in age downhill. Figure 7.5b shows the relation between measured surface flowpath distance from the ridge top divide to the nearest perennial stream and soil water mean residence from a set of suction lysimeters located near the soil–bedrock interface that might more readily respond to upslope contributions of water than rainfall. The downslope aging of water is clearly illustrated. But as we saw in Chapter 6, Maimai is in many ways

[17] To link back to the themes that we looked at in Chapter 6, a plot of discharge vs. change in storage for these California catchments shows classic fill and spill behavior, now at the full catchment scale. Some of them show rather clean fill and spill system behavior; others are more messy. See Sayama et al. (2011) if you are interested in these details.

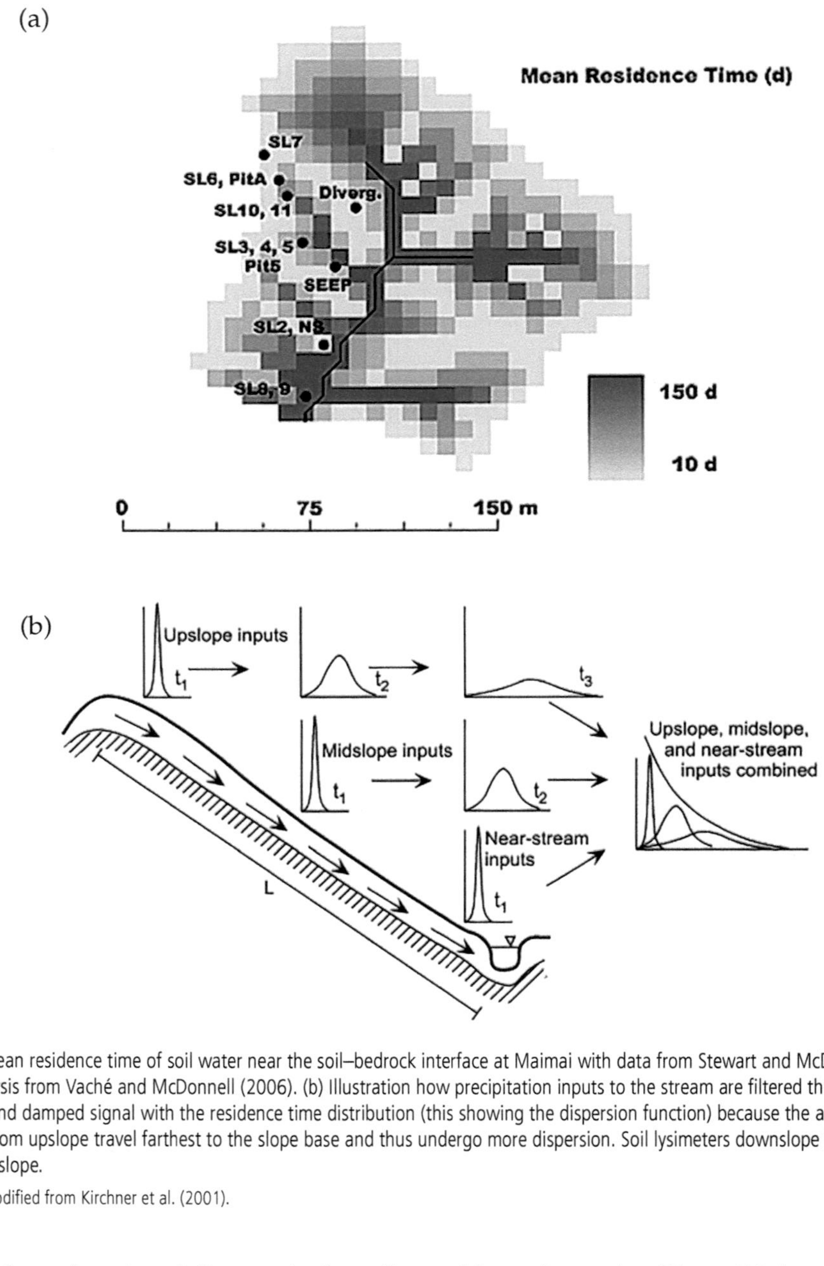

Figure 7.5 (a) Mean residence time of soil water near the soil–bedrock interface at Maimai with data from Stewart and McDonnell (1991). Figure and re-analysis from Vaché and McDonnell (2006). (b) Illustration how precipitation inputs to the stream are filtered through the soil yielding a lagged and damped signal with the residence time distribution (this showing the dispersion function) because the advection of water and tracer pulses from upslope travel farthest to the slope base and thus undergo more dispersion. Soil lysimeters downslope integrate the signal from each point upslope.

Source: Lower plot modified from Kirchner et al. (2001).

an outlier for downslope travel distance in the soil over a bedrock interface. Most hillslope soils are not nearly as "forced" or "aged" laterally.

Follow-up work by Asano et al. (2002) in Japan for the Fudoji catchment reported that soil water ranged in age from 10 to 67 days. But contrary to Maimai, the Fudoji soil water showed no evidence of downslope aging. Figure 7.6 shows how their soil water ages increased exclusively with depth. Here again is a good example of the value of comparative analysis where two sites with very different conductivity differences between soil and what was below the soil: at Maimai, a sharp hydraulic conductivity transition, forcing water during and between events

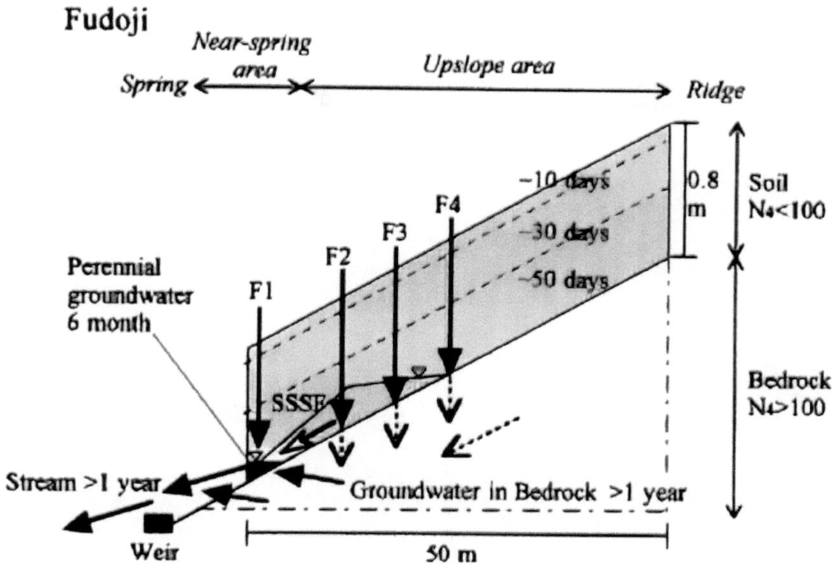

Figure 7.6 A schematic illustration of soil water aging at the Fudoji catchment, Japan.
Source: Asano et al. (2002).

to move laterally downslope. And at Fudoji—likely much more characteristic of slopes generally—a less sharp transition to weathered rock below and hence a continued vertical aging of soil water into and past the interface—similar to what was later observed in the WS10 catchment at the HJ Andrews (McGuire & McDonnell 2010).

Perennial groundwater at a seepage zone at the slope base of the Fudoji catchment showed that the age of the lateral water movement through bedrock was much greater than the movement of water vertically through the soil.[18] Figure 7.6 shows how soil water continues vertically to percolate in the underlying bedrock.[19] The arrows present the dominant flowpath as reported by Asano et al. for their "heavy storms" and subsurface stormflow and the exfiltrating bedrock groundwater at something over 1 year old.

Following Asano et al. (2002), studies would continue to pour in from the USA, Europe, South America, the tropics of Mexico and karst sites in China. Most of these focused on the residence time of soil water up to ~1 m and sometimes to ~2 m depth. But soil can also be deep. In recent years, these deep soils have been investigated—and perhaps at the extreme end of soil water depth and soil water age is the vertical aging of soil water on the Loess Plateau in China. There, soil water ages vertically through 100+ m of unsaturated loess[20] en route to the water table. Figure 7.7a shows a photo of the region and a profile of soil water tritium values. Like we discussed in Chapter 5, the main tritium signal on Earth (well and above the natural background tritium values) is the signal put into our Earth's atmosphere during thermonuclear testing that peaked in 1963 and subsequently reduced dramatically. What do you see in Figure 7.7b? At

[18] As seen also at the WS10 HJ Andrews catchment by McGuire and McDonnell (2010).

[19] Recall from Chapter 6 our examination of losses of subsurface stormflow reported on experimental hillslopes—this is showing that same effect; now from a different perspective.

[20] A modern word derived from German and Swiss to describe wind-deposited, loosely compacted material, where the clay fraction is <20% with the other fraction comprised of approximately equal fractions of sand and silt. We will see an example in Chapter 8 that shows how one can hand-auger 20 m in this material—try that in your home catchment!

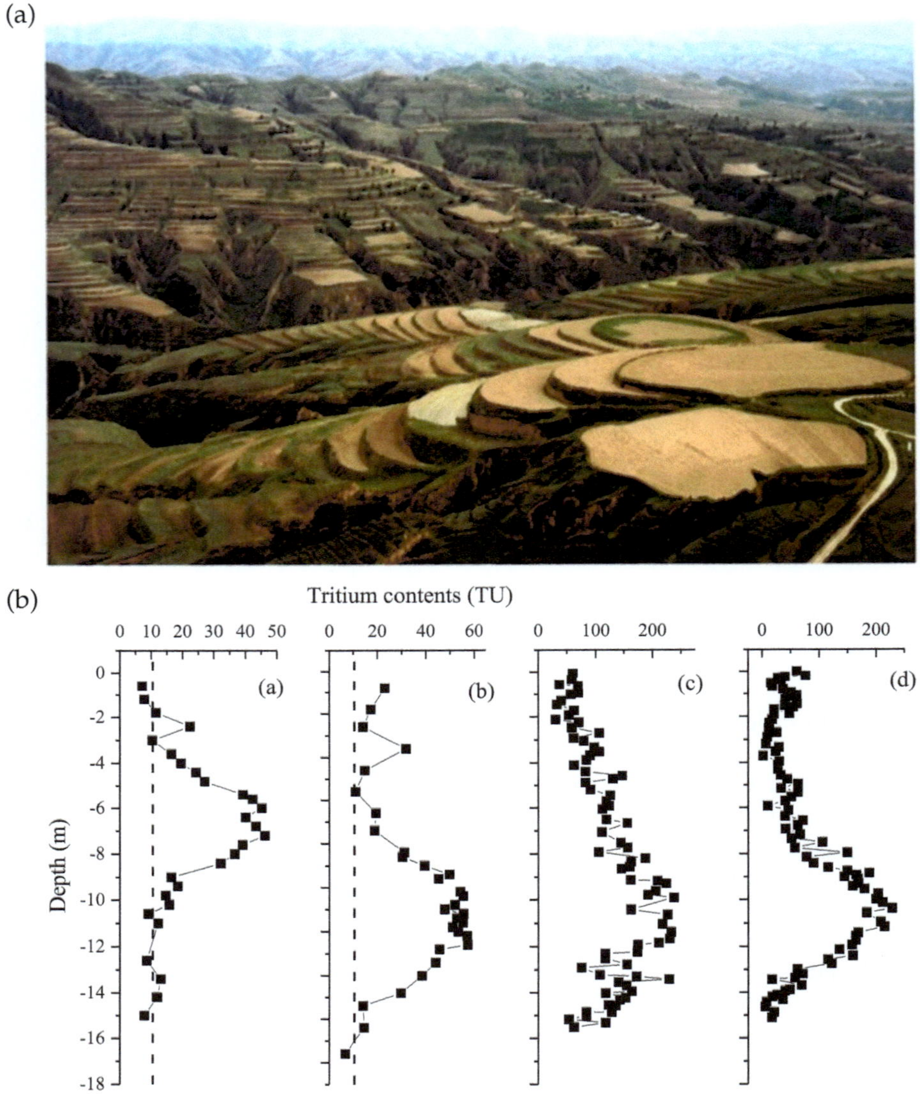

Figure 7.7 (a) The Chinese Loess Plateau, source Shutterstock. (b) Tritium profiles at 4 sites on the Loess Plateau, from Li and Si (2018).

about 8 m depth at this point on the Loess Plateau in Figure 7.7b, you "see" the 1963 bomb peak! Albeit with some dispersion of the signal. It's quite amazing and rare to see this signal today so well preserved in the unsaturated zone.[21] If you do a quick calculation of this bomb peak in the soil water here and consider the time from 1963 to the time when these pore water samples were extracted and analyzed (let's say in 2013), then it takes 50 years to travel 10 m. Indeed, at sites like this on the Loess Plateau, the "soil water" can be up to 1000 years old by the time it reaches the water table at ~100 m (as shown by Wang et al., 2023).

[21] Because in most regions with shallower soil, the bomb signal has been long-ago flushed out.

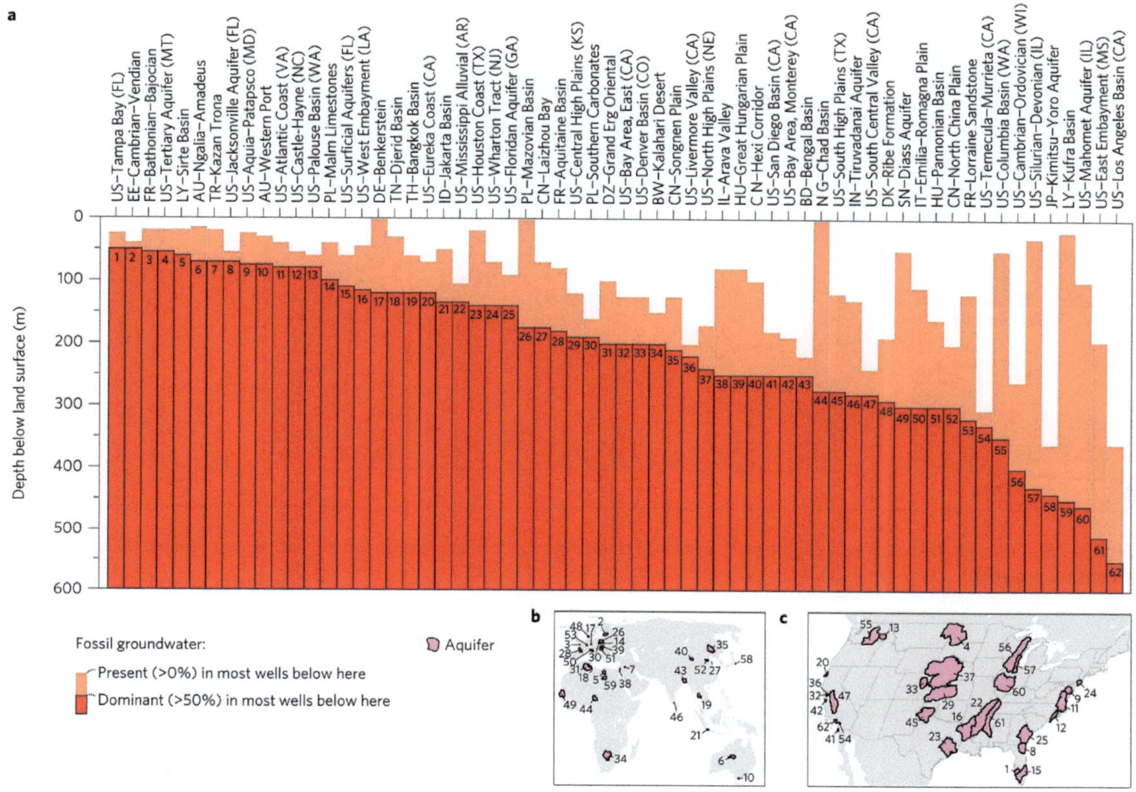

Figure 7.8 Prevalence of fossil groundwater (>12,000 years old) in global aquifers.
Source: Jasechko et al. (2017).

7.3.1 Passive storage and defining where is the bottom of a catchment

Let's now go even deeper into the critical zone and discuss groundwater mean residence time. Recent global syntheses that we have already alluded to in Chapter 2 offer stunning insights into the age of water in storage underground: it gets old *quickly* with depth. Jasechko et al. (2017) analyzed groundwater carbon isotope data (^{12}C, ^{13}C, ^{14}C) from 6455 wells around the globe and found that 2/3 of groundwater below 250 m was 12,000 years old or older. In other words, fossil water. Figure 7.8 shows this prevalence of fossil groundwater for 62 global aquifers. The top of the peach-colored bar represents the depth below which most wells (>50%) contain detectable fossil groundwater. The top of the red-colored bar represents a depth below which most wells (>50%) are dominated by fossil groundwater (minimum fossil groundwater fraction is >50%).[22] Fossil groundwater becomes dominant at a median depth of 200 m. Certainly that would be passive storage!

These observations have sparked much new interest in answering the question: where is the bottom of the catchment? Meaning, how deep does one need to go at a place before what is down there never returns to the surface?[23] As we discussed briefly at the end of Chapter 2, such new work is sparking new discussions of compartmentalization of the catchment water cycle (McDonnell, 2017).

[22] Note that the lower limit of the graph (600 m) does not necessarily represent the lower boundary for the aquifers studied.

[23] Recall earlier our discussion of dead zones.

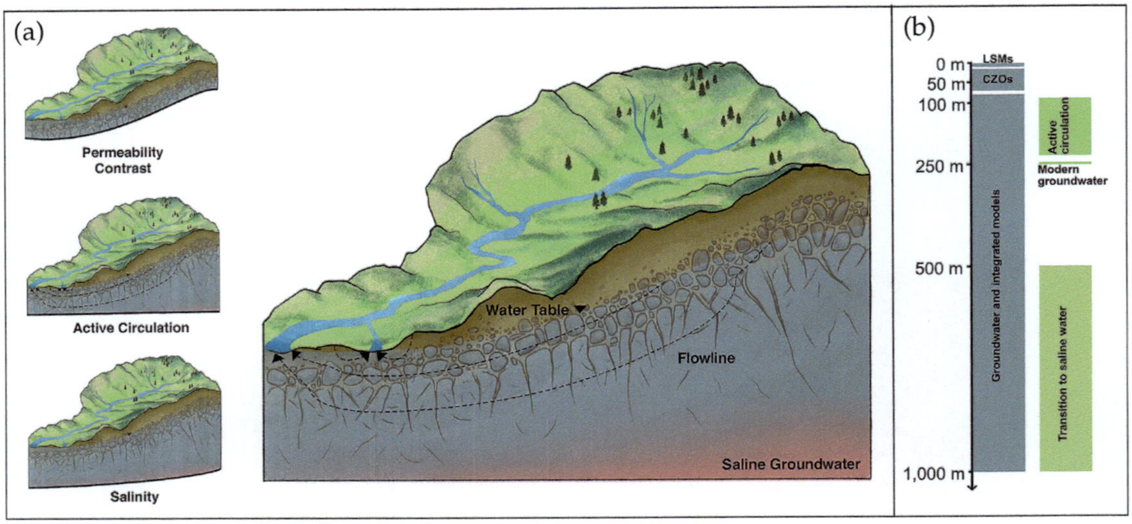

Figure 7.9 "Where is the bottom of a catchment?" as defined by (a) three common approaches and (b) modeling applications (blue) and field observations (green).

Source: Condon et al. (2020).

Condon et al. (2020) have addressed explicitly how to define what the "bottom" of a catchment might be. They note that so-called "active circulation" of groundwater in mountain aquifers has been noted in the upper 100–200 m (Gleeson & Manning, 2008; Markovich et al., 2019). Going deeper, Condon et al. (2020) note that Pleistocene groundwater recharge has been found in crystalline and sedimentary environments up to 1000 m depth (McIntosh et al., 2012).

So where is the bottom of a catchment?

Like most answers: *it depends*. The most recent analysis by Ferguson et al. (2023)[24] shows that less than 0.1% of groundwater below 500 m ever makes its way back to streamflow. So, 500 m might be a good general marker for the bottom of a catchment. Certainly, as Condon et al. (2020, p. 2) note, "below a few thousand meters, most groundwaters are essentially stagnant over periods of millions of years due to a combination of low permeability and negative buoyancy." But place matters. And Figure 7.9 shows that the transition from fresh to saline groundwater often occurs between 500 and 1000 m (Ferguson et al., 2018). But in some mountainous regions with hot springs, so-called meteoric-derived water can circulate to depths of 5000 m (Ferguson & Grasby, 2011). And in other places like the plains of Saskatchewan in Canada, at just 4 m below the surface is groundwater that is 2300 years old; at 7 m depth, that groundwater is over 13,000 years old; and at 16 m it is 20,000 years old (Keller et al., 1988). Not surprisingly, these sites have exceptionally slow groundwater flow velocities in the clay-rich glacial till, of the order of 8 m in 10,000 years. So, in this instance, the bottom of the catchment could be as little as 10 m.

[24] That we mentioned briefly at the end of Chapter 2.

7.4 Catchment transit time and its relation to catchment scale and geology

Location, location, location

—Unknown

We have looked at storage and catchment transit time—but how might other catchment characteristics affect catchment transit time and its scaling relations? Following the catchment mean transit time work and the stored water residence time work,

Table 7.1 Mean transit times and their reported link to catchment characteristics. As per the original paper, the most important local control is highlighted in bold.

Catchment characteristics	R^2	Location	Study
Topographic aspect	0.28	Arizona	Broxton et al. (2009)
Flow paths length/gradient	0.30	Europe, USA	Tetzlaff et al. (2009b)
Flow path gradient	0.08	Europe, USA	Tetzlaff et al. (2009b)
Bedrock infiltration	0.85	Japan	Katsuyama et al. (2010)
Flow paths length	0.98	New Zealand	Dunn et al. (2007)
Distance from divide	0.50	New Zealand	Vaché and McDonnell (2006)
Topographic wetness index	0.92	Oregon	Tetzlaff et al. (2009b)
Flow paths length/gradient	0.91	Oregon	McGuire et al. (2005)
Upslope area	0.88	Oregon	Tetzlaff et al. (2009b)
Flow paths length	0.72	Oregon	McGuire et al. (2005)
Flow path gradient	0.65	Oregon	McGuire et al. (2005)
Distance from stream	0.62	Oregon	Tetzlaff et al. (2009b)
Proportion responsive soil cover	0.94	Scotland	Soulsby et al. (2006)
Downslope index gradient	0.83	Scotland	Tetzlaff et al. (2009b)
Proportion responsive soil cover	0.80	Scotland	Hrachowitz et al. (2009)
Proportion responsive soil cover	0.76	Scotland	Tetzlaff et al. (2009a)
Flow paths length/gradient	0.74	Scotland	Tetzlaff et al. (2009b)
Flow path gradient	0.67	Scotland	Soulsby and Tetzlaff (2008)
Proportion responsive soil cover	0.60	Scotland	Soulsby and Tetzlaff (2008)
Drainage density	0.59	Scotland	Hrachowitz et al. (2009)
Flow path gradient	0.42	Scotland	Tetzlaff et al. (2009a)
Precipitation	0.25	Scotland	Hrachowitz et al. (2009)
Proportion of wetlands	0.59	Sweden	Lyon et al. (2010)
Flow path gradient	0.52	Sweden	Lyon et al. (2010)
Flow path length/gradient	0.43	Sweden	Lyon et al. (2010)
Peclet number	0.40	Sweden	Lyon et al. (2010)
Flow paths length/gradient	0.32	Sweden	Tetzlaff et al. (2009b)

From Tetzlaff et al. (2009a).

more studies started reporting mean transit times across multiple catchments; across a region or across different scales, nested within a larger catchment. Let's now explore how catchment mean transit time is related to catchment characteristics, catchment scale and catchment storage. Table 7.1 shows some of the first studies where such analyses were done; with the most important catchment characteristics affecting the mean transit time shown in bold.[25] Table 7.1 suggests quite a wide range of factors and one where there is no universal control on mean transit time. Let's now think about why this might be.

Soulsby et al. (2006) found that the mean transit times in their catchments in Scotland (of the order of ~2–5 months) were lowest in the catchments with the smallest groundwater contributions and in catchments dominated by peat and/or shallow soils and bedrock. They also showed that the mean transit times were longest (up to ~12–15 months) in catchments with larger groundwater contributions and in catchments dominated by better-drained soils. So, this seems to suggest that transit time is related to storage. Figure 7.10 shows other studies in Scotland that support the importance of soil storage, responsive soils and drainage density—a measure of how dissected a catchment is[26]—on transit time (Tetzlaff et al., 2009 a, b; Hrachowitz et al., 2009).

Most of the catchments in Table 7.1 are from humid regions. There's not too much literature

[25] We will start with only these first papers. At the time of this writing some 2434 papers have cited these collective works so things have grown rapidly in recent years.

[26] Technically total stream length divided by the catchment area.

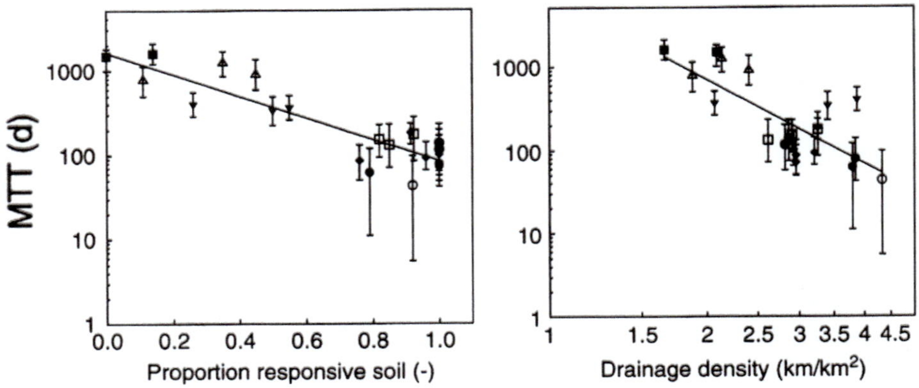

Figure 7.10 Mean streamwater transit times (MTT) vs proportion of responsive soils and drainage density for headwater catchments in Scotland.
Source: Hrachowitz et al. (2009).

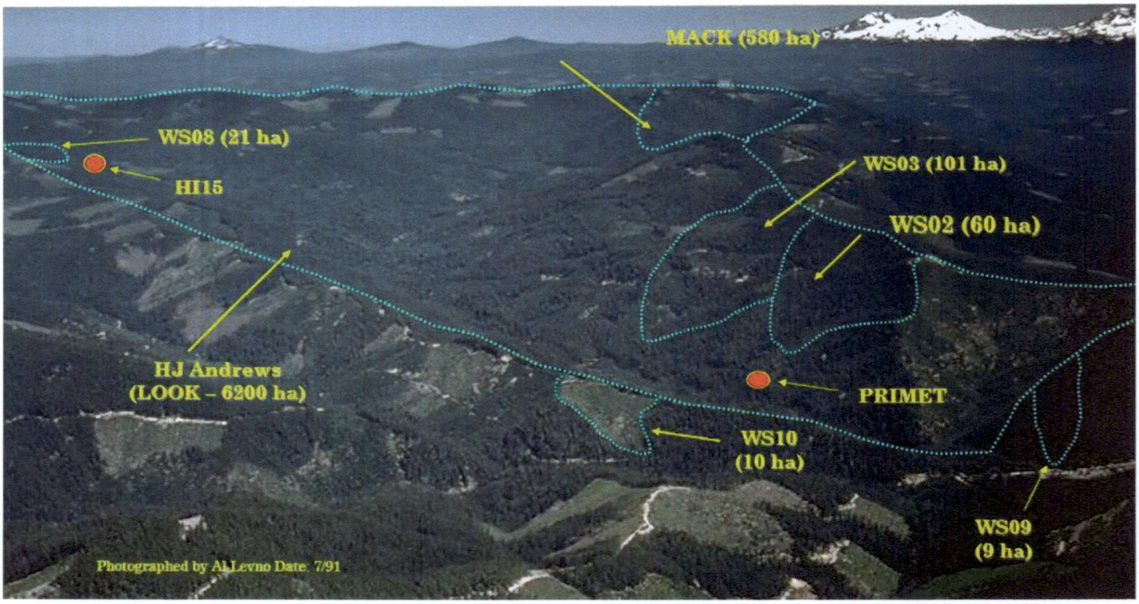

Figure 7.11 The nested catchments of the HJ Andrews (total 62 km^2 area watershed).
Source: Photo from Al Levno, July 1991.

on how catchment characteristics control catchment transit time in arid and semi-arid catchments. That's likely because many have intermittent flows in the headwaters where precipitation events are often few and far between. One rare early example is Broxton et al. (2009) who looked at 15 small (1–15 km^2) catchments that drain different aspects of Redondo Peak, Valles Caldera in New Mexico, USA. They found a strong aspect control on transit time, likely linked to soil depth, where south-facing slopes had deeper soils (due to faster weathering and faster soil production) and hence deeper storage, a longer residence time of stored water and, consequently, longer stream mean transit times.

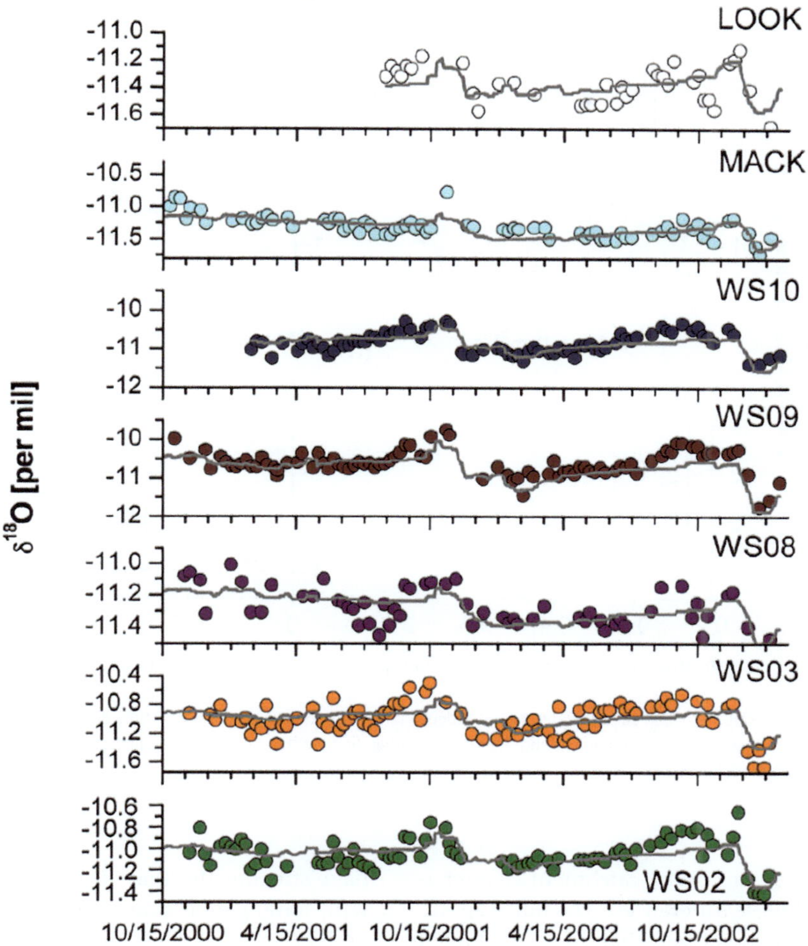

Figure 7.12 The time series of $\delta^{18}O$ for 6 catchments sampled by McGuire et al. (2005). The line running through the time series of isotope data is the best fit line using the convolution integral approach.

But what about catchment area? How might this affect stream transit time? That is a question that puzzled many of us when the first scaling data were published in the early 2000s. Wouldn't catchment mean transit time simply be linearly related to catchment area? Burgman et al. (1987) were perhaps the first to pose this question in a gray literature IAEA publication. But it really wasn't taken up as a research question until the McGlynn et al. (2003) study at Maimai in New Zealand. Let's now look at one study from Table 7.1 that presented the first thorough analysis of transit time scaling relations, this time for the HJ Andrews catchments in Oregon, USA. Figure 7.11 shows six catchments that were sampled by McGuire et al. (2005) that range from ~10 ha to 62 km^2.

Isotope data were collected for the six catchments and Figure 7.12 shows the stream water isotope dataset. Recall that we looked at the WS10 data earlier in this chapter (see Figure 7.2). And like that discussion, the line running through the time series for each of the catchments is the best-fit line using the convolution integral approach with an exponential transit time distribution function. The catchment mean transit times expressing themselves in these streams ranged from 0.8 to 3.3 years.

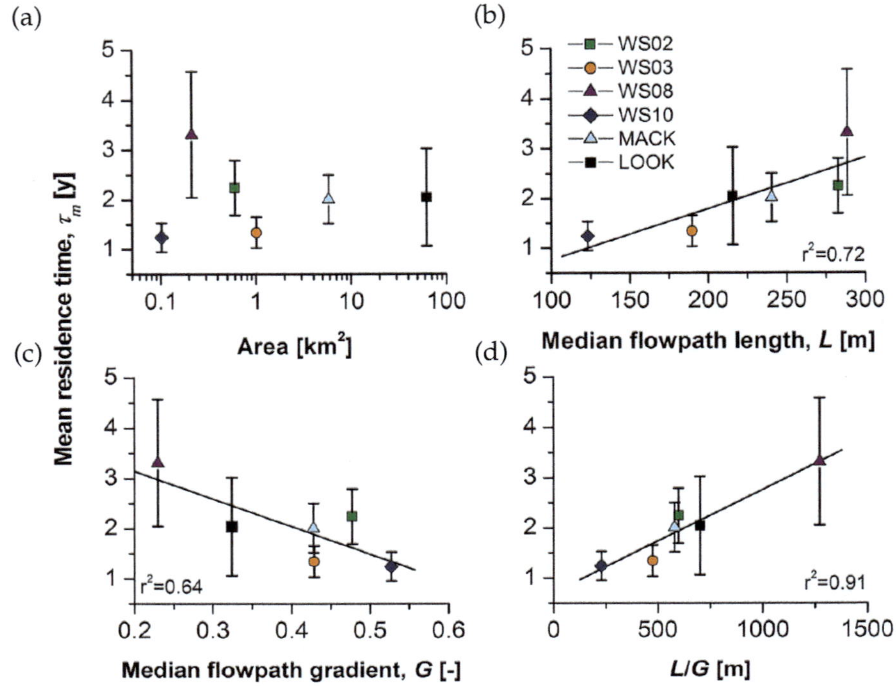

Figure 7.13 The relationships between stream water mean residence time (what we would call transit time or age, today) catchment area, median flow path length, median flow path gradient and the ratio of median flow path length to median flow path gradient—all determined through digital elevation model analysis.
Source: McGuire et al. (2005).

The question then was: what catchment characteristics controlled these differences? The null hypothesis going into the work was that mean catchment transit time would get older with increasing catchment area—just as flow amount in the stream would increase with catchment area. The idea was that as catchments got bigger, perhaps so too would the contribution of longer and deeper flowpaths?

Figure 7.13 shows the results of the McGuire et al. (2005) analysis. Quite surprisingly, there was no relation between catchment mean transit time and catchment area. Meaning that catchment transit time from a 10 ha catchment was not that different to a 62 km^2 catchment—both in the range of ~1–3 years. But look at plots b and c in Figure 7.13. What do you see? As the flowpath length for these catchments increases, so too does the catchment transit time. Conversely, median flowpath gradient showed an inverse relation with catchment transit time. This tells us that the catchments at the HJ Andrews characterized by short, steep flowpaths have streamwater that is younger than catchments with long, gentle slopes. And when combined, as in Figure 7.13d, this length divided by gradient explains over 90% of the variance in streamwater age. That is a very high correlation for environmental variables.

The scaling findings from the HJ Andrews site showed an extremely strong relationship with slope length and gradient but effectively no relation to catchment area. Hale and McDonnell (2016) tested this hypothesis for another set of eight catchments about 200 km to the west of the HJ Andrews site. The new grouping of catchments—the Alsea catchments—had very similar terrain characteristics and strikingly similar rainfall regimes and hydrograph dynamics, as shown in Figure 7.14. But the one thing that was different between the HJ Andrews and Alsea catchments was geology. Rather than very low permeability volcanic rocks at the HJ Andrews, the Alsea catchments were characterized by layered sandstone and mudstone.

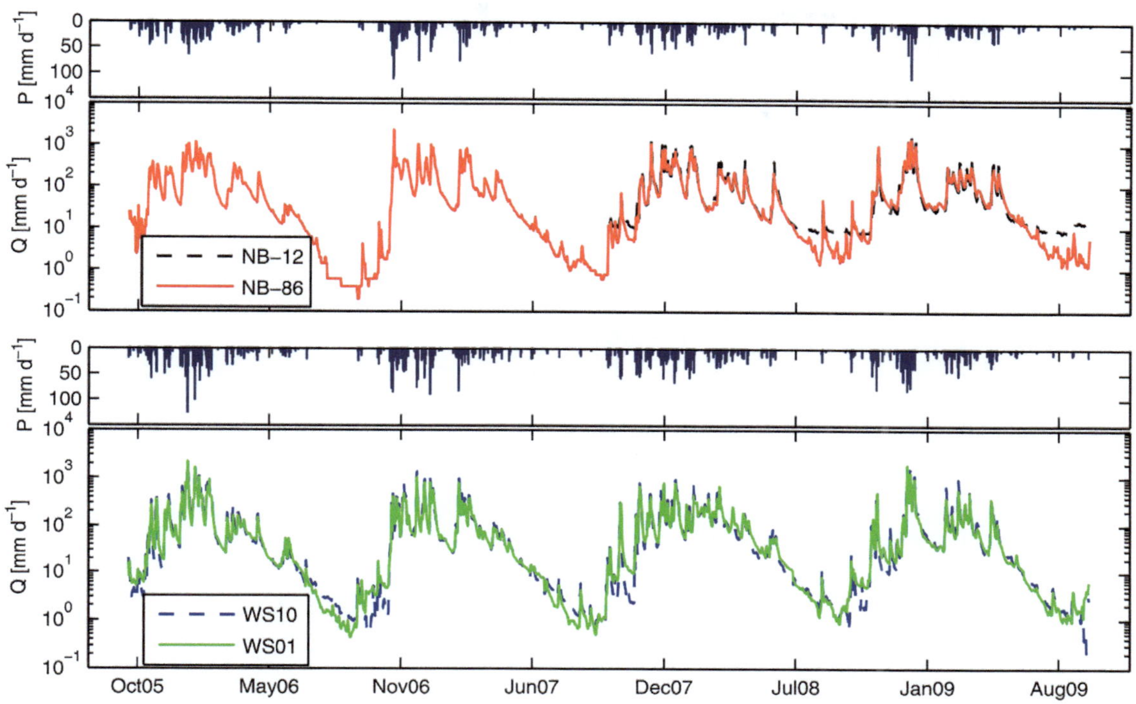

Figure 7.14 Mean daily discharge for the four Oregon catchments. The lower panel relates to the volcanic landscape shown in the earlier photo; the upper panel the new sites some 200 km to the west.
Source: Hale et al. (2016).

While these two catchment systems were very similar in terms of their celerity (as shown by the streamflow response in Figure 7.14), their velocities and scaling relations differed greatly. Hale et al. (2016) found that none of the strong catchment transit time scaling relations from the HJ Andrews catchments applied to the Alsea River catchments; neither slope length nor slope steepness had an effect on streamwater age. There was no relation to *any* of the topographic measures that worked so well at the HJ Andrews site, despite them looking outwardly (and flowing), seemingly identically. And what do you think controlled the scaling relations for catchment transit time in these eight catchments? "Catchment area" was the answer! Figure 7.15 shows that rather than 1–3 years old like the HJ Andrews, the catchment mean transit time at Alsea was of the order of 4–11 years old.

Figure 7.16 shows a perceptual model to explain how these nearly identical stream flow relations hide strikingly different plumbing mechanisms. The greater porosity, permeability and storage offered by the sandstone and mudstone rocks of the Alsea catchment are responsible for the longer, deeper flowpaths and longer catchment mean transit times that scale with area.

Many other studies followed this early Oregon example to show how geology controls catchment transit times. Perhaps the clearest was Pfister et al. (2017) who compared 16 nested catchments (0.45 to 410 km^2) in the Alzette River basin in Luxembourg. Their catchments spanned a large range of clean and mixed expressions of schists, phyllites, sandstones and quartzites. They examined 9 years of precipitation and discharge data, and 6 years of biweekly stable isotope data in streamflow and showed a very strong, positive relationship between bedrock permeability and catchment mean transit time ranging from 0.5 to ~2 years (see Figure 7.17).

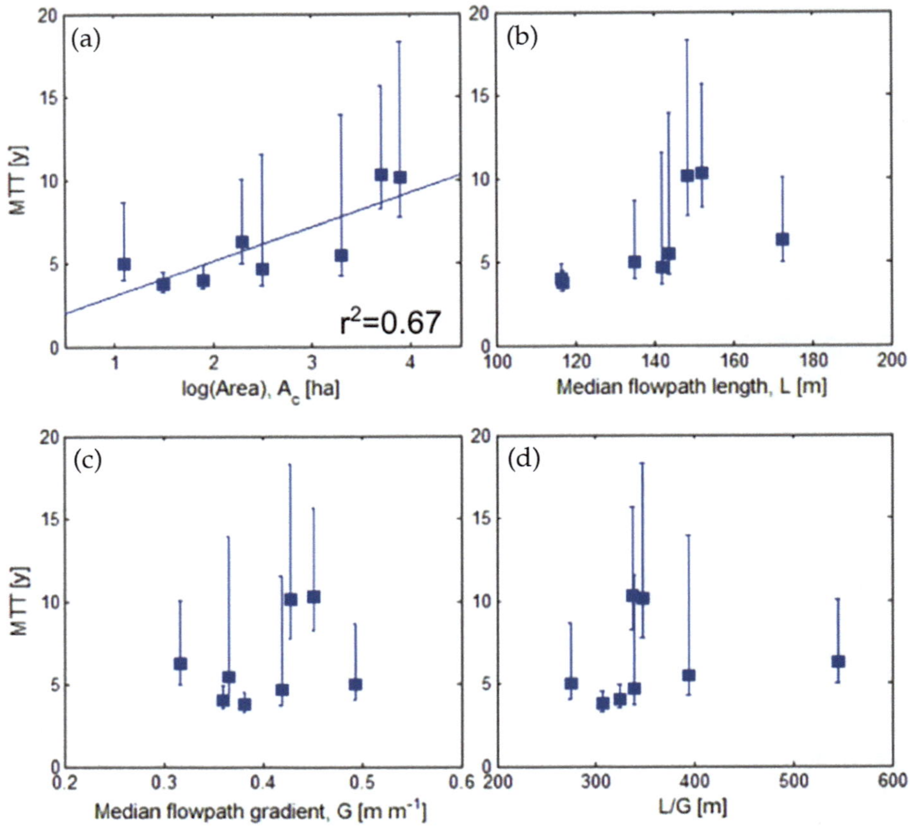

Figure 7.15 Mean transit time (in years) as a function of the (a) log of catchment area, (b) median flowpath length, (c) median flowpath gradient and (d) the ration of median flowpath length and flowpath gradient for the Alsea catchments. Compare the results to those shown for the HJ Andrews catchment in Figure 7.13.

Source: Hale et al. (2016).

7.5 The inverse storage effect

In science the credit goes to the (wo)man who convinces the world, not to whom the idea first occurs
—Francis Darwin

A last key pivot in thinking regarding catchment transit time started in the early 2010s when the idea of "time-varying transit time distribution" functions was born (Botter et al., 2011, followed by the work of van der Velde et al., 2012, Harman, 2015 and others[27]). Figure 7.18 shows how rainfall or snowmelt entering a catchment can spend different amounts of time in the catchment depending on *when* they entered. It shows the importance of changes from season to season, where a subsequent precipitation pulse could experience a very different distribution of transit times—a distribution best described as time-variable. Just to be clear, most of the work up until the mid-2000s assumed a single, *time-invariant* transit time distribution would exist for the catchment (as we outlined at the beginning of this chapter). Figure 7.18 shows how the interaction

[27] Note I am not identifying all the early mathematical and theory papers linked to water age concepts. But these are reviewed in Benettin et al. (2015), Calabrese and Porporato (2017), Rigon and Bancheri (2021) and Rigon et al. (2016) if you want to learn more.

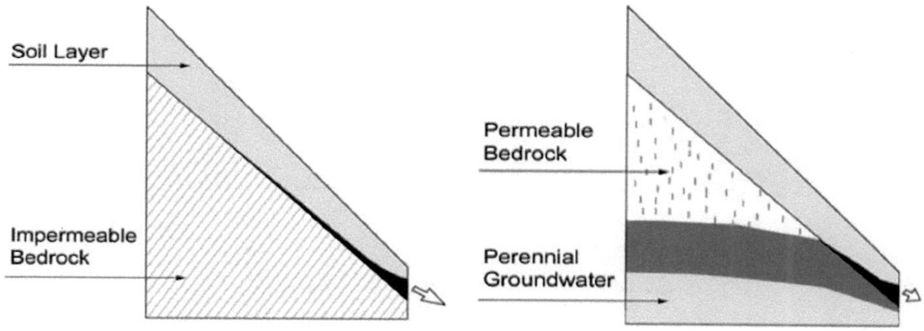

Figure 7.16 The perceptual model to explain seemingly identical celerity and radically different velocities at the HJ Andrews (left) and Alsea (right).

Source: Cody Hale, pers. comm.

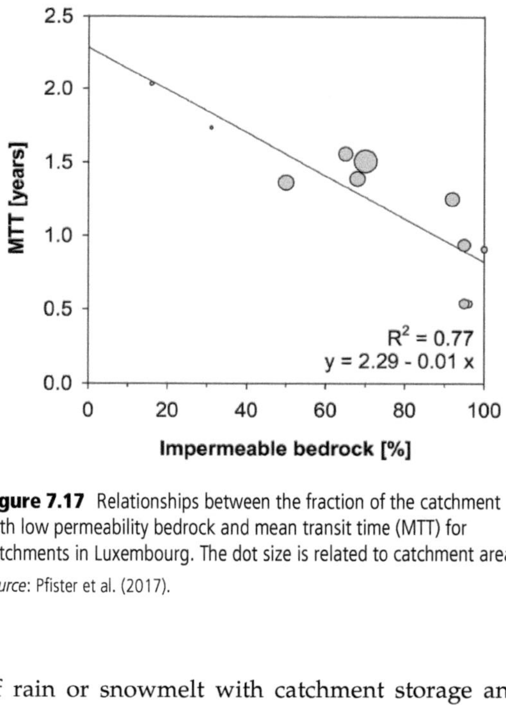

Figure 7.17 Relationships between the fraction of the catchment with low permeability bedrock and mean transit time (MTT) for catchments in Luxembourg. The dot size is related to catchment area.

Source: Pfister et al. (2017).

of rain or snowmelt with catchment storage and evapotranspiration could activate and deactivate flowpaths and change the transit time distribution.

Following the recognition of time-varying transit time distributions, there was increasing recognition that streamflow was not a mixture of *all* the stored water but only some component of that stored water—that at any given point in time might represent only a small component of that storage. Figure 7.19 shows how new methods grew (beyond the simple convolution integral approach in Figure 7.19a) to deal with this, including the use the use of so-called StorAge Selection (SAS) functions for relating dynamically how water residence times in storage related to catchment transit time (e.g., Botter et al., 2011; Harman, 2015; Benettin et al., 2015, 2017; van der Velde et al., 2015; Figure 7.19b) and the use of flux tracking (e.g., Birkel et al., 2011; Fenicia et al., 2010; Hrachowitz et al., 2013; Figure 7.19c), as well as particle tracking (e.g., Pangle et al., 2017; Figure 7.19d)[28] in hydrological models for understanding time-varying catchment transit time dynamics.

The outcome of all this work and the new methods was to explore what has become known as the 'inverse storage effect' (Harman, 2015)—the idea that streamwater transit times are usually shorter with wetter catchment conditions.[29] The only field example that I am aware of that shows this inverse storage effect using empirical data and independent transit time estimates is shown in Figure 7.20. It is a plot of streamwater transit time[30] vs. monthly

[28] If interested in these techniques, the papers by Sprenger et al. (2019) and Benettin et al. (2022) offer useful reviews on each of these approaches.

[29] We knew that catchment systems were not steady-state with an invariant transit time distribution. But what we did not have prior to ~2010 were any theoretical approaches to represent the dynamic processes.

[30] This plot was constructed by taking samples of $^3H/^3He$ and CFC-12 from Burns et al. (2003) (that allowed calculation of water age) and then plotting them against the much easier-to-measure Si concentrations in well water and stream water. Using this strong relation, Peters et al. (2014) could then use monthly Si data from the site to craft this flow-age plot. It's

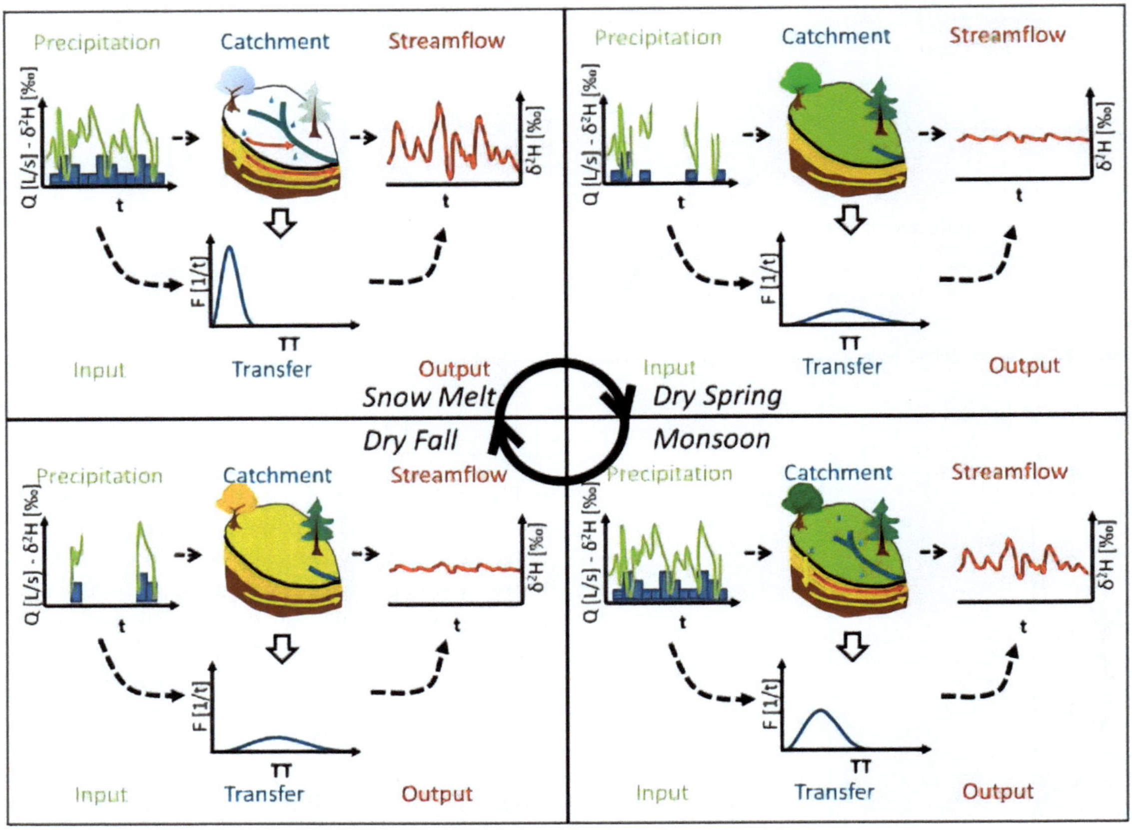

Figure 7.18 Time-varying transit time distributions. Rather than one single function for a catchment, different wetness states in different seasons generate different transit time functions.

Source: Heidbuchel et al. (2012).

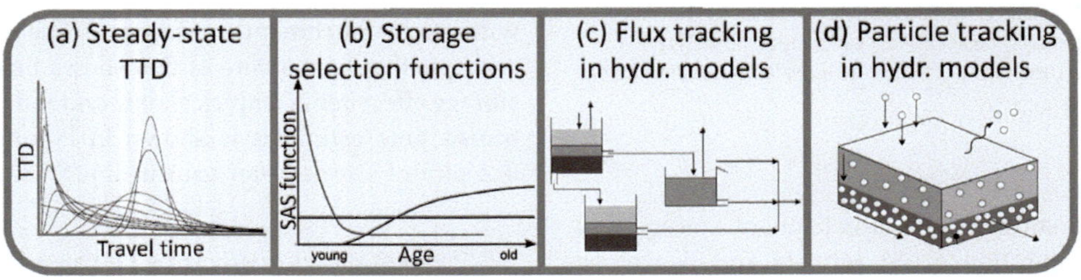

Figure 7.19 The development of new approaches to get time-varying transit times following the early transit time work (a), now linked to storage selection functions (b), flux tracking in hydrological models (c) and particle tracking in hydrological models (d).

Source: Sprenger et al. (2019).

beautiful, really. A top-down description of a catchment that incorporates deep insight into the groundwater–subsurface stormflow–streamflow connection and its changes with time through the hydrologic year.

flow from the Panola catchment in Georgia, USA. When the flow is low, the water in the stream is old; recall we discussed this in a mean transit time sense earlier in comparison to the Maimai and HJ Andrews catchments. As the flow increases, so does

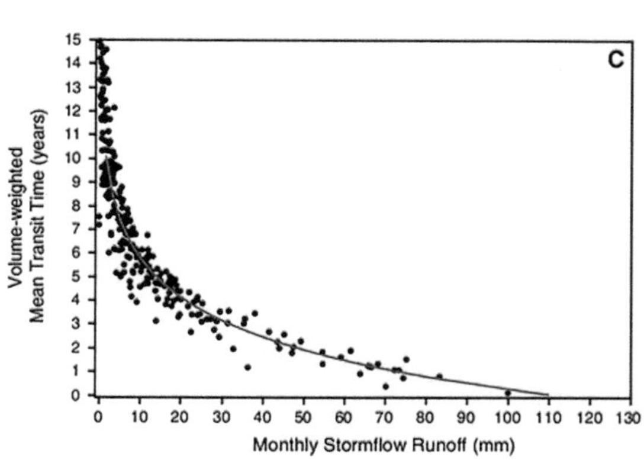

Figure 7.20 Scatter plot showing a strong non-linear relation between catchment mean transit time and monthly storm runoff at the Panola Mountain experimental catchment (for water years 1986 to 2008).

Source: Peters et al. (2014).

the activation of faster and shallower flow pathways. These data are uncommon, sadly, in our field. But hopefully they will become more common as tracers are used more extensively across many study catchments.

While the figure above is a unique relation for one catchment, we might expect in the years ahead that other catchments will show a similar, non-linear stream age behavior[31]—toggling back and forth between young and old water. With the amount of bimodality perhaps related mostly to the seasonality of the flow regime. Meaning, if there was little seasonality, we may see little expression of the very old water; with extreme seasonality, we would see some of the oldest water appear in the lowest of flows in the stream. Of course, the hydrogeology of the catchment will also have a strong bearing on all of this. Unlike the situation in the mid-1960s when Hewlett and Hibbert (1963) wondered about what sustained streams during periods of no precipitation (see Chapter 3)[32] and so too the philosophers many centuries before them (see Chapter 2), we know now that the source of stream baseflow (once subsurface stormflow has shut off) is comprised largely of groundwater. Recent estimates suggest that groundwater-sustained stream baseflow is 59% ± 7% globally (Xie et al., 2024).

Lastly, how can we visualize the link between groundwater and stream water and the activation of processes that deliver shallower and younger waters to the stream—especially when that catchment is wet. Figure 7.21 shows what is often hidden from view—here, in a road cut. If you look carefully at the photo, the pipes are showing groundwater draining perhaps 50 m from below the slope ridge. That's the water that keeps streams going in extended dry periods and what contributes the oldest waters to the stream. Shallower flowpaths activate and contribute younger water to the stream—often independent of the deeper groundwater system—via subsurface stormflow or via other even shallower, faster young-water delivery processes associated with saturation excess and infiltration excess overland flow. Much theoretical and modeling work is now underway to quantify how streams are affected age-wise from these compartments that vary vertically through the catchment critical zone (e.g., Marçais et al., 2022 who show how different runoff processes contribute to the age distribution). Chapter 8 will address the new and growing awareness of how vegetation affects this storage too, in the competition for water between the streams

[31] Here I mean from over such a large range of ages; otherwise there are some examples since Peters et al. (2014) that at least show similar relations for dynamics in the younger age range (although these are largely model results and hence influenced by model structure, etc.): e.g., Fig 4b in Kirchner (2016a, b); Fig 3 in Benettin et al. (2017); Fig 5 in Torres & Baronas (2021).

[32] An astute reader might immediately say here, hey wait a minute, didn't Hewlett and Hibbert (1963) tell us that *soil water* sustained baseflow?! I have polled many hillslope hydrologists in the writing of this book to get their take on the idea of soil water drainage as the source of stream baseflow. In retrospect, the famous Hewlett and Hibbert (1963) hillslope concrete trench study that we discussed in Chapter 3 likely led us astray for many decades. Lee et al. (2020) have gone back to the original concrete model (recall Figure 3.2 from Chapter 3) and repeated the experiment on what is now a cracked and leaky cement slab. Lee et al. (2020, p. 14) noted that "the leak in our model was analogous to a realistic leak to bedrock fractures commonly found in many natural catchments." And with leaks, that slope now does *not* in fact show that soil water sustains baseflow because without the artificially impermeable base, deeper percolation occurs. As we saw in Chapter 6, hillslopes pretty much everywhere are leaky. And as we will see in this chapter, headwater catchments are leaky; as Fan (2019) has discussed. And of course many headwater streams are themselves intermittent (see van Meerveld et al. (2019) for exciting new work on that). This is another side of the same groundwater–streamflow coin.

Figure 7.21 Groundwater seeping out of a roadcut (and contributing to stream baseflow) in the Oregon coast range where layers of sandstone and mudstone underlie responsive forested catchments.
Source: Photo by McDonnell.

draining the catchment and the vegetation draining the catchment into the atmosphere.

7.6 Summary and looking ahead

> *In all science, error precedes the truth, and it is better it should go first than last*
>
> ***—Horace Walpole***

What do we take away from this chapter? The first perhaps is that critical zone storage and streamwater age are hot areas of research. And perhaps two useful catchphrases to summarize this chapter: that "geology is destiny" for catchment transit time; and that in many ways, "streamflow is simply a catastrophic failure of groundwater storage"[33] (outside of rainfall and snowmelt events). Figure 7.22 summarizes another key factor we learned in this chapter—that *topography conceals more than it reveals*. This was true for much of the shallow fill and spill behavior linked to thresholds and connectivity in Chapter 6. And it is even more apt a phrase for this chapter where storage—active and passive—are largely hidden from view. As we saw for the two Oregon catchments, nearly identical outward forms

[33] Both of these are catchphrases that I love; I first heard them from Gordon Grant and Adam Ward, respectively.

Figure 7.22 A perhaps useful analogy for the headwater catchment: similar outward forms can hide radically different plumbing, largely linked to geology.
Source: Chouard (2008).

and nearly identical rainfall–runoff behavior can have multifold differences in catchment mean transit time; and also fundamentally different transit time scaling behavior. It is a cautionary tale for places elsewhere that a catchment could be indistinguishable from another in terms of topography and flow dynamics and yet have radically different plumbing.

Many of the themes identified in this chapter show where the field is now—and what is driving current research. But before we finish this chapter we should at least say where we are now with hydrograph separation—something that we saw in Chapter 5 so radically influenced the perceptual models of streamflow generation. Burns (2002) was among the first to question the continued value of the standard two-component hydrograph separation. He noted that while the "thrill is gone" for doing isotope-based hydrograph separations in forested, humid catchments and that "new studies in catchments with different climatic and human disturbance regimes, and studies that combine water-isotope and solute isotope measurements" (Burns, 2002, p. 1516) could provide more thrills and surprises for years to come. And while this certainly happened between then and now,[34] perhaps the biggest development is the calculation of something called the "young water fraction."[35]

The young water fraction was developed by Kirchner (2016a, b). It is not a hydrograph separation technique per se, but a way to quantify "the proportion of the transit-time distribution younger than a threshold age"; or roughly 0.2 years or 2–3 months old. The Kirchner papers show how it can be inferred from the amplitude of seasonal periodic isotope tracer patterns in precipitation and

[34] For a review of this see Klaus and McDonnell (2013).

[35] With both of these developments stimulated by Jim Kirchner at ETH Zurich; with a vision for high frequency analysis of storm event tracer signatures (Kirchner et al., 2004). And Kirchner (2016a, b) with the new "young water fraction" approach that we will examine below.

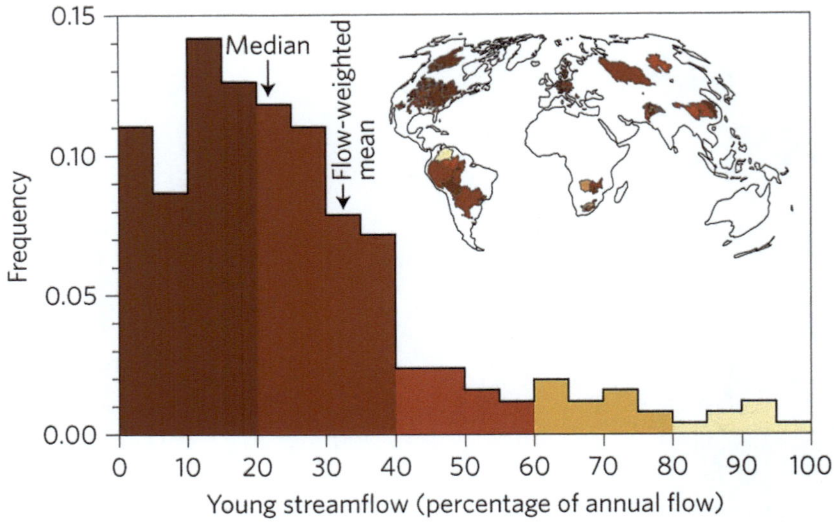

Figure 7.23 Histogram showing the young streamflow fractions, with the median value of 21%. The inset shows the catchment locations.
Source: Jasechko et al. (2016).

streamflow.[36] The first application of this technique after the Kirchner proof-of-concept papers was the work of Jasechko et al. (2016) who computed the young water fraction from stable isotope data from ~250 streams. They found that the flow-weighted young streamflow fraction was 34% of annual flow (Figure 7.23). Of course, this varies greatly from catchment to catchment. But the hunt now is for the varying ages that make up the remainder; the blend of older waters[37] that combine to form the balance of the stream hydrograph—the focus of much active research today.

[36] Fractions of new water—defined as the fraction of water that is "new" since the last sampling, can be also inferred from so-called "ensemble hydrograph separation" of a tracer time series (see these methods in Kirchner, 2019; Knapp et al., 2019; Kirchner and Knapp, 2020).

[37] See new work by Visser et al. (2019) that combines cosmogenic radioactive and stable isotopes (i.e., tritium, ^{22}Na, ^{35}S and ^{18}O) to determine the so-called hydrochronology of water storage dynamics of catchments.

CHAPTER 8

Ecohydrological impacts on streamflow generation (~2000–present)

8.1 Catchment ecohydrology

And there hold intercourse, with roots of trees and stones, with rivers at their source

—Elinor Wylie

Starting in the early 2000s, the question of how plants affect streamflow generation became increasingly a focus of study—largely through the lens of ecohydrology: the interconnectedness of water, plants and the catchment. Of course, the effect of plants on interception loss and evapotranspiration is well-covered in textbooks. And we discussed in Chapter 6 the effects of plant roots on soil macroporosity. Interestingly, most of the catchments that we mentioned in the earlier chapters started out as paired catchment experiments (something we will discuss later in this chapter) to determine the effects of forests on streamflow. But these paired catchment studies did not focus so much on the details of the hydrological processes. A turning point came ~2000 with a paper by Rodriguez-Iturbe (2000) that made the case for the process study of the interconnectedness of ecology and hydrology linked to the soil- and catchment water balance. The journal *Ecohydrology* also began in 2008 and with it came a flood of papers getting at exciting new questions: where do plants get their water? And do plants use water that would otherwise become streamflow?

If you have had an undergraduate hydrology course, then you will know about things like interception and stemflow and their influence on storm runoff. Overall, vegetation reduces the input to the soil surface compared to open rainfall.[1] Throughfall, its drip points around the vegetation canopy and stemflow can arrive to the soil pre-funneled from above (Spencer and van Meerveld, 2016). If you've stood in a forest during rainfall, then you will know that under a canopy, the water that comes through the canopy (and thus is added to the soil surface) is rarely spatially uniform (Keim et al.2005; Klamerus-Iwan et al., 2020). Similarly, you know how roots can impact the subsurface via preferential flow as we examined in Chapters 2, 4 and 6. Recently, ideas about soil as a porous medium have evolved (Gao et al., 2023) and been hotly debated (Zhao et al., 2024). Perhaps a good recent ecohydrologically oriented definition of soil to kick off this chapter is "an interconnected network of interacting entities, micro- and macrobiota, existing, growing, and communicating over a pore network" (Hunt et al., 2024, p. 1).[2] And these interconnected and interacting entities can include biota that is big and small. To first show the big, Figure 8.1 shows an image from the North Island of New Zealand where a high-pressure hose was used to remove soil and to expose the tree root network. Now you really

[1] And compared to bare soil evaporation, evapotranspiration from trees and shrubs can far exceed bare soil evaporation. We will address evapotranspiration later in this chapter.

[2] In hydrology, we "dumb down" the complexity of soil to deal with it based on pure physics. A recent Special Issue of the *Vadose Zone Journal* examines this (edited by Najm et al., 2019) and new work shows how they can evolve over time (Hartmann et al., 2020).

Streamflow Generation. Jeffrey J. McDonnell, Oxford University Press. © Jeffrey J. McDonnell (2026). DOI: 10.1093/9780191953736.003.0008

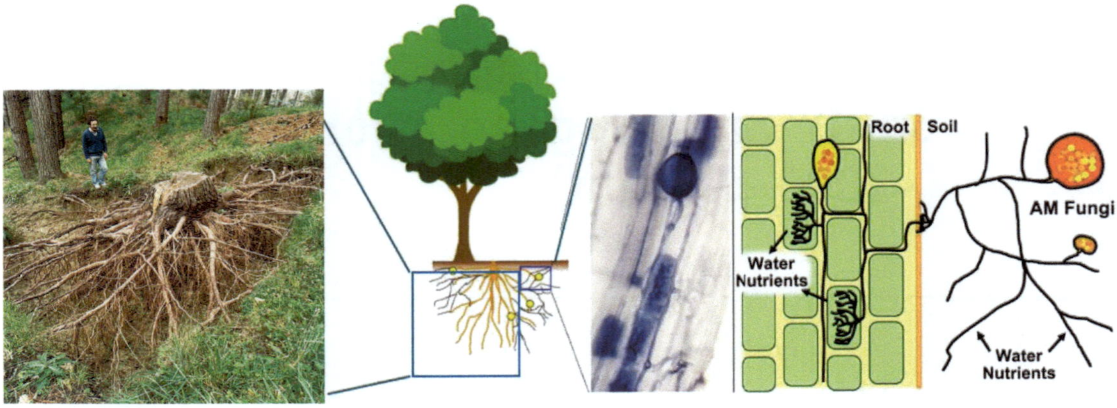

Figure 8.1 Photo from the North Island of New Zealand showing the extensive root network of a tree following high pressure hose removal of the mineral soil. (Soil, defined by my PhD examiner and soil physicist extraordinaire, Brent Clothier, as "cement with a bunch of pipes in it." He said this to me in a December 1988 PhD defense. I'm still processing these words and realizing the full extent of these remarks after almost 40 years of further digging.) The right-hand side shows the rhizosphere; the region of soil that is influenced by things like root secretions and soil microorganisms, now with an Elsevier journal called *Rhizosphere* as a repository for the interdisciplinary study of this narrow zone.

Source: Photo courtesy of Chris Phillips, Landcare NZ and used with permission. Right-hand plots from Ezawa et al. (2023).

start to appreciate that soil is not just physics: it is also biology. And this photo does not include the other extreme of the preferential flow distribution: the rhizosphere with mycorrhizal fungi and other living matter, as shown on the right-hand side of Figure 8.1.

The question now is: how does water extraction by these root systems impact streamflow? Of course, the clearest plant signal we "see" in headwater streams is the often-observed diurnal fluctuation in lowflow. These diurnal fluctuations often manifest themselves at the outlet of the headwater catchment in summer when vegetation is most active. Figure 8.2 shows two perceptual models for the controls on this diurnal signal for a forested headwater catchment at the HJ Andrews in Oregon, USA. These plant linkages to streamflow are often expressed in the water generating baseflow, depending on the nature of the soil, subsoil, unsaturated saprolite and fractured rock access by plant roots. As Figure 8.2 shows, topographic position can influence the plant's connection to the stream.[3]

Many studies have shown that such effects by trees in headwater catchments are limited spatially to a few meters on each side of the stream channel (Bond et al., 2002; Tian and Xu, 2024). We can sometimes explain these process controls in experimental headwater catchments—with the work of Harmon et al. (2020) being the most extensive thus far in a single catchment. However, we have great difficulty with generalization to any new, unstudied catchment. It's such a basic and fundamental connection—but like the streamflow generation processes we examined in the previous chapters, it is largely related to stored water we cannot see underground, now made all-the-more complex by the biological interactions and how and where plants uptake water through their root systems.

So how have the effects of vegetation on streamflow been quantified at the catchment scale beyond the ups and downs of diurnal low flow? Let's now look at the traditional approach—the paired catchment approach—that has been used to quantify forest influences on annual and seasonal streamflow in headwater catchments. We'll then move on to isotope tracing and the latest thinking on ecohydrological impacts on streamflow generation.

[3] Of course, trees use water from across the whole slope but further upslope the signals are either smeared out and cancel each other or there is no connection to the stream from the further upslope places (as flow is vertical there).

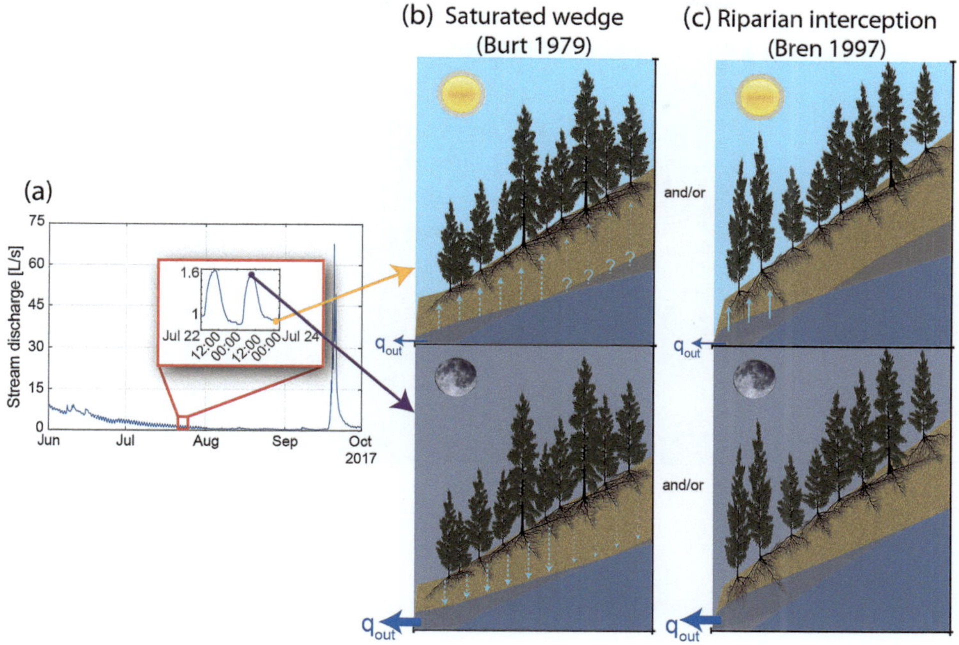

Figure 8.2 Diurnal changes in streamflow due to plant transpiration showing two alternative perceptual models on the flow pathways to cause these diurnal changes.

Source: Harmon et al. (2020).

8.2 The paired catchment approach

> *Knowledge is a big subject; Ignorance is a bigger one*
>
> ***—Unknown***

Our understanding of plant water uptake impacts on streamflow and the catchment water balance have come mostly from paired catchment studies. Bates and Henry (1928) were among the first to conduct a paired catchment experiment (in Colorado, USA). Since then, some 150+ such studies have been completed. The paired catchment approach is a simple, albeit slow and expensive approach, that includes a "control" and "treatment" catchment as shown in Figure 8.3. It leverages the annual catchment water balance to examine how changes in vegetation impact streamflow.

Figure 8.3a shows the paired catchment approach visually for the Hubbard Brook experiment forest in New Hampshire, USA. Neighboring headwater catchments are gauged for precipitation input and streamflow output for a number of years. Once a long enough record has been obtained—to make sure that the control catchment and the would-be treatment catchments behave similarly—catchment(s) are manipulated via clearcut tree removal and/or disturbances like thinning or patch cutting. Before–after statistical tests identify changes to the water balance linked to vegetation removal and re-growth. So, in effect, the paired catchment experiment is a paired water balance experiment.

Perhaps the best and most concise summary of what one might expect from a paired catchment study was made a long time ago by Hibbert (1967). He summarized the three main effects of paired catchment forest cover manipulation on streamflow via three simple, declarative statements that still define our perception of the annual-scale vegetation impacts on streamflow:

1. Reduction of forest cover increases water yield.
2. Establishment of forest cover on sparsely vegetated land decreases water yield.

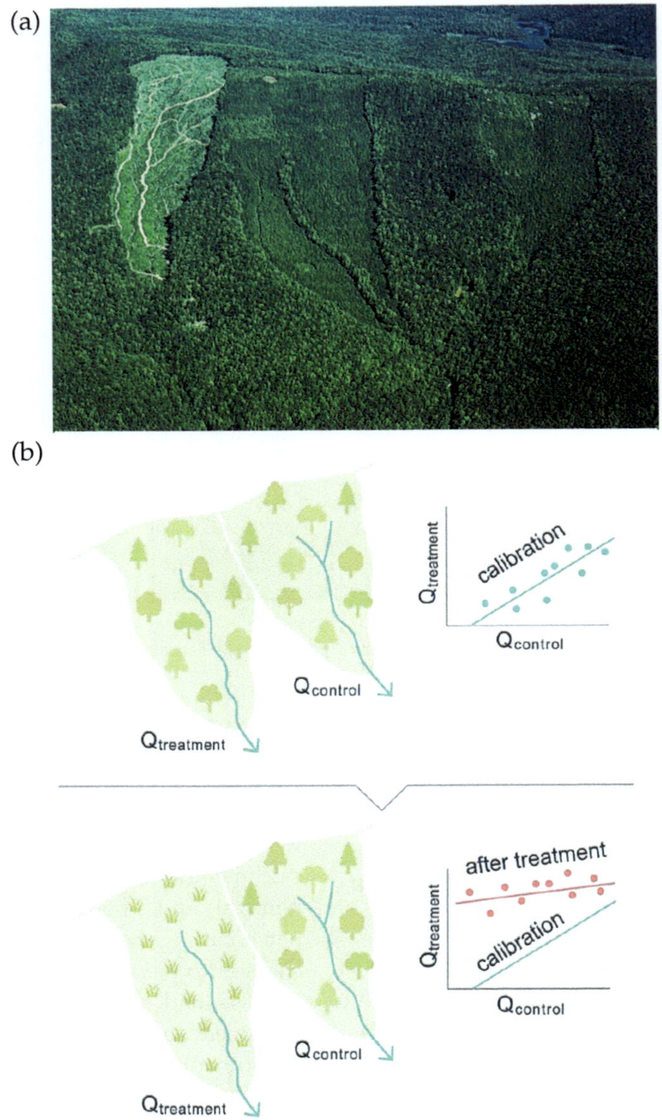

Figure 8.3 (a) Examples of paired catchments in the Hubbard Brook experimental catchment in New Hampshire, USA showing a control (photo right) and two treatment catchments (photo center left). From https://lternet.edu/findings/watersheds-as-laboratories/. (b) A schematic showing the before-after approach to quantify change using the paired catchment approach. (A crisp, recent definition of the paired catchment experiment is: "Streamflow is measured for two (in this case forested) catchments for a period and the streamflow signatures (e.g., annual water yield, peak flow or seven-day minimum flow) for the two catchments are related to each other (top panel). After several years of measurements, one catchment is treated (in this case, the catchment to the left is deforested). Streamflow measurements continue for several more years (or decades) and the signatures for the two catchments are again related to each other (bottom panel). The difference in the regression line for the pre-treatment (i.e., calibration period) and the post-treatment period is attributed to the effect of the treatment." (Van Meerveld and Seibert, 2025, p. 6)).

Source: Van Meerveld and Seibert (2025).

3. Response to treatment is highly variable, and, for the most part, unpredictable (Hibbert, 1967, p. 535).

The term water yield that Hibbert refers to is simply the streamflow per unit area (i.e., discharge divided by catchment area) delivered to the catchment outlet (usually over a year, in the context of such work). The main follow-on reviews of paired catchment studies since the 1960s have each supported the rather dismal viewpoint of Hibbert's third point that predictability was low. For instance, in their review, Bosch and Hewlett (1982) examined the relationship between percent forest area cut and increases in water yield for 94 paired catchment studies. They found that the coefficient of determination (r^2) for the linear relationship between percent cut and water yield increase was 0.2. Just think about that number for a tick. It means that the predictive ability of forest removal observed via these experiments—through perhaps the most expensive and lengthy (often more than a decade) type of catchment experiment—is less than 20%. You wouldn't want to gamble knowing those odds! And like the Bosch and Hewlett (1982) review, continued follow-on paired catchment reviews with ever-more studies (e.g., by Stednick, 1996; Brown et al., 2005) showed similarly poor relationships between percent forest area cut and water yield response. Jones et al. (2009, p. 2700) basically repeated the words of Hibbert (1967) when they wrote "the principles of forest hydrologic response are well established but prediction is still a challenge." A more recent review of the paired catchment literature by Goeking and Tarboton (2020, p. 172) weighs in on the Hibbert declarative statements and notes that "decades of subsequent research have supported the [third Hibbert] hypotheses." But even more bleakly, they note that in drier climates:

> recent studies suggest that the variability of water yield response is a fundamental characteristic of semiarid western watersheds and raise questions about the universality of the first hypothesis regarding the relation between forest cover and water yield (Goeking and Tarboton, 2020, p. 172)

Here you might ask: why has this paired catchment approach not yielded clearer results? The main reason is that paired catchment studies are a black box approach.[4] And trying to summarize and intercompare paired catchment findings is a fraught exercise given all the differences in the studies and study designs: e.g., record lengths of the different studies vary considerably; what is replanted may differ to what vegetation existed before the study; post-treatment recovery varies from years in some studies to decades in others. Some studies do vegetation manipulations from grasslands and scrublands to bamboos and other non-hardwood species; some studies involve clearing (i.e., vegetation removal) prior to planting; others do not. Even the simple grouping of studies into planting or harvesting categories is difficult since many published paired catchment studies include harvesting *and* planting. But above all, paired catchment studies in many ways ignore the biggest thing we have learned in the past eight chapters; that the catchments *themselves* are different—different subsurface structures, belowground storages, different geology and topography. In many ways, these differences can trump the aboveground alterations (McDonnell et al., 2018). And if we start to look at how forest cover affects peakflows or lowflows, things become even *less* clear.[5]

So how can we sidestep all the issues of paired catchment studies and open up the black box to develop a mechanistic understanding of how vegetation influences streamflow? One way is to look at all the things that we learned about runoff generation processes in the previous chapters and apply that sort of mechanistic thinking to where the water goes, its flowpaths and residence times.[6] And how vegetation changes the physical processes of catchment storage and release that we have examined in Chapters 6 and 7 that connect to transpiration. And

4 Like we discussed for catchment water balance studies in general, in Chapter 2 (recall the black box diagram).

5 And often also, more hotly debated!

6 And studies outside temperate catchments have shown that changes in landcover can alter the dominant runoff processes. Forest clearing and the development of an extensive road networks, can enhance the occurrence of infiltration excess overland flow and a fast routing of subsurface flow via the drainage networks associated with roads. Reforestation of degraded land can lead to improvements in microporosity and infiltration, and reduce the occurrence of overland flow (whether due to saturation above a shallow low permeability layer, or pure infiltration excess overland flow). See much recent work on all this by van Meerveld et al. (2021).

to include the use of stable isotope tracing to illuminate the black box of catchment processes. Let's now look at stable isotope tracing of plant water and how that has enabled new understanding of plant water impacts on streamflow in headwater catchments.

8.3 Isotope tracing of plant water

"Curiouser and curiouser" cried Alice

—Lewis Carroll

Interestingly, plant water uptake studies in recent years have followed a similar trajectory to streamflow source studies; where stable isotope tracing has been used to quantify water sources, water residence times and water transit times—now of plant transpiration from the roots, through the xylem (i.e., the most extensive water conducting pathway of the tree[7]) and out through the needles or leaves. Essentially, precipitation input signals through the year are recorded in the soil and plants extract water from different depths and storage units, linked to their root distribution. Figure 8.4 shows an image of this with many pieces that we will unpack later in this chapter.

Isotope tracing in plant studies dates back to the 1930s when Washburn and Smith (1934) examined the isotopic composition of xylem water in willow tree branches and leaves and related those signatures to streamwater. A few years later, Guntz and Beltran (1939) noted there was a difference in isotope values between water from xylem and leaves vs. sap, revealing how these differences in isotope composition in plants could be attributed to the difference in the rate of evaporation of heavy water and potentially to the different diffusion of these waters "through the cellular complexes." Fast forward ~45 years and White et al. (1985) already had a thoroughly modern take on how to determine the source water for water in the tree xylem where they note that:

trees can be separated into three categories: those which rely solely on summertime rainfalls, those which rely solely on groundwaters, and those which utilize both water sources. For trees in the last category, D/H ratios of sap can be used to quantify the relative contributions to the tree's source water of summertime rainfalls and groundwaters (White et al., 1985, p. 237)

What White et al. (1985) are saying is that by tracing a stable isotope ratio like D/H (meaning ^2H to ^1H), "where a tree gets its water" can be determined.[8] The study that *really* kicked off the whole notion of tracing plant water sources and that showed the most clear and powerful promise for others to follow was Dawson and Ehleringer (1991). They went straight to the point to ask "do streamside trees use streamwater"? And they showed that, contrary to expectation, they did not:

mature streamside trees growing in or directly next to a perennial stream used little or none of the surface stream water. The deuterium to hydrogen content of both source and xylem waters indicated that mature trees were using waters from deeper strata (Dawson and Ehleringer, 1991, p. 335)

They went on to comment on the effect of tree age where they noted that:

Although adult trees maybe have roots distributed continuously through a profile, it seemed that the most active sites of water adsorption were limited to deeper soil layers. In contrast, small streamside individuals appeared to use streamwater, whereas small non-streamside individuals used recent precipitation as their primary water source. (Dawson and Ehleringer, 1991, p. 335)

Figure 8.5 shows their classic results where the smallest (youngest) trees quickly shifted source waters from an isotopic signature of precipitation to that of groundwater. This makes sense as the plant begins to grow and root profiles deepen. Impressively, they also did potted tree tests with labelled water that showed that no apparent fractionation[9] occurred during uptake. This was consistent with

[7] The definitive textbook by Taiz and Zeiger *Plant Physiology* (2010, 5th Edition) notes that "In most plants, the xylem constitutes the longest part of the pathway of water transport. In a plant 1 m tall, more than 99.5% of the water transport pathway through the plant is within the xylem, and in tall trees the xylem represents an even greater fraction of the pathway. Compared with the complex pathway across the root tissue, the xylem is a simple pathway of low resistance." (p. 52)

[8] A bit later, we will look at the seasonal origin of plant water uptake (and streamflow)—but note now the link to this much earlier study.

[9] Fractionation, very simply, is the sorting of light and heavy isotopes of O and H; in other words a "discrimination" against the heavy one over the light during the phase change from water to vapor, as we saw in that hand-drawn figure from Horton in Chapter 2. Many useful review papers can be

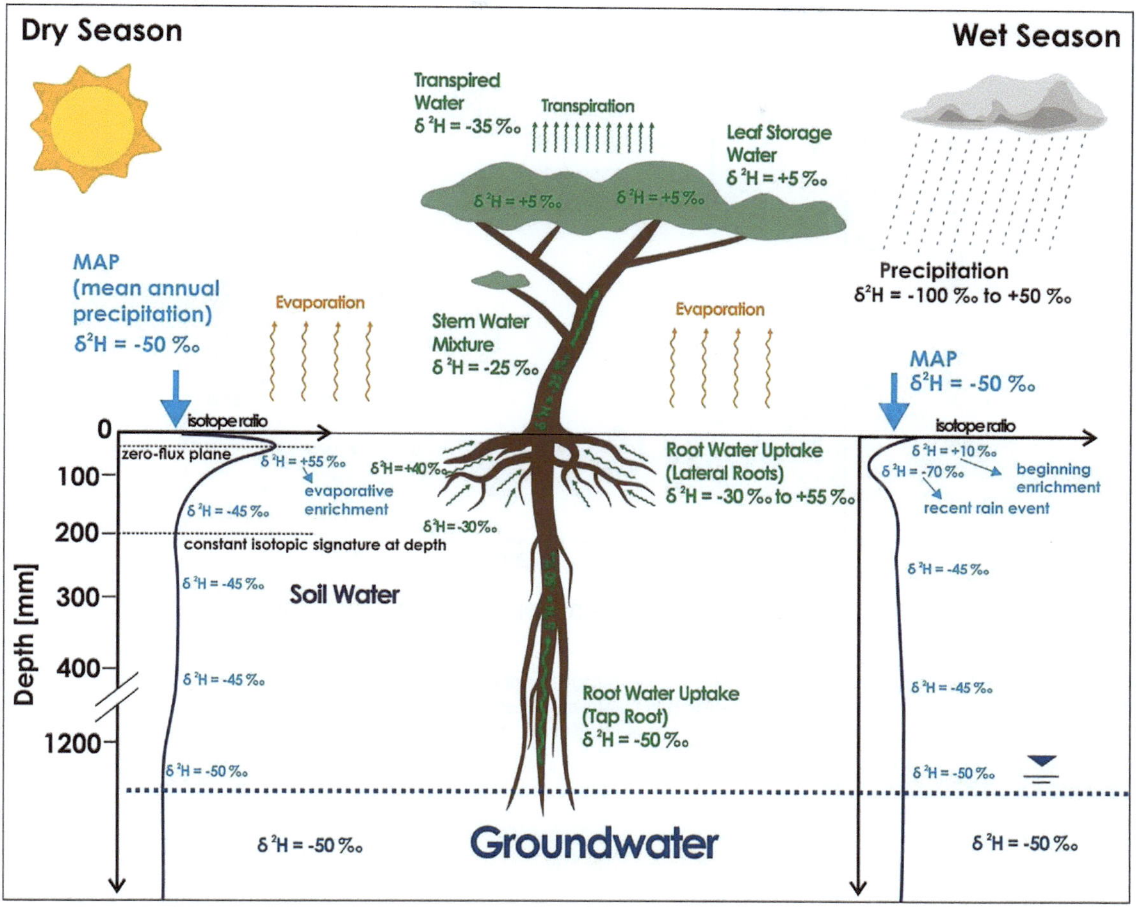

Figure 8.4 Variations in the isotopic composition of the soil with depth below the surface and in different parts of the plant that are used for isotope tracing of plant water uptake. We will cover other elements of this figure later in the chapter that will make it more understandable than it may be now for some.

Source: Beyer (2018, https://www.isodrones.com/post/2018-08-31-isotopes-deep/).

several early studies like Wershaw et al. (1966), Allison et al. (1985) and White et al. (1985) who showed that water movement in the xylem was non-fractionating. We'll get into the details of what all this means in a little bit.

Up until about the late 2000s, the consensus view was that the subsurface system was well mixed. Meaning that, despite the appreciation of preferential flow (like we examined in Chapters 1–6), we somehow still assumed that the structured heterogeneity of soils would result in a set of well-mixed sources that would supply plant transpiration, stream flow, groundwater recharge, etc.; in other words, all would draw from the same pool of water in the subsurface. This view was largely based on single-isotope findings; meaning that studies used either ^{18}O *or* ^2H but not both together.[10]

consulted for this information (e.g., Sprenger et al., 2016; von Freyberg et al., 2020).

[10] Perhaps more accurately though is the fact that pre-2000 or so, ecologists and hydrologists talked together much less. There were indeed some ecology studies that used dual isotopes for plant water uptake studies pre-2000 (e.g., Busch et al., 1992).

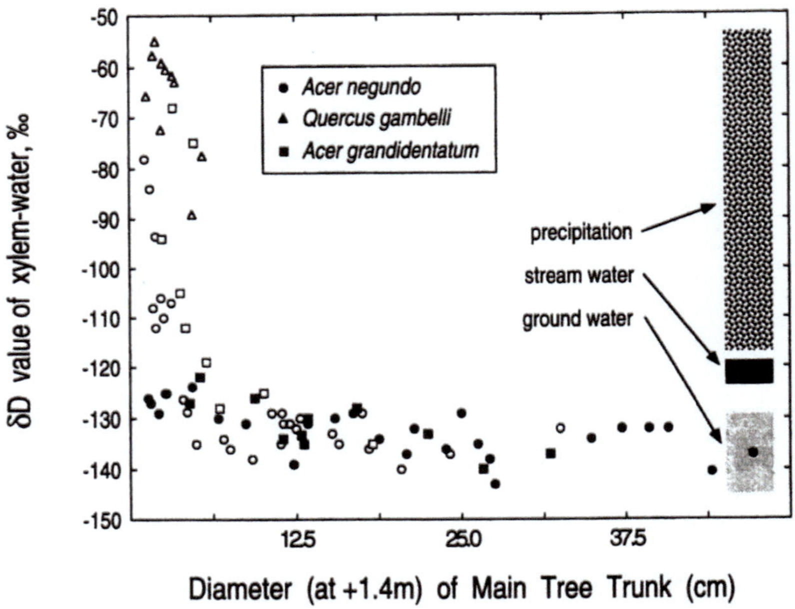

Figure 8.5 The work of Dawson and Ehleringer 1991 where the hydrogen isotope ratio of extracted xylem water is shown as a function of tree size. Open circles denote samples from trees growing nearest to the stream. Closed circles denote those from trees further away from the stream. The right-hand vertical bar shows the range of long-term values of precipitation, streamwater and groundwater.

Source: Dawson and Ehleringer (1991) (This figure version from a textbook chapter in a book I co-edited with Carol Kendall titled "Isotope Tracers in Catchment Hydrology").

Dual isotope analysis (^{18}O *and* ^{2}H) enabled something different to be observed—namely, that perhaps the subsurface was *not* so well mixed. We will unpack this new understanding shortly. But for now, let's see how most catchment rainfall, streamflow, groundwater and soil water data plot in this dual isotope space—at least in a humid location. Figure 8.6 shows data for a humid headwater catchment in Eastern Mexico. There is a lot we could discuss here with respect to the solid line, the so-called Global Meteoric Water Line.[11] Whether you realize it or not, this plot is showing the phenomenon that we saw in Chapter 7 in Figure 7.1, where a sinusoidal input signal for precipitation is lagged and damped when one looks at stream water. See the extremes of the wet season and dry season precipitation in Figure 8.6? See how the soil water samples stretch a bit less along this line compared to precipitation? Or how deeper soil water is spread even less along the line? And see too how the stream data is very compressed compared to all the other values? That is the lagging and damping we have already looked at in Figure 7.1, but now through the lens of both isotopes: i.e., the $\delta^{18}O$ and $\delta\ ^{2}H$ values together.[12] As individual isotope signals, they

[11] Very simply, there is an average annual relationship between ^{2}H and ^{18}O in precipitation (meteoric waters) across the globe which was first discovered by Craig (1961). The Local Meteoric Water Line is simply that: a line derived from precipitation samples collected at a local place.

[12] Note that when we speak of the values for ^{18}O and ^{2}H we use the δ notation, but when we speak of ^{18}O and ^{2}H generally, we do not use any δ notation. And while I am discussing terms, there is no "water isotope" as sometimes seen in the literature! Only the isotopes of hydrogen and oxygen that *make up* the water molecule. A clever reader may note that I have used the term "water isotope" in one or two of my own papers over the years (and sometimes in this book!). Mea culpa!

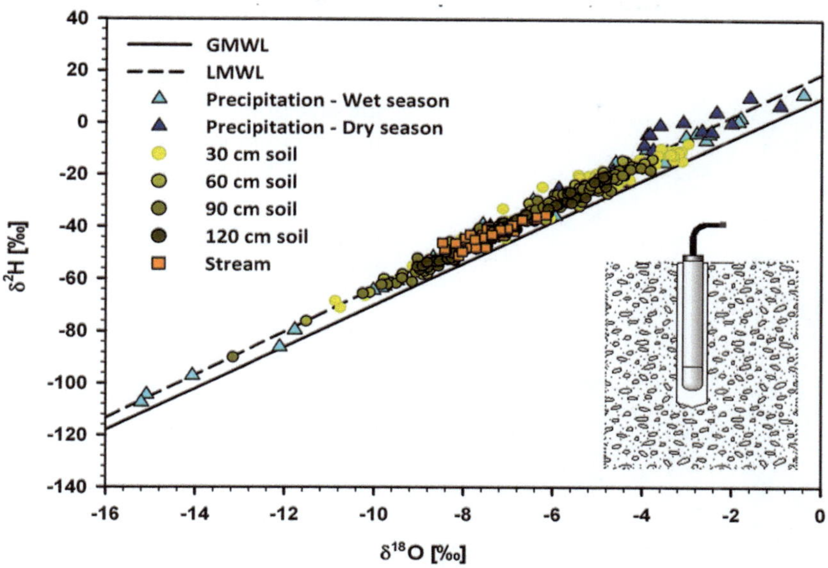

Figure 8.6 Precipitation, soil water and streamflow from a headwater catchment in Veracruz State, Mexico showing typical patterns for humid catchment with less variability in the isotopic composition for the deeper soil water and streamflow than the precipitation or shallow soil water. Soil water was extracted from a suction lysimeter.

Source: Muñoz-Villers and McDonnell (2012).

would all follow a sinusoidal pattern that we have seen in Chapter 7.

Figure 8.6 shows soil water extracted from a suction lysimeter—shown in the inset of the figure. A suction lysimeter is a porous cup sampler that uses a hand pump to apply a vacuum—and it enables extraction of the mobile water.[13] But soil water held under greater tension cannot be extracted with suction lysimeters. That soil water requires a lab process called cryogenic vacuum distillation, of which there are many different types of lab set-ups. Nevertheless, this so-called cryogenic extraction typically involves placing a frozen soil sample, housed in a small glass tube (e.g., like in Figure 8.7), into a heating block and then connecting it to another glass tube in a cold liquid nitrogen chamber. Depending on how much the soil sample is heated and how long the extraction goes on for, all of the water comes out of the soil and into the collection vial; up to high tensions >1000 kPa and in some cases to 5000 kPa of tension. Everything! That extracted and collected water is then run through a mass spectrometer or laser spectrometer to obtain the stable isotope values.[14]

The first study to compare the soil water samples from suction lysimeters and cryogenic extraction with stream water and plant water was Brooks et al. (2010). That study was conducted in the same catchment we looked at in Chapters 6 and 7: Watershed 10 at the HJ Andrews Experimental catchment in Oregon, USA. Figure 8.8 shows

[13] Recall that the suction lysimeter is the same tool used to obtain the soil water samples for many of the soil water residence time studies that we looked at in Chapter 7. The water that is held under low tension in the soil is thus fairly mobile. A suction lysimeter can only sample soil water when the soil is wet—or more precisely, when the soil suction is not greater than the vacuum applied to it; these hand-pump suction lysimeters typically apply < 60 kPa of suction and water is slowly drawn into the cup over some hours. It is an inexact science. And beyond that tension range, the cup sits dry. No water can enter. But there *is* still water in the soil, of course, held under greater tension.

[14] Please bear with this methods discussion. I know that I have not discussed methods anywhere else in the book. But understanding the things we will explore later in this chapter hinge on knowing how the methods influence the interpretations.

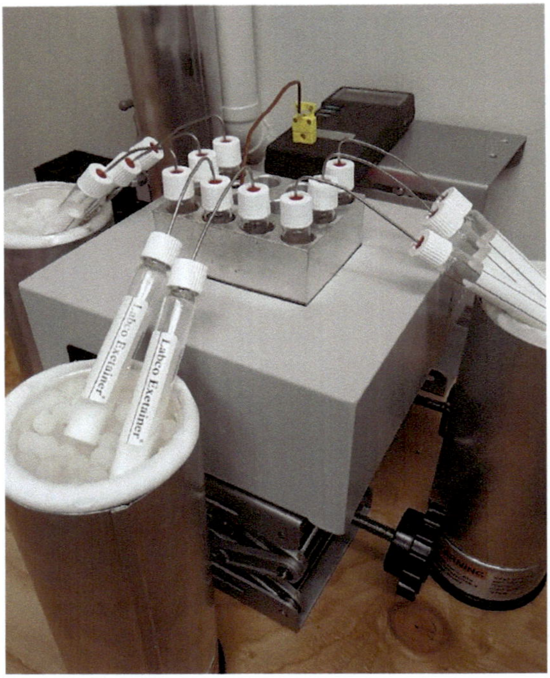

Figure 8.7 The cryogenic extraction system. This one is designed to the specifications of Koeniger et al. (2011)—a system now widely used today (although, there is considerable debate about such methods and how best to extract water from soil and plant material, e.g., Duvert et al., 2024).

Source: Photo by Magali Nehemy.

the meteoric water line for the site. Many years of sampling for soil water, subsurface stormflow, streamflow and groundwater showed that all of these waters plotted along the meteoric water line. Renee Brooks, an ecologist, started sampling the 50-yr old Douglas fir forest for their xylem water isotope values. Her data—shown in the green triangles in Figure 8.8, plotted well *below* the local meteoric water line, far from where all the mobile water samples had plotted.

The reason the soil water plotted below the meteoric water line in Figure 8.8 was due to evaporation. Evaporation is very systematic. If we think about the simplest example to explain this, consider evaporation from a saturated surface as in Figure 8.9. During evaporation, the heavier water molecules ($^{1}H\,^{2}H\,^{16}O$, $^{1}H_2{}^{18}O$) have their bonds broken and evaporate less easily than the lighter water molecules ($^{1}H_2{}^{16}O$). In other words, during evaporation, lighter water molecules vaporize faster than heavier water molecules. Figure 8.9b shows progressive isotopic enrichment of the remaining water in this example liquid that plots on an "evaporation line" that typically lies *below* the local meteoric water line on a shallower slope. The angle of that evaporation line is linked to the aridity and the isotopic composition of the air.

If we then think about soil water, evaporation can penetrate down into the soil profile. And during soil water evaporation, the mass differences are such that the remaining soil water is enriched in heavy isotopes along a trend line, based on how long it has been exposed to the evaporative effect. These "evaporated soil waters" are more complex than that shown for open water in Figure 8.9. They combine the effects of source water variations, mixing and fractionation, as Benettin et al. (2018, p. 2887) note where "all three of these effects jointly determine the isotopic patterns in the evaporated soil samples." Nevertheless, the cryogenically extracted soil water (that shows signs of evaporation and therefore, fractionation) plots on a line of a lower slope (the so-called soil water evaporation line). You can imagine then an imaginary line in Figure 8.8 now running through the soil water data.

We will explore all this in more detail in the next section. But for now, just know that the evaporative front in soils typically extends down into the profile some many tens of centimeters so what this means is that there is a disproportionate amount of the heavier isotopes in this "evaporation zone" (i.e., the shallow soil profile). The upper soil profile, perhaps down to 50 cm[15] is typically isotopically "enriched" (values closer to 0 per mil). But deeper down and when away from the evaporation effect, the cryogenically extracted soil water "ties back" to the meteoric water line. Bottom line: most ecological theory suggested at the time of the Brooks et al. (2010) work that there was no fractionation during transpiration, so there must be an evaporated soil

[15] We could spend a lot of time discussing this; how soil texture impacts this depth of evaporative penetration, etc. There is also the change from direct evaporation to the so-called Stage 2 evaporation where vapor diffusion dominates. See Or et al. (2013) for a good review paper on all this.

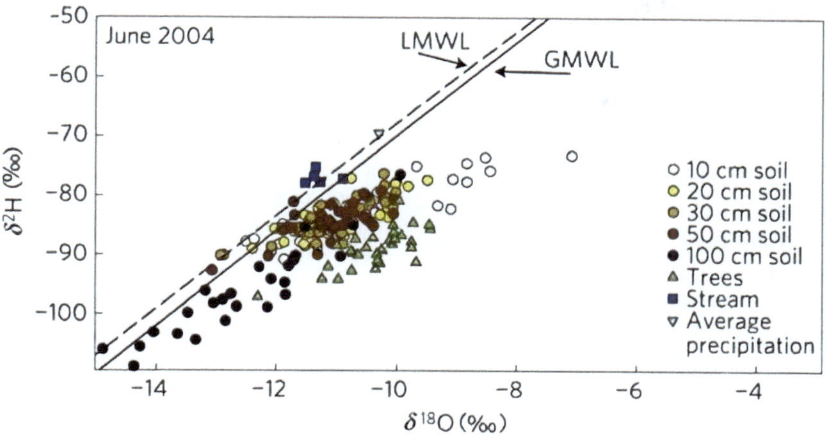

Figure 8.8 Data from the Brooks et al. (2010) paper showing ^{18}O and ^{2}H of xylem water, streamwater, estimated annual average precipitation and "bulk soil water" from different depths below the soil surface. These samples were collected during 3 different sampling periods at 32 randomly distributed sampling points in the 10-ha catchment. The two lines are the local meteoric water line (dashed) and global meteoric water line (solid).

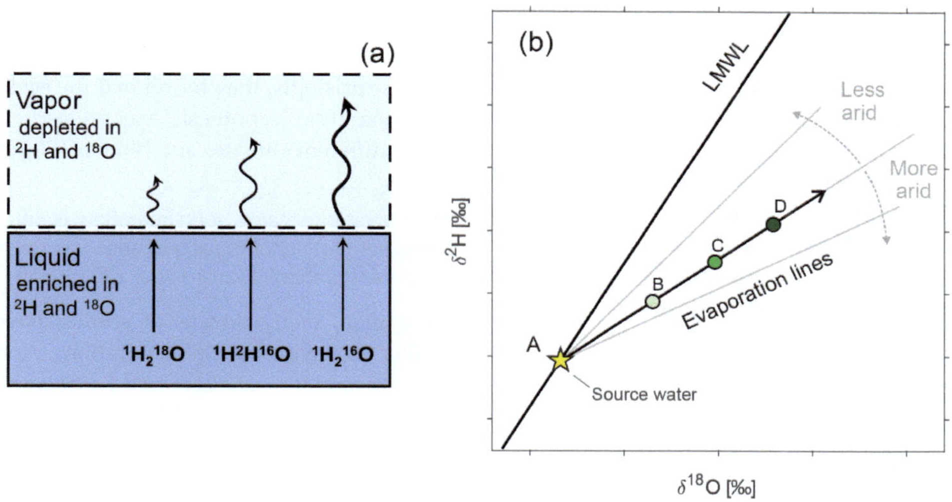

Figure 8.9 (a) Fractionation during evaporation from a saturated surface (open body or surface-saturated soil) where the heavier water molecules have their bonds broken and evaporate less easily than the lighter water molecules. (b) Progressive enrichment of the remaining water, B–D of a water source (yellow star A) subject to evaporation. These samples plot on an "evaporation line" that typically lies below the local meteoric water line on a shallower slope.

Source: Benettin et al. (2018).

water source water that the trees were using. The question then became: how?[16]

[16] Of course, as we will see below, the answer would be the ecohydrological separation hypothesis. And this comparison of lysimeter-derived soil water vs. cryogenically extracted soil water would then challenge the consensus view that the subsurface system in a catchment is well mixed.

8.4 The ecohydrological separation hypothesis

Strong opinions but held lightly

—Charles Darwin

The Brooks et al. (2010) work demanded some sort of process explanation. Again, almost all of the work up until ~2010 had assumed that the subsurface was well mixed isotopically,[17] that trees extract water from that mixed reservoir and that mixed reservoir drains to the stream. The Brooks et al. (2010) study advanced a new perceptual model—the ecohydrological separation hypothesis—whereby:

a pool of tightly bound water that is retained in the soil and used by trees does not participate in translatory flow, mix with mobile water or enter the stream. Instead, water from initial rainfall events after rainless summers is locked into small pores with low matric potential until transpiration empties these pores during following dry summers. Winter rainfall does not displace this tightly bound water (Brooks et al., 2010, p. 100)

This means that in a sense, there are two water worlds—one for the mobile water that forms preferential flow and streamflow generation—and a lower mobility water world that supports transpiration. With the two acutely *not* well mixed.

The ecohydrological separation hypothesis posited that the trees were using a fundamentally different water source than we see in streams; different from the waters we have discussed in Chapters 5 and 7 from hydrological studies in humid catchments. Figure 8.10 shows a simplification of this. Mobile water plots *on* the meteoric water line with the stream water as a damped version of the precipitation input (see Figure 8.10a) and the less mobile water (meaning, the cryogenically extracted water) plotting on a soil water evaporation line with the plant xylem water as a damped version of that end member (as shown in Figure 8.10b).

[17] But as Sprenger and Allen (2020, p. 1) would note a decade later after more and more studies came in, "based on extensively observed heterogeneities in soils and watersheds, we posit that differences in isotopic compositions of water in plants, streams, and other subsurface pools are expected in most scenarios."

If we return again to Brooks et al. (2010), they further noted that:

As transpiration and stormflow are out of phase in the Mediterranean climate of our study site, two separate sets of water bodies with different isotopic characteristics exist in trees and streams. We conclude that complete mixing of water within the soil cannot be assumed for similar hydroclimatic regimes as has been done in the past. (Brooks et al., 2010, p. 100)

So, behind these words, Brooks et al. (2010) are wondering if things plotted the way they did because of their Mediterranean climate and out-of-phase nature between the hydrology (wet winters) and ecology (summer growing season). When reports of similar behavior started coming in a few years later from other sites (e.g., Goldsmith et al., 2012), it still seemed like this on–off the line behavior and its link to specific climates (where flow and grow were out of phase) could be tested. But how? Evaristo et al. (2015) were able to find 48 studies in the literature that indeed had precipitation, extracted soil water, streamflow and plant water data to test this hypothesis. Surprisingly, they found that the ecohydrological separation hypothesis was widespread across many different climates and biomes (Figure 8.11):

at 80% of the sites, the precipitation that supplies groundwater recharge and streamflow is different from the water that supplies parts of soil water recharge and plant transpiration. (Evaristo et al., 2015, p. 91)

Since then, many dozens of studies have shown similar ecohydrological separation. And in an interesting plot twist, there was a retrospective analysis of the Dawson and Ehleringer (1991) "Streamside trees" paper we examined earlier; this time incorporating dual isotope analysis. In that work, Bowling et al. (2017, p. 1) found that "neither groundwater nor stream water matched the isotope composition of xylem water during two growing seasons"[18] and that:

[18] Not to get too carried away with this description, but this work has also been re-examined by Nehemy et al. (2022) who noted that: "in contrast to Bowling et al. (2017), our hydrometric observations support the notion that trees use snowmelt water during the snowmelt season (Figures 1 and 2). Melt infiltration increased soil water availability providing a water source for stem rehydration and transpiration in these otherwise dry soils (i.e., following fall and early winter conditions), with low soil water content (Figures 1b and 2c).

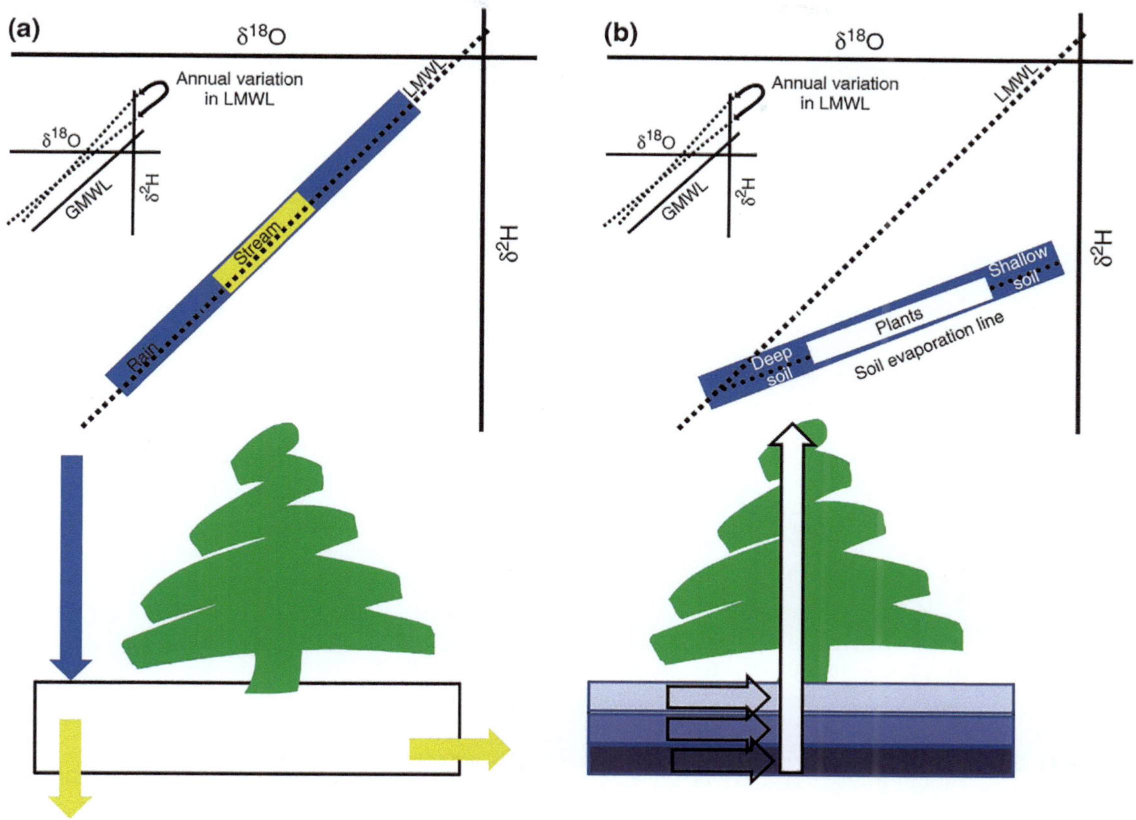

Figure 8.10 The perceptual model of ecohydrological separation. Streamwater follows the precipitation water generating streamflow. The plants follow the soil water signature that shows signs of evaporative fractionation—more near the soil surface and progressively less with depth.
Source: McDonnell (2014).

soil water (0–1 m depth) was closest to and periodically overlapped with xylem water isotope composition, but overall, xylem water was isotopically enriched compared to all measured water sources. The "two water worlds" hypothesis postulates that soil water comprises isotopically distinct mobile and less mobile pools that do not mix, potentially explaining this disparity (Bowling et al., 2017, p. 1)

Our tracer data shows evidence of snowmelt water use." They further note that "it is possible that because of the 'snapshot' sampling approach, Bowling et al. (2017) might have missed the possibility of observing the potential overlap between trees and soil by sampling a bit too early in May when leaves were only beginning to emerge." So too, Nehemy et al. (2022) note that "their monthly sampling frequency may have reduced their ability to detect tree water sources, where trees might have been lagging in relation to soil water δ-values. We would have faced similar challenges had we not sampled weekly in combination with continuous phenological monitoring with stem radius change and phenocams."

They further found that the "depth trends in water isotopes following snowmelt (as shown also in Figure 8.4) were consistent with the two water worlds hypothesis, but snow metamorphic isotope effects could not explain the highly enriched xylem water." In effect, their dual isotope analysis completely changed the story of the Dawson and Ehleringer (1991) paper—from a perceptual model of young plant use of rainfall and old plant use of groundwater to an interpretation of soil water use and ecohydrological separation.

But how was ecohydrological separation imagined?

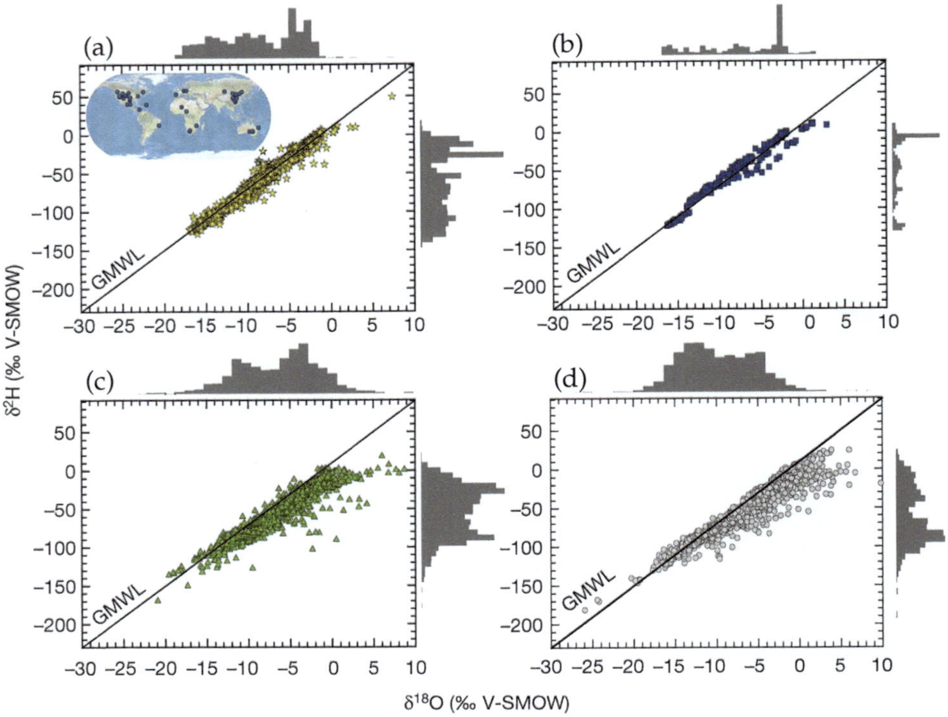

Figure 8.11 Global evidence of ecohydrological separation, with more of the plant and soil water samples than groundwater and streamwater samples plotting below the global meteoric water line (GMWL). The individual plots show (a) groundwater, (b) streamwater, (c) plant xylem water and (d) soil water each plotted against the global meteoric water line.
Source: Evaristo et al. (2015).

What is lost perhaps a bit in the subsequent studies since Brooks et al. (2010) is that knowing *both* the suction lysimeter (mobile water) and the cryogenically extracted (bulk water) together helps to identify these small pore/big pore end members. A good part of the "story" of Brooks et al. (2010) hinged on Figure 8.12. It shows the soil water lysimeter data and the bulk soil water extracted with cryogenic extraction. Bulk soil was always more isotopically depleted than the lysimeter water. In other words, the two pools were isotopically different, and the water thus did not appear to mix.

This difference between the lysimeter water and the bulk water was later seen in nitrate biogeochemistry studies by Hall et al. (2016) who used the same approaches to look at the nitrate composition of cryogenically extracted water vs. suction lysimeter water and found "two nitrate worlds" where:

> ...soil extractions sampled the anaerobic microsites inside soil aggregates where denitrification can occur, whereas lysimeters largely reflected the transport of N that was nitrified in the aerobic macropore environment or in surface litter (Hall et al., 2016, p. 1228)

Similarly, in an isotope remote sensing study, Good et al. (2015, p. 175) noted even more generally that their observed limited connectivity between soil and surface water "fundamentally structures the physical and biogeochemical interactions of water transiting through catchments." But the question still was: how and when is the bulk water recharged? The way that Brooks et al. (2010) described it was that:

> the observed depletion in heavy isotopes of water with depth is caused by soil–water recharge during the first large autumn rain event, whereby increasingly isotopically depleted precipitation through the event recharges deeper and deeper soil. (Brooks et al., 2010, p. 102)

So, what Brooks et al. (2010) are saying here is that the water that trees transpire starting in the spring, is water that recharged the previous fall,

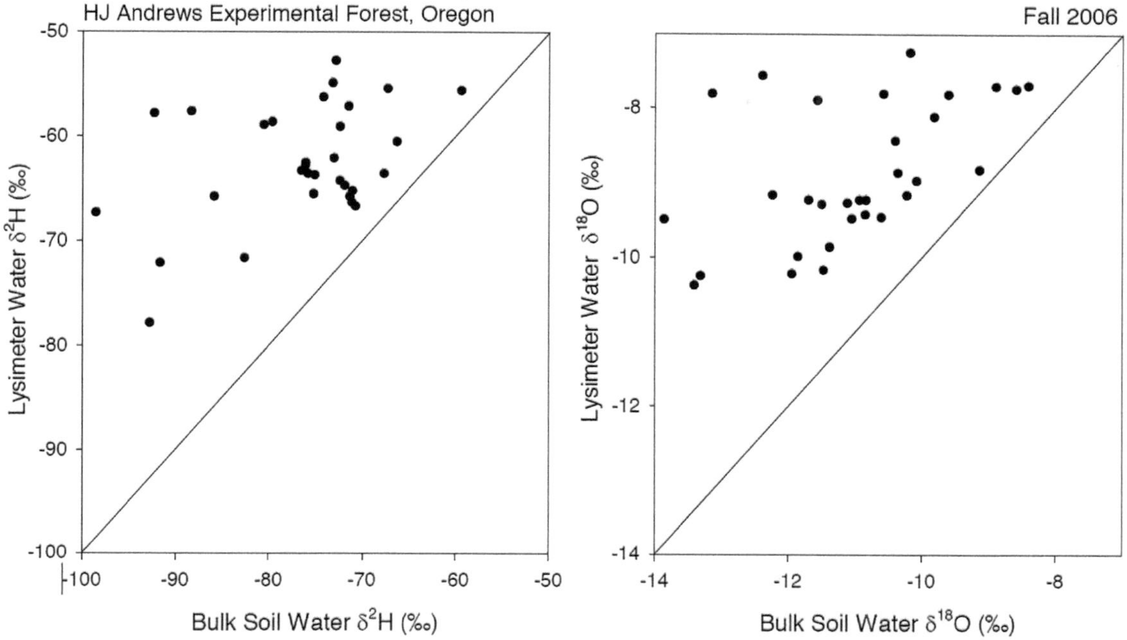

Figure 8.12 A comparison of lysimeter water isotope data to bulk water (i.e., cryogenically extracted water) isotope data for samples collected at the same depth and time. Bulk samples were taken at the same time that the vacuum was applied to the lysimeter. Lysimeter samples were collected a week later. However, pairing based on collection dates produced the same result.

Source: Brooks et al. (2010).

as shown in the perceptual model in Figure 8.15. In other words, careful tracing of the rainfall signatures of the first fall rains (and monitoring of soil moisture recording devices at depth) showed that the first one or two events following the multi-month summer drought recharged the water in the small pores. Figure 8.13 shows then how the many rain events through the winter—totaling over 2000 mm of input—seemingly did not mix with the water in the smaller pore spaces; that "tightly bound water" then is what was used the following spring and summer when the trees reactivated their water use.

But it was not until the work of Allen et al. (2019) who did extensive sampling throughout Switzerland (at 182 sampling sites) that the seasonal origins of plant water uptake came into better focus. Recall that at the heart of the ecohydrological separation was the fact that summer transpiration was comprised of soil water recharged the previous fall. Allen et al. (2019, p. 1199) looked at three tree species and found that "xylem water isotopic signatures show that summer rain was not the predominant water source for midsummer transpiration in any of the three sampled tree species." Figure 8.14 shows how the beech and oak "mostly used winter precipitation," whereas their sampled spruce "used water of more diverse seasonal origins." They noted too that even in the same plots, "beech consistently used more winter precipitation than spruce." So, in many ways, this was similar to the Brooks et al. (2010) finding but now showing some additional nuance with additional species, for more sites, and more complex behavior.

Ecohydrological separation is not a perceptual model panacea. And of course, there are not *literally* two water worlds. It is *one* soil water balance—albeit one we now realize that is exceptionally poorly mixed. Berry et al. (2018) and Sprenger and Allen (2020) argue that we need to move beyond ecohydrological separation and I agree. We have seen in the previous chapters just how heterogeneous soils

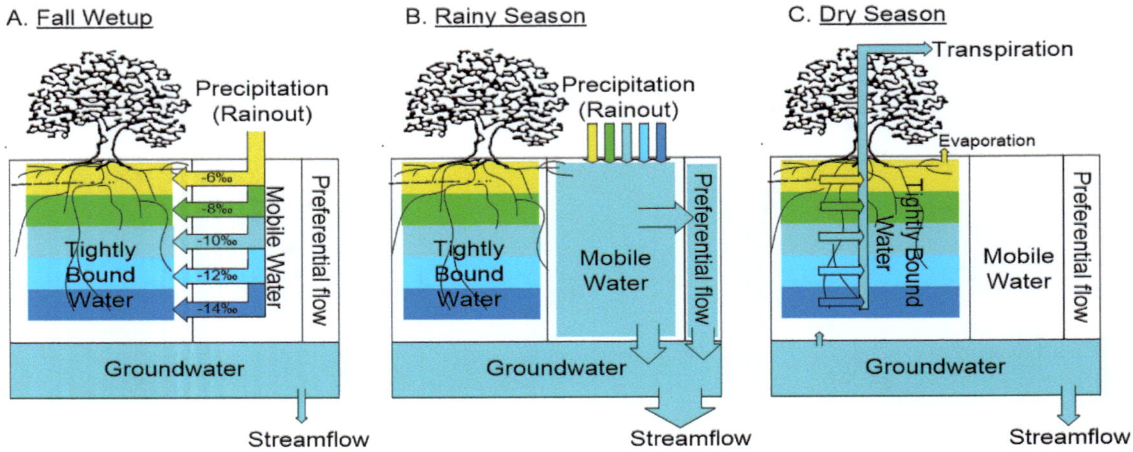

Figure 8.13 Perceptual model of ecohydrological separation linked to the time of recharge of the bulk soil water, the bypassing of winter rainfall through larger pores in the soil profile and then subsequent use of the water whose seasonal origin was the previous fall.
Source: Brooks et al. (2010).

and subsoils can be[19] and that we now can *expect* differences in the isotopic values of water in streams and soils and plants. The seasonal origin of plant water is likely also linked to the seasonal origins of streamflow, where, in many places, most streamflow comes from winter precipitation since most of the summer water often doesn't get very far down into the soil profile. Recent work has shown how often the stream runoff composition is insensitive to summer rain contributions (Bush et al., 2025). While in other areas—like northern boreal forests in Canada—during the main hydrological event of the year (i.e., snowmelt), trees and streams can and *do* share the same sources (Nehemy et al., 2022).

8.5 Dynamic and deep water sourcing by plants

> *Between every two pines is a doorway to a new world*
>
> ***—John Muir***

Let's now look at recent work that shows new evidence for dynamic water sourcing by plants. And later, we'll explore bedrock moisture use by plants.

We will start at a semi-arid site in Spain of del Castillo et al. (2016)—drier comparatively than the places we have looked at thus far in this chapter. The site has two tree species: Aleppo pine and holm oak. These mixed forest conditions are typical across large swaths of Spain and elsewhere in the Mediterranean. del Castillo et al. (2016) measured stable isotope values of xylem water covering an exceptionally long summer drought and subsequent recovery. They went beyond the snapshot sampling that we have seen so far and focused on sampling at different times throughout the year. Figure 8.15 shows that for the first ~2 months of drought, the xylem water isotope composition of oaks tracked the soil water changes but then started to differ from the soil water values. What this showed was that during drought conditions the oaks were able to remain ecologically active for longer periods by using different soil water sources that were not accessible to the pines. What del Castillo et al. (2016) showed was that their oak trees used deeper water during extreme water stress (down to as much as 4–6 m) and well beyond the root depth of the shallow-rooted pines. They further noted that through the:

> dynamics of interaction patterns, we conclude that interspecific differences in pine–oak mixed forests hold two components: a static, spatial component determined

[19] Still most studies take samples at only one or a handful of sites, and basically ignore the spatial variation (see a review by Penna and van Meerveld (2019)).

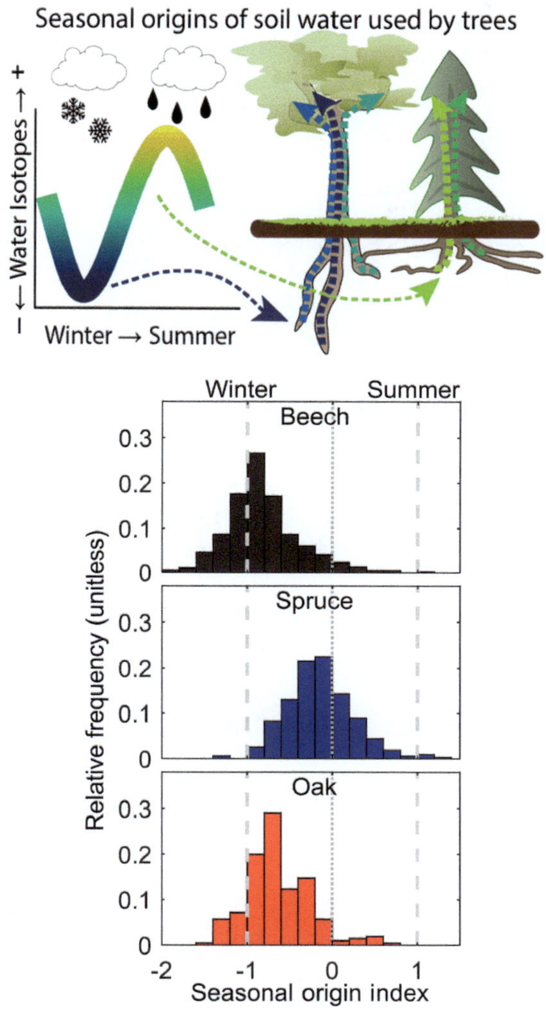

Figure 8.14 The seasonal origin of transpiration for the 3 tree species sampled across Switzerland. An origin index of −1 would mean winter precipitation, 1 summer precipitation, and 0 a mixture of summer and winter precipitation.

Source: Allen et al. (2019).

by root distribution, and a dynamic, physiological component related to water uptake capacity within the soil profile (del Castillo et al., 2016, p. 214)

These results are interesting because they suggest that these two species behave—hydrologically at least—quite similarly during the wet season, in terms of their water source extraction; but that during summer drought, they shift and differ from one another. It is an example of "niche segregation."[20]

In much wetter, tropical sites, the sort of niche segregation shown in the example from Spain has also been seen—but now across dozens of co-existing species together on a single hillslope. One example of this is work by Sohel et al. (2021, p. 1) who sampled 46 tropical hardwood tree species (of the more than 200 that existed on their hillslope) in a small plot in Northeast Queensland, Australia. They noted the "remarkable variation covering the full range of soil composition, suggesting strong sorting and niche segregation across the small plot." Their results suggested that wood density, tree size and mean basal area increment together explained more than half the variance of xylem water isotope composition and source water uptake depth. More recent work in the Amazon has shown that depth of uptake is driven by a key tree hydraulic trait: *embolism resistance*. Nehemy et al. (2025) showed how embolism resistance can explain the sizeable use of shallow soil water and recent rainfall, suggesting that in some tropical sites, trees might quickly use current season rainfall; the opposite to the seasonal origin work we looked at earlier.

Going deeper, a classic example of deep water use by trees and its link to streamflow comes from southwestern Australia and its Jarrah forest, a type of deep-rooted eucalypt species. Accounts of the effect of forest clearing and streamflow quantity and quality started in the early 1900s. Burvill (1947) wrote one of the earliest formal reports on how cutting Jarrah and replacing it with wheat and other crops did two things: it increased stream baseflow and significantly increased stream salinity. Before clearing, the forest decreased the amount of rainfall entering the soil due to interception and the deep roots took water from the soil for transpiration. When the trees were removed, the winter wheat and other crops used much less water and more water percolated deeply and increased groundwater recharge. But as the water table rose, contributing to stream baseflow, it flushed out the salt that had built up in the soil profile in this semi-arid

[20] a popular ecological notion defined often as a "process by which competing species use the environment differently in a way that helps them to coexist."

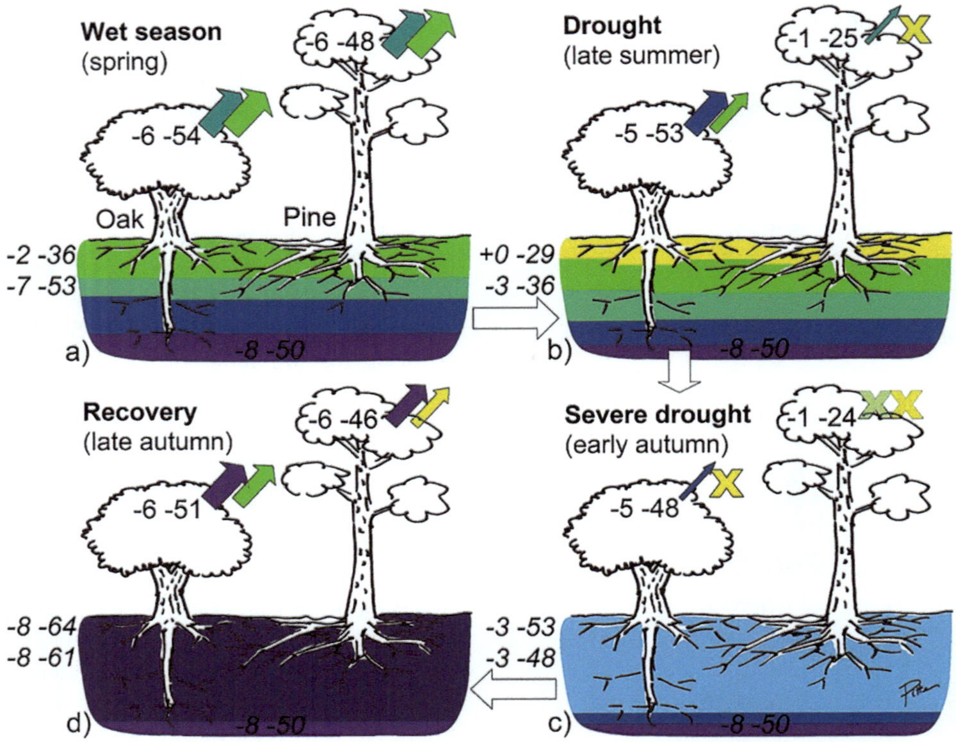

Figure 8.15 The seasonal changes in the isotope composition in soil and xylem water in holm oak and pine where the arrows on the tree top show the transpiration rates (the cross indicates stomatal closure). The numbers are the $\delta^{18}O$ (always the leftmost number) and $\delta^{2}H$ (the rightmost) values (in per mil, or parts per thousand; technically, per mil relative to VSMOW).
Source: del Castillo et al. (2016).

area over 1000+ years (as described by Peck and Williamson 1987). Stream baseflow increased 2–3× the values pre-clearing, but the salt output from the catchments increased 100–1000-fold.

Of course, deep water use can go the other way too—with the planting of forest as opposed to the removal of it. If we zoom in on one site with deep soils and deep roots in China, we can see how deep roots develop in the Loess Plateau—an area we examined in Chapter 6. Li et al. (2023) showed how soil water and root distributions ranged from 6 to 25 m across 72 paired sites: farmlands typical of pre-afforestation conditions and neighboring tree plantations linked to afforestation across the Loess Plateau of China. Figure 8.16 shows soil moisture data below trees ranging from 1 to 25 years of age. What is striking is the distinct "water mining." Figure 8.18b shows the evolution of the decrease of soil water content following afforestation in shallow-rooting farmlands and grasslands. And Figure 8.18a shows the maximum rooting depth of trees across the stands' ages. Following afforestation, the mean soil water decline was about 75 mm/yr linked to a root deepening rate of about 1 m/yr.

Other new perceptual models for plant water sourcing of deep water are being developed. One of the most exciting areas of work is in plant water use of rock water. Anecdotal evidence of rock water use dates back many decades. Dye (1996) saw >8 m depth water use by trees in the Frankfort State Forest, in the Eastern Transvaal province of South Africa. And even Dawson and Ehleringer (1991, p. 336) anticipated this in their work where they noted that "studies of plant interaction and competition for water should include the possibility that

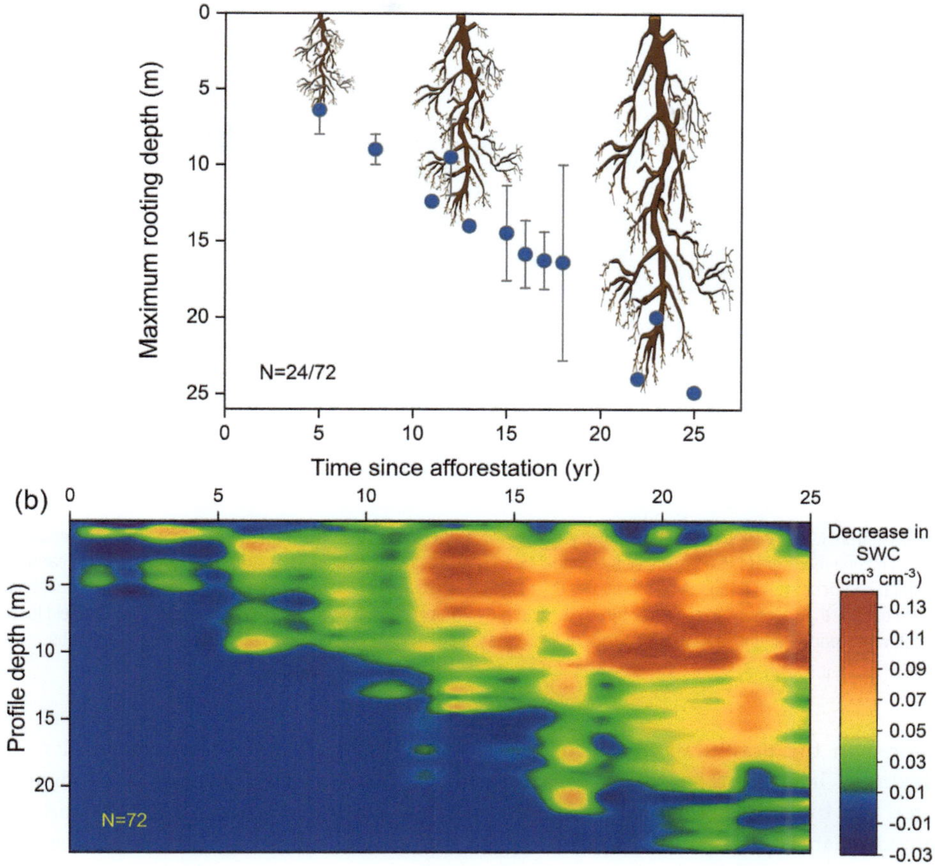

Figure 8.16 (a) The maximum rooting depth of trees across stands ages. (b) The decrease of soil water content (SWC) following afforestation in shallow-rooting farmlands and grasslands.
Source: Li et al. (2023).

some trees might use deeper water sources." But it was not until the isotope tracing work of Salve et al. (2012) and Oshun et al. (2016), that showed how plant water uptake of moisture in weathered rock could be a mechanism to explain plant water behavior during drought, that this was fully appreciated. In follow-up work, Rempe and Dietrich (2018) outlined this perceptual model of plant water use of rock moisture where:

> Recent theory and field observations suggest that [a] systematically varying weathering zone, that can be tens of meters thick, commonly develops in the bedrock underlying hillslopes. Weathering turns otherwise poorly conductive bedrock into a dynamic water storage reservoir. Infiltrating precipitation typically will pass through unsaturated weathered bedrock before reaching groundwater and running off to streams (Rempe and Dietrich, 2018, p. 2664)

What is remarkable from this quote is that often difficult-to-see, difficult-to-understand and difficult-to-access unsaturated zone, was, until this 2018 work, "virtually unexplored" compared with the sorts of soil water studies we have examined thus far.[21] The Rempe and Dietrich paper proposed the term "rock moisture" to describe

[21] Remember from Chapters 5 and 6 that the importance of bedrock or saprock water for runoff generation remained largely unexplored until the landmark work by Bill Dietrich's

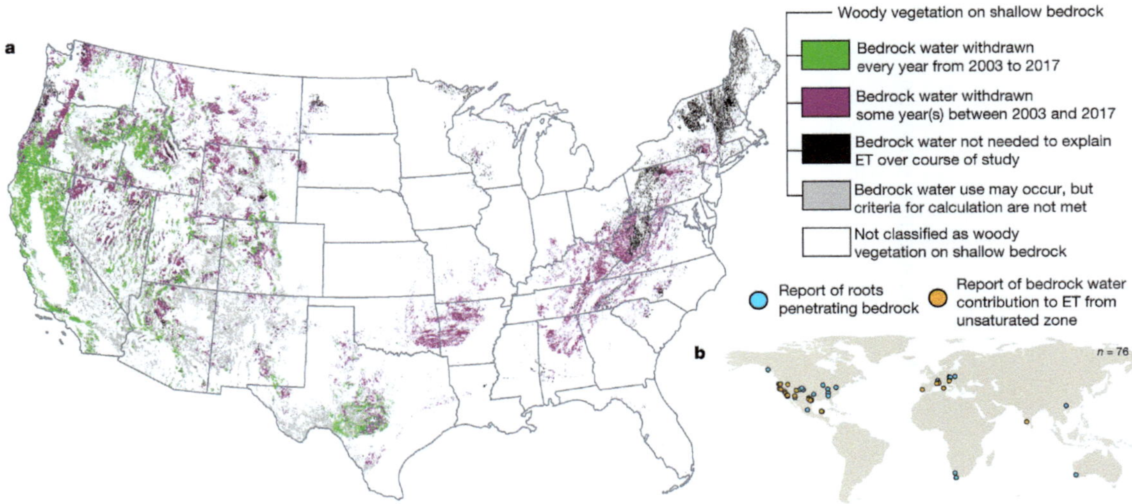

Figure 8.17 The map from McCormick et al. (2021) showing (to quote) (a) the extent of woody vegetation where bedrock is encountered within the upper 1.5m. Green zone represents the area where the annual bedrock water storage deficit is greater than zero for every year of the study; pink where the annual bedrock water storage deficit is greater than zero for at least one year of the study; black where the annual bedrock water storage deficit is not greater than zero for any years of the study; and gray = not reported. The figure also shows, in blue, the field locations where rooting into bedrock was reported and in orange where rock moisture has been quantified as a contribution to evapotranspiration.

Source: McCormick et al. (2021, p. 227).

this water stored in the unsaturated zone of the weathered bedrock. In a more recent analysis of bedrock water use by plants, McCormick et al. (2021) presented what they called a "lower-bound estimate" of the contribution of bedrock water storage to transpiration across the continental United States" (Figure 8.17) where they showed that:

> woody plants extensively access bedrock water for transpiration. Plants across diverse climates and biomes access bedrock water routinely and not just during extreme drought conditions. . .our findings indicate that plants commonly access rock moisture, as opposed to groundwater, from bedrock and that, like soil moisture, rock moisture is a critical component of terrestrial water and carbon cycling (McCormick et al., 2021, p. 225)

A final rock water sourcing example worth looking at is from trees growing on seemingly no soil at all in dryland karst environments. I've marveled at such limestone areas while in the field—wondering, where the heck these trees get their water; where soil, if it does exist, is patchy, stony and often very shallow. Figure 8.18 shows the results of Carrière et al. (2020) who traced water sources in three tree species in France (fir, beech and holm oak) and showed how they each used deep karst rock moisture more intensively during drier years. They noted that the holm oak, which is a species that is very well adapted to water stress "uses the deep water resource very early in the summer season" (Carrière et al., 2020, p. 1). Conversely, they found that "the two other species exploit the deep water resource only during severe drought" as shown in the histogram plots in Figure 8.18. Similar work is now exploding in China in karst and epikarst where isotope tracing of plant water is revealing new perceptual models of plant water extraction from soil-filled fractures in limestone (e.g., Liu et al., 2025).

How all this rock moisture work will link to streamflow is an open question. Certainly, water can be mined from deep cracks and fissures but that water may not be in communication with the stream. So, in effect, this could be a separate closed system that plants are tied into, unlinked to streamflow generation. Another set of water worlds

group at Metman Ridge in Oregon (recall papers we discussed by Montgomery et al. (1997) and Torres et al. (1998)).

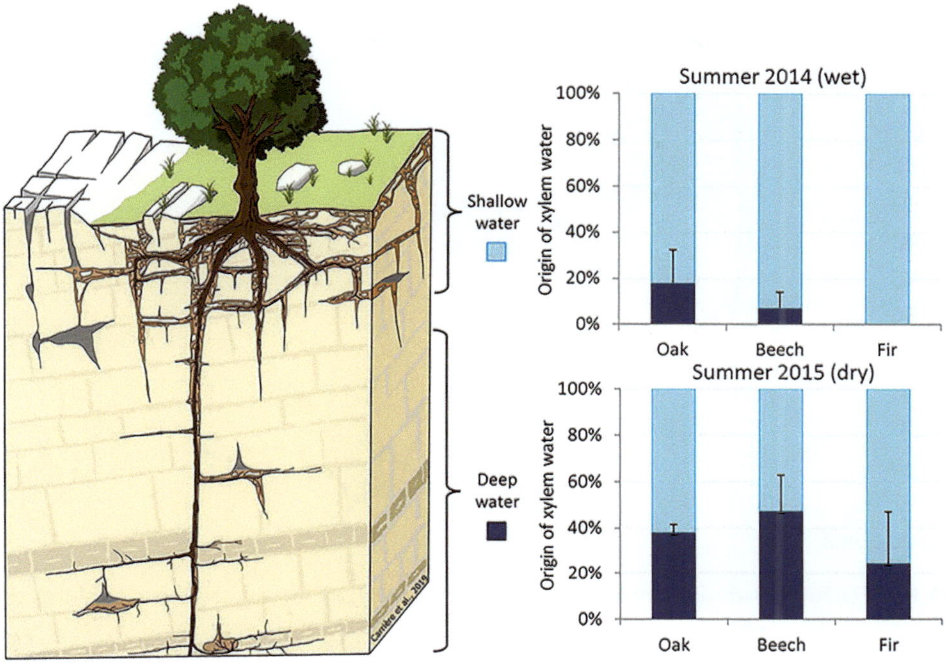

Figure 8.18 Plant water extraction from karst with portions of shallow and deep water quantified for oak, beech and fir.
Source: Carrière et al. (2020).

perhaps? Yet, refilling the moisture storage may affect runoff generation during the first storms after a drought. The coming years will prove very exciting in this regard.[22]

8.6 Tree water transit times

The best time to plant a tree was 20 years ago. The second-best time is now

—Chinese Proverb

We have examined the source water for plant uptake. What about questions of residence time of water in the plant itself and the transit time of water through the plant? We're still in the very early stages of process understanding of this "transpiration water age" (as noted by Soulsby et al., 2016 and Sprenger et al., 2019). And our perceptual models are even more nascent.

But what do we know?

[22] And new work is now getting at the age of water stored in cracks that trees are mining in some areas (see Liu et al., 2025).

Early work that performed tritiated water injections into old growth trees in the Pacific Northwest USA forests showed us that transit times through stems were on the order of 20–40 days (Kline et al., 1976). But these were huge trees. More contemporary work with deuterium irrigations of jack pine in Canada has shown transit times on the order of 11 days—i.e., the lag for irrigation at the roots to be detected at the canopy level (Hamilton, 2019). More recently, Nehemy et al. (2022) quantified transit times during snowmelt water uptake in a Saskatchewan boreal forest stand of the order of 9–18 days, for jack pine and black spruce, respectively.[23] Unlike the dispersion we discussed in Chapter 7 linked to catchment transit times, recent model exercises indicate more "piston-flow"-like water movement of tracers within the trees (e.g., Knighton et al., 2020). Gaines et al. (2016)

[23] Technically it was 12 days based on a summer estimate of velocity data, but for the snowmelt period, it was 9–18 days (1–2 weeks), for black spruce and jack pine, respectively.

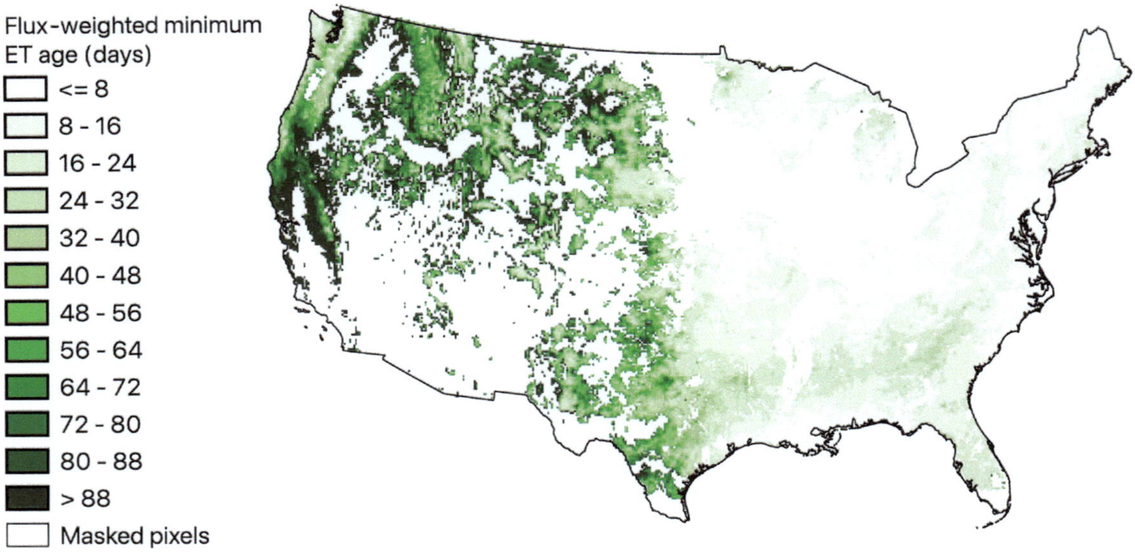

Figure 8.19 Map of (a) flux-weighted, last-in first-out inferred evapotranspiration (ET) age, indicating that ET must be relatively old across much of the western continental United States.
Source: Hahm et al. (2022).

combined a deuterium tracer, sap flux and anatomical techniques to study tree water transport on a forested ridge top in central Pennsylvania, USA. They hypothesized that tracer velocity would be higher in so-called ring-porous tree species because of their greater vessel diameters and water transport efficiency. They also hypothesized that tracer residence time would be largely dependent on tree size. They found tracer velocities in the xylem between 1 and 18 m/d with maximum deuterium concentration in the tree crowns found 1–12 days after tracer injection. The tracer residence times in the tree were 5–22 days with "no evidence of longer residence times in larger trees" (Gaines et al., 2016, p. 1554). Interestingly, they found no evidence of a relationship between tracer velocity and calculated xylem-specific conductivity, which varied by nearly an order of magnitude between species.

All these examples have been for single-site studies. What generally might we know about tree water transit time? Hahm et al. (2022, p. 1) developed what they call a "lower-bound constraint" on what they call "evapotranspiration age" where they calculate this age by assuming that evapotranspiration is "preferentially sourced from the most recent precipitation through a last-in, first-out algorithm." Their work for the western USA (Figure 8.19) shows that such ages are weeks to months old. But they acknowledge that "true ET fluxes are composed of distributions of ages and may be composed of much older water." Follow-on work by Felton et al. (2025) used a blend of satellite remote sensing products to reveal a global average transit times of water through "aboveground vegetation" that vary from "~5 days in croplands to ~18 days in evergreen needleleaf forests, with a global median of 8.1 days" (Felton et al., 2025, p. 59).

There are indeed examples of much older water comprising plant transpired water. Again, we need to distinguish between the transit time of water revealed from an injection in the trunk or root vs. a rainfall or snowmelt application. The latter includes the soil water residence time and this can greatly skew the calculation of tree water age. For instance, the Loess Plateau site that we looked at in Chapter 7 showed that mature apple trees planted in the deep unsaturated zone, with roots extending down some 20+ meters, had transpiration water leaving the plant on the order of 29 years (Zhang et al., 2017). By linking the apple water tritium concentration to the

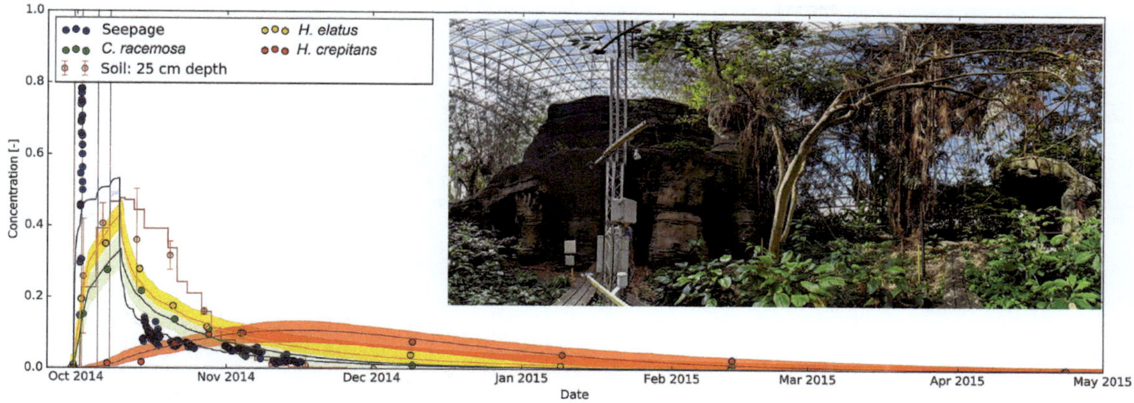

Figure 8.20 Breakthrough curves from the tracer additions at the Biosphere 2 Tropical Rainforest biome. Data for 3 tree species are shown, together with the groundwater and soil water tracer concentrations.

Source: Modified from Evaristo et al. (2019) with the photo from https://biosphere2.org/research/research-initiatives/tropical-rain-forest.

soil water tritium concentration, Zhang et al. (2017) used a mixing analysis to show that ~40% of fruit tissue water came from depths between 4 and 9 m within the soil profile.

Unpacking the components of soil water residence time and transpiration water age has been achieved mostly in large lab settings. For instance, Evaristo et al. (2019) performed a 9-month drought and rewetting experiment at the Biosphere 2 Tropical Rainforest biome. Figure 8.20 shows how after an imposed 10-week drought, they added labeled rainfall with 152‰ δ^2H distributed over four events, followed by background rainfall (−60‰ δ^2H) distributed over 13 events. Their results showed that mean transit times through groundwater recharge and plant transpiration were markedly different: groundwater recharge was 2–7 times faster (~9 days) than transpired water (range 17–62 days). The "age" of transpired water showed strong dependence on species and was linked to the difference between midday leaf water potential and soil matric potential. Linking back to earlier themes in this chapter, their results also showed that the trees used water with the extracted soil water signature (~90%) and not the "more mobile" seepage water (~10%).

Other recent controlled experiments have addressed transit time distributions of tree water and soil water and groundwater recharge. Benettin et al. (2021) performed a labeled tracer experiment on a 2.5 m^3 lysimeter planted with a willow tree (Figure 8.21) where they tracked the fate of the labeled water addition for 43 days through the soil, the bottom drainage and in the plant xylem. Unlike the Evaristo et al. (2019) work, they found tracer in the plant xylem water within a day or two of application, with a very long tail in the transit time distribution of tree xylem water (over 40 days later) and a slower breakthrough through the bottom drainage of the lysimeter.[24] Again, such studies are in their infancy and hold great promise for new perceptual models of how plant water uptake affects not only the streamflow signal, but also the components of that flow and its transit time distribution (see Chapter 7). We'll touch on the modeling work in this regard in Chapter 9 when we discuss the needed dialog between experimentalists and modelers—especially for determining the full

[24] Just to add a caveat that this stuff is made all the more complex by the plants where, as Benettin et al. (2021) note a little farther down in the paper: "The tracer breakthrough also had irregular fluctuations that likely reflected some progressive shift toward shallower/deeper sources depending on tree water status as outlined in Nehemy et al. (2021). This behavior is consistent with the observed root architecture: fine roots (<2 mm) were observed throughout the entire soil column, up to 2 m depth (Nehemy et al., 2021). The relative uniform and deep presence of fine roots suggests that water uptake was not constrained to any depth or sector."

Figure 8.21 The EPFL lysimeter experiments.
Source: Photo by Paolo Benettin.

exit time distributions for vegetated catchments (meaning the stream transit time distribution and the plant transpiration transit time distribution).

8.7 Summary and looking ahead

Science does not know its debt to imagination
—Ralph Waldo Emerson

We have learned in this chapter the complex relationships between vegetation, transpiration and streamflow. Catchment ecohydrology is a hot research area and including plants in the transit time distribution of streamflow is *the thing* right now in research studies—meaning the combined *exit time distribution* for water out of a catchment. This work will no doubt alter our perceptual understanding of water movement through catchments. Interestingly, the disruption of conventional thinking in ecohydrology by isotope tracing mirrors the disruption caused by the use of isotopes in understanding streamflow generation processes that we reviewed in Chapter 5. But that pace of change is much faster now, with *Science*, *PNAS* and *Nature* papers appearing sub-annually, focused on plant water uptake in relation to the critical zone and streamflow (e.g., McCormick et al., 2021; Miguez-Macho and Fan, 2021; Radolinski et al., 2025; Nehemy et al., 2025).

But while research and understanding of ecohydrological impacts on streamflow generation is exploding, one major challenge remains: how to quantify and understand mixing in the subsurface. One very important caveat in all the isotope tracing work—that could hold many keys to process interpretation and re-interpretation—is the *methodology* for water extraction from soils and plants and the pressing need for improvement (as outlined in Rothfuss and Javaux, 2017; Sprenger et al., 2016). We only discussed cryogenic extraction in this chapter—and it has issues in terms of the many options for extraction time and temperature but also consistency from lab to lab (Orlowski et al., 2018). Standard operating procedures are desperately needed (Millar et al., 2022; Ceperley et al., 2024). But other methods also exist like

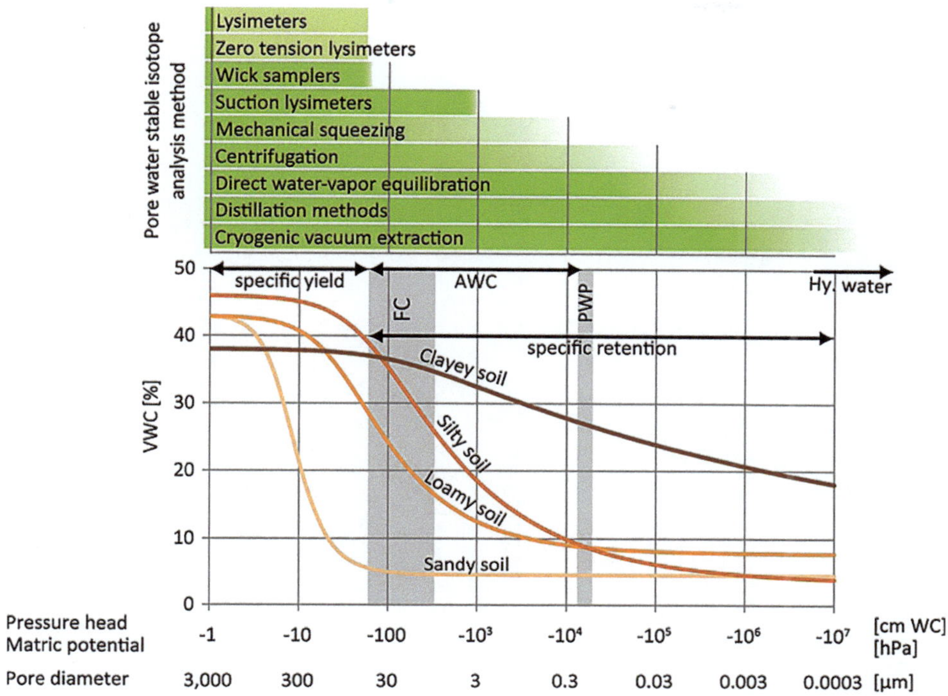

Figure 8.22 The extraction methods get water held under different tensions, where VWC = volumetric water content, FC = field capacity (gray area), AWC = plant available water capacity, PWP = permanent wilting point (gray area), Hy. water = hygroscopic water. The upper part of the figure shows the most widely used extraction methods linked to their range of matric potentials and pore diameters along the bottom of the plot. *Source*: Sprenger et al. (2015).

centrifugation and vapor extraction. Some of the most recent literature suggests that for plant water, a special type of centrifuge technique called a cavitron may be a way to circumvent some of the sensitivity of cryogenic extraction of plant waters to the extraction temperature and extraction time (Barbeta et al., 2022). In the end, each of the extraction techniques sample different waters, as shown in Figure 8.22, where ultimately, a good portion of the difference between extraction techniques could simply be related to the *type* of water extracted. We desperately need ways to extract water sequentially along the moisture release curve, to compare isotope ratios of plant water at particular tensions and the soil water at the same tension. Some work has begun in this area (Orlowski and Breuer, 2020).

Lastly, what things might usefully point the way to new and better understanding of plant water connections to the stream? Figure 8.23 shows depth functions of many of the things we looked at in previous chapters: soil water potential, soil water content and soil hydraulic conductivity. But it also shows perhaps the things we are now trying to understand mechanistically in our quest to define mixing in the subsurface: the role of biological factors like mycorrhizal fungi, hydraulic redistribution, root conductivity and root embolism, and how they interact with soil physical factors like soil water potential and hydraulic conductivity to control root water uptake with depth. Here, Bachofen et al. (2024) note how root conductivity is linked to root surface area, that itself, decreases exponentially with depth. New work is suggesting how mycorrhizae (that for many plants have a symbiotic relationship with their roots) can discriminate against heavy isotopes (Poca et al., 2019). And new work in the Amazon is showing how resistance to

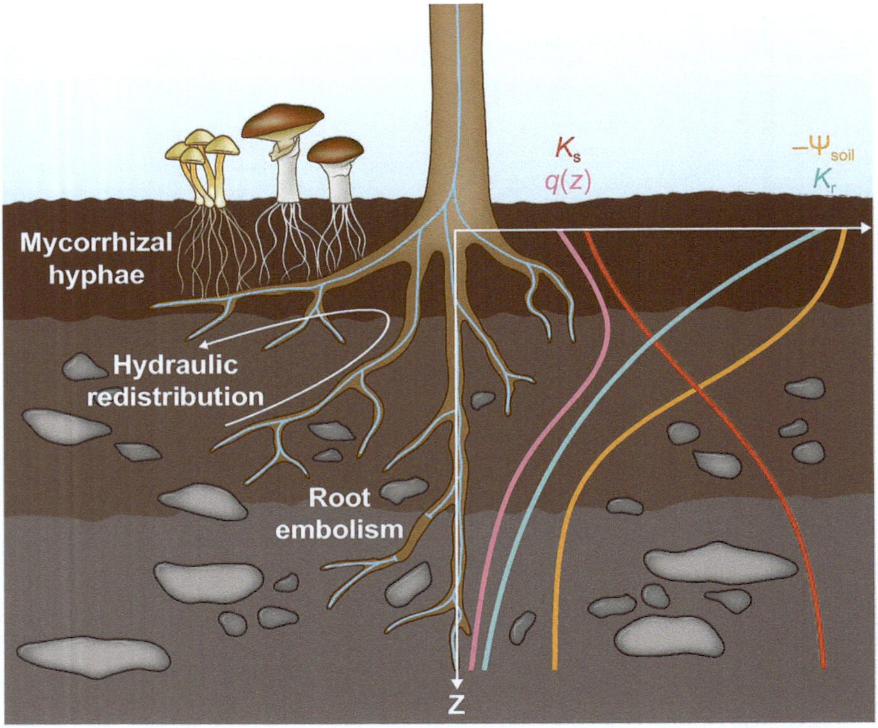

Figure 8.23 Three factors: mycorrhizal fungi, hydraulic redistribution and root embolism, and how they link to root water uptake, where soil water potential (Ψ_{soil}), soil hydraulic conductivity (K_S) and root water flux (q(z)) change with depth and are linked to root conductivity (K_r). All variables increase from left to right.

Source: Bachofen et al. (2024).

embolism can explain differences in shallow–deep water uptake across different topographic positions (Nehemy et al., 2025). All this points to a need for dialog between plant ecologists and hydrologists on how we develop new ways of understanding and quantifying mixing in our numerical catchment models, something we will explore in the next chapter.

PART 3

What Do We Need To Know?

CHAPTER 9

Opportunities and challenges ahead

9.1 Introduction

No one ever notices what has been done; one can only see what remains to be done

—Marie Curie

This book is my attempt to provide a narrative of streamflow generation processes and how perceptual models have developed and evolved over time. It has led from specific catchment examples—well-chosen particulars, I hope. And rather than being comprehensive in summarizing all studies, I've tried my best to show where ideas originated and how we have moved through the squiggle we discussed in Chapter 1. Of course, this is merely *my* narrative—one drawn from my own experiences as a field hydrologist. So, it is biased. Nevertheless, I hope that it is of *some* value to the beginning graduate student as they navigate the literature, proceed to their own field studies and learn all the relevant quantitative approaches used in catchment hydrology. I similarly hope that it is of use to aspiring hydrological modelers; to see what perceptual models their mathematical models are based upon, what important processes are excluded from the models; to provide an appreciation of the importance (and difficulties) of doing fieldwork. As such, this book is meant to complement and not replace the main equation-led textbooks in our field.

How can we summarize the book in a single phrase? Maybe: streamflow happens when storage says no. If we reflect on some of the things we discussed in Chapter 1, I hope that you can see that streamflow generation is more than the 1970s era process diagrams of infiltration excess overland flow, saturation excess overland flow and subsurface stormflow that dominate most undergraduate textbooks. While these process descriptions are a necessary part of perceptual models of streamflow generation, they are not in themselves a sufficient perceptual model. A perceptual model is about the aggregate behavior of all the interacting components of streamflow generation. We know now that catchments store and release water to streams in often highly non-linear ways; ways linked to patterns of connectivity. That there is much loss along all flowpaths. That thresholds and other characteristic forms of non-linearity (e.g., hysteresis, feedbacks, etc.) abound in headwater catchments. And that the residence time of water within catchments and the transit time of water leaving catchments are long—especially at low flow; with progressively younger waters with increasing flow rates. I hope that you can appreciate that celerity and velocity are two different and equally important considerations for streamflow generation. And that between events, streamflow is affected by plants that plumb the critical zone and wick water from soil and weathered rock in ways that impact subsurface mixing and the streamflow signal. And that they do this in ways that we are only now beginning to understand—with the selection of water for streamflow and plant transpiration from storage being a key open question in our field.

The throughline of this book is that field discovery should be our north star in catchment hydrology. To make modeling the sole focus—as is often the case today—is like young kids who want to have candy for dinner. It is a huge problem in our field now, as noted most recently by Gleason and Brown (2025) who present perhaps the harshest critique yet

Streamflow Generation. Jeffrey J. McDonnell, Oxford University Press. © Jeffrey J. McDonnell (2026). DOI: 10.1093/9780191953736.003.0009

of our model-obsession in hydrology. It may seem like we know a lot. But, arriving at a new headwater catchment, I would challenge the best in our field to be able to make a non-trivial prediction about the age, origin and pathways for streamflow generation without any field measurements; even with geological maps, a good DEM and known climate information. Water flowpaths—especially belowground where most of the action happens—are still the dark matter of streamflow generation. We remain at the point where we can particularize streamflow generation at a given catchment (after sufficient field work) but we have great difficulty generalizing (or predicting) streamflow generation behavior.[1]

This book advocates that we need to redress the balance between field work and modeling. And for the health of our discipline, we need to return to field evidence and discovery science as our ultimate goal. Modelers, politicians, funders and the general public often ask whether we still need more field experiments "when we have not made full use of the process understanding we have gained from previous field studies?"[2] But we are dealing with a trillion-fold range of scales in headwater catchments. There is no genie-in-a-bottle cure for this problem. There is a medley of processes and networks that remain hidden from us. We've noted throughout the book that the boundary conditions are the science.[3] And knowing that boundary conditions are the science means that contingencies (the soil depth, subsurface interfaces, preferential flow networks, inexplicable mixing during and between events) are at least equally important to what we learn as undergraduates with Darcy's Law and the Richards equation.

We need to be voracious in our appetite to describe streamflow generation across different environments; to do what the famous geologist WM Davis told us in Davis (1926) (paraphrased by Kennedy, 1983, p. 329): "we need to revert to the creation, testing and discarding of genuine hypotheses" and to spend less time and effort "congratulating ourselves that some concept or other explains a significant proportion of the observed variation." In our case, we need to spend less time congratulating ourselves that we have fitted a storm hydrograph with a rainfall–runoff model and get back to the field to improve our process understanding. Because, the beginning of knowledge is the discovery of something we do not understand.[4] And the amount we currently understand is minuscule to what remains to be understood!

My crystal ball is no better than anyone else's, so I will not speculate too far into the future in this final chapter. But I will leave you with some parting thoughts on measurements, the dialog between experimentalist and modeler, the challenge that heterogeneity poses and the urgency for perceptual model development in a changing world. And then lastly, some ideas on navigating the early stages of a field hydrology career.

9.2 Streamflow generation: Grouping, diagnosis and discovery

Science consists of grouping facts so that general laws or conclusions may be drawn from them

—Charles Darwin

Grouping of streamflow generation behaviors is something dreamed about for decades (McDonnell and Woods, 2004).[5] There have been some useful attempts (e.g., Wagener et al., 2007; Gharari, et al., 2011; Barthold and Woods, 2015; Berghuijs et al., 2016) and some recent work getting at signatures of streamflow generation across large regions. Some recent work has started to link geographical grouping and classification of catchments to model structural development, in the UK (Wagener et al., 2021). And these are great developments. To fully accomplish it, we will need to accumulate and house our perceptual knowledge somehow, as noted by Wagener et al. (2021). I hope this book is one small

[1] Paraphrasing Pearce (1990).

[2] In fact, this quote comes from Sivapalan (2003, p. 1037).

[3] As we noted earlier, this was first clearly stated by Beven (2006).

[4] To quote Frank Herbert, author of the book "Dune."

[5] Some textbooks have commented that every catchment is a "law unto itself" (Brutsaert, 2005). I strongly reject that notion and hope that this book has convinced you otherwise. While there is always some uniqueness of place, I firmly believe in grouping of dominant behaviors; behaviors that help show that there are indeed similar perceptual models that can explain the streamflow response at multiple places.

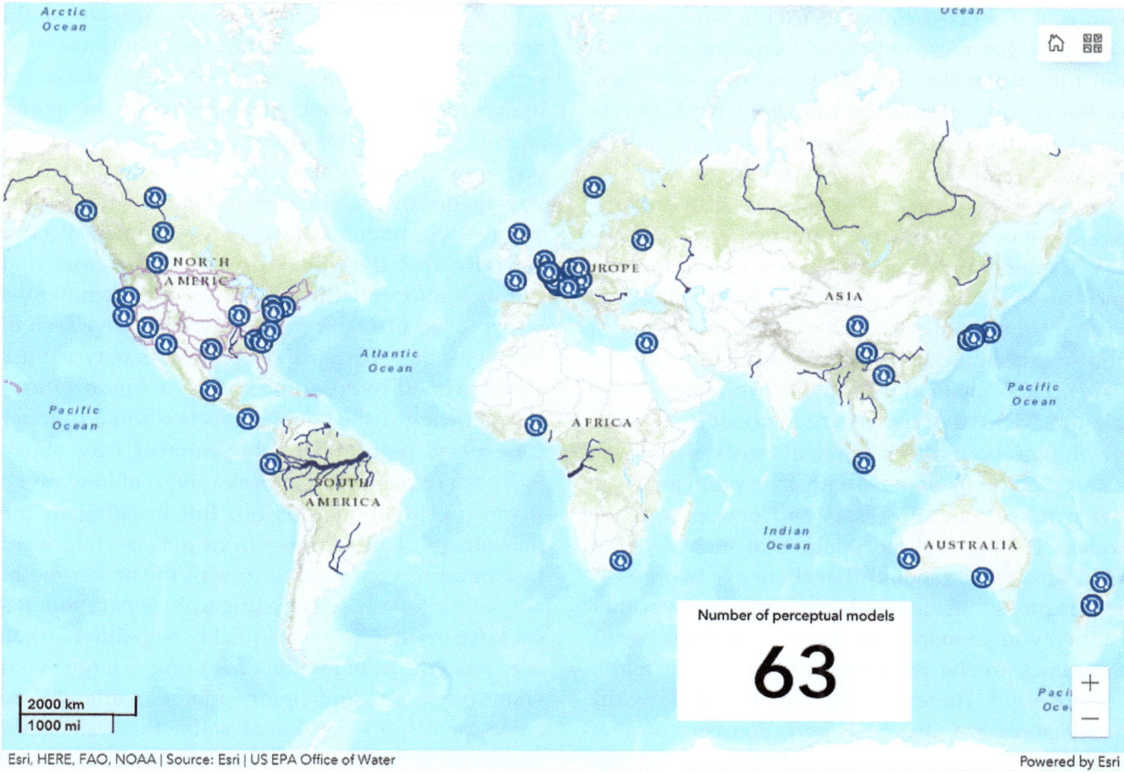

Figure 9.1 Perceptual Model Database of McMillan et al. (2023) that includes many of the research catchments covered in this book. McMillan et al. (2025) have extended this and the GIS dashboard is available via https://github.com/RY4GIT/perceptual-models.

step in this regard. Housing perceptual models has begun by others (Figure 9.1): 63 perceptual model figures with taxonomic attributes, including vegetation, soils, topography and topographic characteristics were assembled by McMillan et al. (2023). This has recently been updated to include 131 perceptual model figures by McMillan et al. (2025) who combed through 400 published catchment process studies.

There are new large datasets that will no doubt lead to new grouping of streamflow generation behavior and catchment functioning. These may be based on things like the Catchment Attributes and MEteorology for Large-Sample SPATially distributed analysis (CAMELS-SPAT) dataset (Knoben et al., 2025) or the multiple other CAMELS datasets[6] or the new EStreams integrated dataset and catalogue of streamflow, hydro-climatic and landscape variables for Europe (Do Nascimento et al., 2024). Indeed, some exciting new work has already begun in this regard, grouping the three main processes classes across the USA with the CAMELS dataset (Wu et al., 2021). They found that climatic and physiographic characteristics (like climate aridity, median event rainfall volume, topographic slope, soil permeability, rock type and vegetation density) were the main factors controlling the classification of infiltration excess and saturation excess overland flow and subsurface stormflow. Even more recent functional grouping of streamflow has been done with this and additional datasets extending into Canada (Sharif and Ameli, 2025).

But while assembling these large datasets and existing perceptual models and grouping the three

[6] See a compilation at: https://essd.copernicus.org/articles/collection6.html

main runoff process classes is exciting and useful, my fear is that these efforts will give the new student the impression that we know more than we do. We need to incentivize new field work (Allen and Berghuijs (2018). The community is calling for this (Blume et al., 2017). Almost all our discoveries of key processes in catchment hydrology have come from field work (Kirchner et al., 2023)—measurements and observations—since our models (and perhaps even our groupings) lock in our preexisting conceptualizations.

So where should field work be headed?

We go into the field usually for two reasons[7]: for obtaining data to parameterize a model or to learn new things about how catchments work. This book has focused on the latter—with an express focus on new patterns, new processes and new perceptual models. It's like the old adage that in hydrology (as in science in general) that there are two possible outcomes: "if the result confirms the hypothesis, then you've made a measurement. If the result is contrary to the hypothesis, then you've made a discovery."[8] There are many pathways forward for measurement, diagnosis and discovery of new behaviors. There is new discussion about hypothesis testing (Pfister and Kirchner, 2017). There are new, inexpensive sensors (Tauro et al., 2018). There are new visions for hydrological observatories (Nasta et al., 2025). There are new recipes for how to set up and sustain an experimental catchment (Penna, 2024). These are all useful points and discussions.

But still the questions remain for streamflow generation research going forward: What should we measure? In what order? And why? For this, I will offer two thoughts; one a simple top-down way to start research at a new site. And the other, a simple bottom-up way to follow the initial top-down work. I like to think of this as a diagnosis; a systematic way to interrogate a headwater catchment—a way to get started; an on-ramp to developing a new perceptual model of a site, toward developing a respectable first draft of hydrological reality.

Figure 9.2 shows three top-down measures that I find very helpful to get started: the runoff response signature, the percentage of event water in the storm hydrograph and some measure of stream water age[9]—descriptors that tell you what sort of beast you might be dealing with. These measures are largely orthogonal to each other. Meaning, they are non-redundant and each reveal some aspect of behavior that the others do not; descriptors that can then inform field work where the catchment is later probed in a bottom-up way.[10] We have seen in this book that many runoff pathways may yield a similar runoff hydrograph shape or runoff ratio.[11] But knowing if the runoff ratio is a small number (say, a few percent) or a big number (say, above 50%) can immediately inform a sense of how much storm precipitation runs off. But bringing in the percentage of new water from a two-component hydrograph separation (or one of the newer methods of isotope-based hydrograph separation we spoke of in Chapter 7) can quickly help differentiate one behavior from another. Meaning, a large event water fraction would signify a large overland flow component[12]; a small event water fraction would signify a small overland flow contribution. While there is really only one way for event water to reach the stream, we have seen in this book that there are countless ways to generate the pre-event water component.

[7] There is a great early discussion of this in Eberhardt and Thomas (1991) related to ecological field studies.

[8] This quote originally from the Nobel prize winning physicist, Enrico Fermi.

[9] Returning to our discussions of grouping and thinking even more inclusively with the role of vegetation and the other key outflows from the catchment, Gibson et al. (2021) sampled 103 catchments in Canada and grouped their vapour and runoff partitioning. They found that 19 of their 103 catchments were runoff-dominated; 5 were evaporation-dominated and 54 were transpiration-dominated (with 27 catchments showing more than one dominant water loss mechanism). This is yet another top-down way to group catchments.

[10] A great example of this approach is the work at the Bruntland Burn catchment in Scotland where these topdown measures were made by Tetzlaff et al. (2007) to then inform bottom-up ground-based field studies (Tetzlaff et al., 2014) and then finally, model analysis (Kuppel et al., 2018a, b).

[11] This is the equifinality thesis of early work by Keith Beven—borrowed from earlier geomorphological thinking that different geomorphic processes could create similar-looking landforms.

[12] Albeit sometimes with considerable mixing with exfiltrating groundwater as shown by Lapides et al. (2022) and Leuteritz et al. (2025).

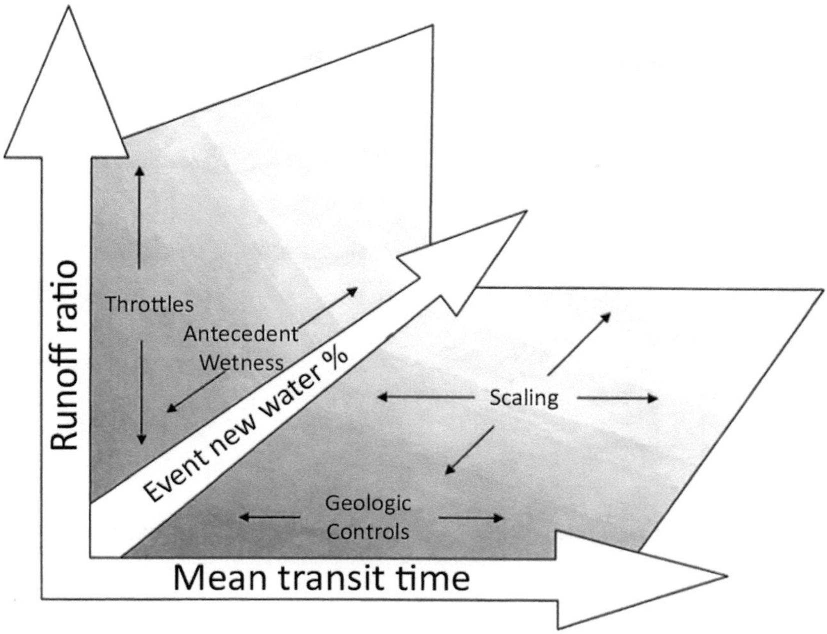

Figure 9.2 The three top-down diagnostic measures that have worked best for me to diagnose how a catchment stores and release water: the runoff ratio, the % event water and the streamwater mean transit time.

Source: Figure drawn by Cody Millar.

A third top-down diagnostic to help understand the subsurface control on flow and transport—and a measure of storage in a catchment—is based on some estimate of catchment transit time. Again, this could be done in a rudimentary way (e.g., a sine wave-fitting exercise of stable isotopes or tritium sampling and analysis—depending on where you are) or one of the many newer techniques outlined in Chapter 7. In this case, a high mean transit time (decades or more) would suggest a sizeable subsurface storage; a low number (months to a year or so) would suggest a small subsurface storage.

Following this simple top-down diagnosis, bottom-up field work can help go beyond the initial diagnosis and to understand *how* the system works. I like to think of what is filling and spilling as the most rudimentary form of bottom-up interrogation when I visit a new catchment; to start to ponder *how* a catchment stores and releases water and drives streamflow generation. This is guided by age-old advice that "all observation should be for or against some theory if it is to be of use."[13] For me, thinking what is filling and spilling is an initial, rudimentary "theory"[14] from which one can structure observations in a bottom-up way. Recall from Chapters 6 and 7 that the core idea behind fill-and-spill is that water first accumulates in various storage units (e.g., soil pores, depressions, hillslopes) until reaching a threshold, at which point connectivity is established, and flow is triggered. So, I like to use fill-and-spill thinking to consider how, where, when and why water is stored and subsequently connected to the stream. Knowing that all runoff processes are highly scale-dependent, one first determines the scale of interest before embarking on this examination (Figure 9.3). Adopting the fill-and-spill approach to an initial bottom-up diagnosis allows me to structure my observations

[13] This quote is attributed to Charles Darwin as noted in in Costa (2017).

[14] Of course, fill-and-spill thinking is not a theory per se as it cannot make falsifiable predictions.

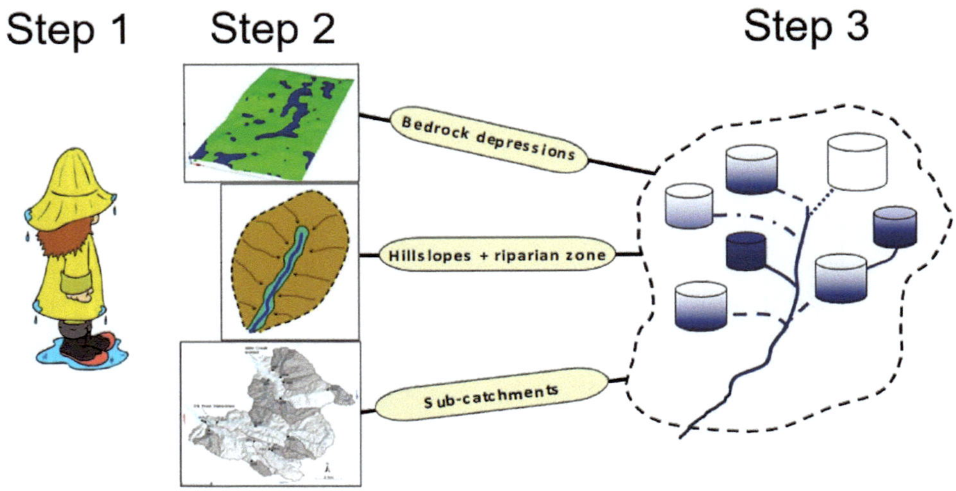

Figure 9.3 A bottom-up approach that can work well for starting a new investigation. Step 1: go to the field. Step 2: choose your scale of interest. Step 3: Take a storage-based view of your scale of interest and see how it fills and spills, what the thresholds are that are linked to connectivity. And then how these align with top-down metrics.

Source: Modified from Taka Sayama.

more effectively with the aim of ultimately determining key thresholds, identifying storage units and understanding connectivity patterns.

Finally, like a medical doctor, your skill at diagnosis will vary with your experience: the more catchments that you work in, the better you are able to draw upon your other "clinical experience" to recognize patterns and behaviors. And I believe that the goal for a field hydrologist is to, over time, develop perceptual models of many places.[15] To develop a latticework of perceptual models. Because often the quality of our hydrological thinking is proportional to the perceptual models that we carry around in our heads. The bigger your field bag of mental models, the better your ability to design experiments to interrogate and diagnose new places and develop better predictive models.

Lastly, perhaps the best use of one's time at a new catchment is to walk it, to dig a soil pit or two and to see the rooting depth of vegetation, what the soil depth might be and what the soil–bedrock interface might look like. And especially to *see it*[16] during rain and/or snowmelt events to get an intuitive sense of how precipitation inputs are handled. It's like visiting Saskatoon, Canada in summertime. If that is your only visit to the city you might believe that it is a hot, dry place. But anyone who has visited Saskatoon in January can tell you that it can be wildly different with −40°C and blowing snow. So, there is no substitute for the field knowledge you gather on site and under a range of conditions. And above all, to think. Think about where the water goes when it enters the soil. Can all the snowmelt or rainfall infiltrate into the soil? How much water can the soil and subsoil store? By what means and how fast can the soil drain? These are great starter questions. And tractable with simple experimental methods as outlined by Scherrer et al. (2007). Building on these initial questions: What flow pathways seem to occur at different catchment positions? How flashy is the stream? Can you see any bedrock springs that sustain low flow? What water are plants returning to the atmosphere? How is that water linked to streamflow generation (or

[15] Of course, there is also value in becoming an expert for one catchment, especially if it allows you to interact with other disciplines at the same site to push the frontiers of interdisciplinary knowledge.

[16] van Stan et al. (2023) is a useful discussion of the importance of being out in the rain.

not)? Such questioning and research visits can pay enormous dividends in ultimately how you attack the subsequent instrumentation types and locations at the site for longer-term, more detailed analysis.

9.3 Getting to grips with heterogeneity in the critical zone

> *To doubt everything and to believe everything are two equally convenient solutions; each saves us from thinking*
>
> ***—Henri Poincaré***

Understanding streamflow generation in headwater catchments is thwarted by two factors: the lack of data and the extraordinary levels of heterogeneity (Figure 9.4). We discussed in Chapter 1 the fact that headwater streams are largely ungauged. In the USA alone, there are >14 million first- and second-order streams (Jay Christensen, EPA, pers. comm., 2025). How we have come to understand them functionally is via the field-based process work and perceptual model development in a small number of research catchments. Countries around the world have extensive plot networks to measure infiltration (and erosion).[17] Many countries around the world have extensive stream gauging for their large catchments and river basins (and as we noted in Chapter 1, ~70% of gauging stations are on fourth- to sixth-order streams). And now there are even remote sensing products to give us flows on larger-order rivers from the Surface Water and Ocean Topography (SWOT) mission with a 21-day observation interval for rivers >50 m width (Getirana et al., 2024). But the headwaters and their hillslopes remain the terra incognita since they suffer the dual problem of low data availability and high process heterogeneity. That is why this book has focused on providing a narrative of the hydrology of headwater catchments from the various fragments of information that we now have.

As we discussed in the previous section, fill-and-spill offers some hope for dealing with the challenge of heterogeneity; we don't need to measure everything everywhere. We can instead focus on measuring or finding ways to estimate the hotspots for connectivity and the key thresholds at the scale of the problem that we are interested in. To go into the field knowing that the larger-scale behavior is not controlled by averages, but often by critical "choke points" that must be exceeded. These choke points are the critical features to be found and identified. But still, heterogeneity is a major issue.

Going forward, we need to understand this heterogeneity.[18] And to know that a hillslope or headwater catchment is not a linear superposition of well characterized soil cores![19] McDonnell et al. (2007) have argued that in order to bring greater coherence to catchment science and streamflow generation research we might benefit from exploring the organizing principles that might *underlie* the heterogeneity and complexity, including the use of comparative analysis, classification, optimality principles and network theory—all with the intent of defining, understanding and predicting catchment function and enunciating important catchment functional traits. Indeed, such work is now being started; binning catchments into simple, intermediate and complex based on their functional traits (e.g., Sharif and Ameli, 2025). But much more needs to be done.

[17] Many countries now have networks of overland flow plots to characterize point-scale overland flow—often reported in "plot years"; meaning the number of plots times the number of years of record. The USA has >10,000 plot years of data (Wischmeier and Smith, 1978; Flanagan et al., 2007); Brazil some 3500 plot years (Anache et al., 2017).; China 2800 plot years (Guo et al., 2015); and Germany over 1000 plot years (Auerswald et al., 2009). So, unlike the hillslope scale and headwater catchment scale there is rather abundant data. Consequently, our understanding of where and when infiltration excess overland flow is produced, at least at the plot scale, has advanced considerably.

[18] And perhaps say what we mean by the word heterogeneity. The dictionary might say that it means "diverse in character or content." For hillslopes and catchments, this relates to things like soil depth, storage distribution, subsurface porosity and permeability. And I'd be remiss not to mention too the range of hydroclimate conditions that exist: from the northern Atacama of Chile with ~1 mm per year of annual rainfall to the Western Ghats of India with over 12,000 mm of precipitation per year. Or the Canadian Precambrian shield where unweathered rock is largely exposed at the surface (totaling an area the size of Brazil) to the Chinese Loess Plateau where 100 m of unsaturated zone overlies rock (covering an area the size of France). One could go on with examples related to topography, soils, drainage density and other geomorphic factors. The combinations and permutations are staggering.

[19] Sadly, often the model-based view.

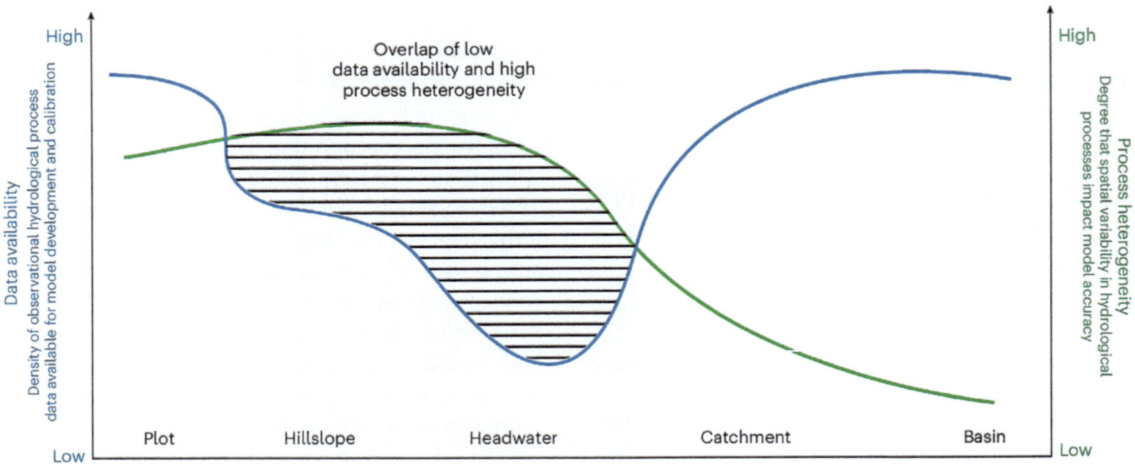

Figure 9.4 The overlap of poor data availability and high process heterogeneity: hillslopes and headwater catchments.
Source: Modified from Golden et al. (2025).

Perhaps the biggest opportunity going forward is to reframe heterogeneity as structure. And get to the root of the heterogeneity problem: the permeability architecture of the full subsurface—the critical zone. This gets at the co-evolution of climate and geology and topography; how plants plumb the critical zone (Sidle et al., 2000; Brantley et al., 2017) and how bedrock weathering produces saprolite and soil (Heimsath et al., 1997). It also gets at why heterogeneity exists. I'm a huge fan of Adrian Bejan and his constructal theory (Bejan, 2007; Bejan and Lorente, 2011). And his take on why heterogeneity and diversity of water flowpaths exist; and his theory that heterogeneity is one way that "nature tells us how she works." In his words, "the unbounded freedom, from which the superposition of numerous flow designs, all observable and evolving toward easier access, keeping what works, and going with the flow" (Bejan, 2024, p. 9).

We've talked a lot about preferential flow. Soil is a living entity. We discussed this in Chapter 8 where at the macro end of the preferential flow spectrum are soil pipes; and the micro end is the rhizosphere. How can we bring biological and geomorphological thinking into what defines the permeability architecture? You have seen references to preferential flow in pretty much every chapter of this book. And yet, here we are, already in 2026 with no real way (beyond dual porosity/permeability models that are really are just tweaks to the soil hydraulic conductivity) to deal with preferential flow in any sort of theoretical way. We have seen in Chapters 2, 4 and 6 how preferential flow can enhance subsurface stormflow, and of course, vertical infiltration. Indeed, Beven has argued that (despite the early recognition of the importance of preferential flow[20]) we've had a century of denial of the existence and impact of preferential flow (Beven, 2021). And yet preferential flow into the soil, within the soil and weathered rock, is now almost axiomatic. Old ideas about soil are rapidly changing (Gao et al., 2023) and hotly debated (Zhao et al., 2024). As are ways to conceptualize and model preferential flow (Jarvis et al., 2016 vs. Germann, 2017). Recall from Chapter 8 the good, recent definition of soil as "an interconnected network of interacting entities, micro- and macrobiota, existing, growing, and communicating over a pore network" (Hunt et al., 2024, p. 1)[21] I think that we need to lean into these new definitions of soil; and into new biologically based thinking.

[20] See for example the early drawing by Engler (1919) in Chapter 2.

[21] And as we noted in Chapter 8, in hydrology, we "dumb down" the complexity of what soil really is to deal with it based on pure physics. A recent Special Issue of the *Vadose Zone Journal* examines this (edited by Najm et al., 2019) and new work shows how they can evolve over time (Hartmann et al., 2020).

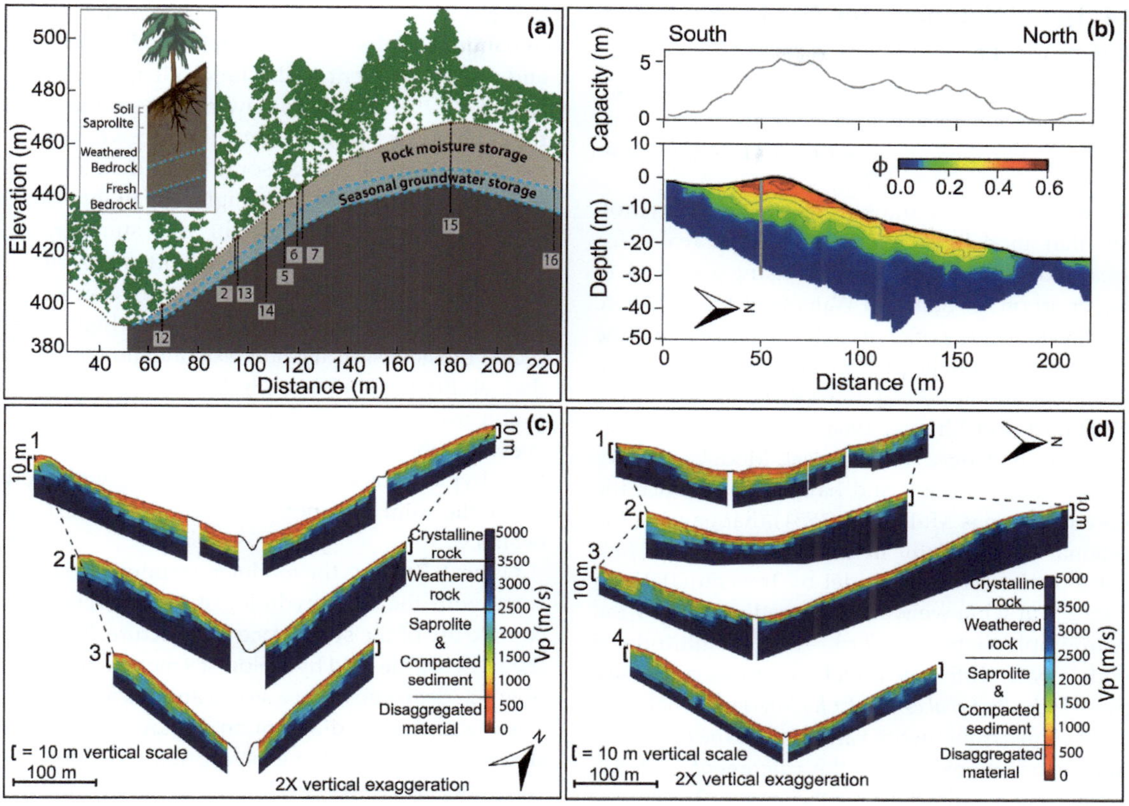

Figure 9.5 New geophysics techniques to describe the heterogeneity in the geophysical structure of our hillslopes and catchments. Increasingly, these are becoming standard bottom-up measures for subsurface storage and how is it distributed across the catchment. The four examples here are from the USA Critical Zone Observatory (CZO) network from (a) interpretations at the Eel River CZO, (b) seismic surveys at Southern Sierra CZO), and (c) and (d) seismic velocity surveys across different transects at Boulder Creek CZO.
Source: Fan et al. (2019).

Beyond getting to grips with heterogeneity at the pore scale, we need to get to grips with heterogeneity in the geophysical structure of our hillslopes and catchments. New work is getting at different testable hypotheses about the deep critical zone and its evolution. Riebe et al. (2017) have laid out new ways of thinking of how tectonic and topographic stresses control the distribution of say, open fractures, in the subsurface. Or how pore fluids from the bedrock drain and activate chemical weathering at the bottom of the critical zone. And how aspect- and climate-driven differences in freeze–thaw may control the patterns of subsurface rock "damage." And how erosion rates and water residence times control the degree of weathering and the thickness of the critical zone. Geophysical measurements like ground penetrating radar, electrical resistance tomography, etc., are helping us test different hypotheses for subsurface structure and heterogeneity, as shown in Figure 9.5. Such measurements help us with basic questions like: Are the subsurface interfaces sharp or gradational? How much rock moisture might there be and how much does the water table in the weathered bedrock vary seasonally? What is the total vs. dynamic water storage and how might this relate to subsurface porosity?

9.4 The dialog between field hydrologist and modeler

> *What is important is seldom urgent and what is urgent is seldom important*
>
> ***—Dwight D. Eisenhower***

There is a pervasive thinking in catchment hydrology that modeling is the coin of the realm for streamflow generation.[22] This thinking goes back to the report on the Stanford Watershed Model where the developers are often noted to espouse that "the ability to accurately predict behavior is a severe test of the adequacy of knowledge in any subject" (Crawford and Linsley, 1966).[23] This book fundamentally challenges that mindset. Models can give us estimates of runoff and have done so since the work of Thomas Mulvaney (1851) that gave us the Rational Method (still in use today) and the first distributed catchment model by Imbeau (1892) in France. But what we want and need is not only the hydrograph shape and form—that a multitude of models with different model structures can easily give—but what *makes up the hydrograph*: where the water comes from (in time and space), the residence time distributions within the catchment and the transit time distributions for the stream. To achieve this, we need better dialog between the field hydrologist and modeler[24] to get perceptual models at the center of the modeling efforts, as advocated by Wagener et al. (2021) and Beven and Chappell (2021) and as shown in Figure 9.6.

A visual example of how field work can influence the model structure is shown in Figure 9.7. Here, a field hydrologist is seeking to develop a perceptual model of the critical zone in a headwater catchment of the Doki river in Japan. The challenge at this site is to develop a bespoke or tailored catchment model that can predict streamflow generation and stream dynamics to inform a new flash flood forecasting and risk assessment tool. The challenge—and the reason for the need for a new model—is that with rapid global change, this area now receives unusually large rainfall events (some more than 300 mm in 24 hr) with exceptionally high rainfall intensities (some on the order of 80 mm/hr) each summer. The streams respond to these events, with staggeringly high runoff ratios of ~90%(!); and with *no* evidence of *any* surface runoff on these steep and highly dissected hillslopes. As a result, existing models fail in their ability to simulate both the "normal" rainfall–runoff events and the few unusually high rainfall events with the extraordinary runoff ratios. The team hypothesizes deep bedrock groundwater may affect the rainfall–runoff behavior for these large events.[25]

Figure 9.7 shows the modeler examining the rock cores with the field hydrologist to appreciate the complexity of the subsurface architecture. Together they discuss the field hydrologist's mental picture of how the deep subsurface stores and releases water. The hand-drawn diagram in Figure 9.7b is the sharpening and evolving perceptual model of this site. After discussion and refinement, the modeler then develops a model structure that honors this perceptual model and decides on the equations to use to get the code running on the computer, before moving on to model calibration and validation.

But beyond the dialog between field hydrologist and modeler as it relates to model structure, we need a new dialog on how to test our models; to know if they are working for the right reasons or not. Hewlett and Troendle (1975, pp. 21–22) said 50 years ago that "accurate prediction of the headwater hydrograph implies adequate modeling of sources, flowpaths and residence time of water and solutes." Recent calls by McGuire et al. (2024) echo this where they note that "to build better-informed and more realistic hydrologic models" we need to collect simultaneously hydrometric and

[22] Analogous to the great quote by Edwin Thompson Jaynes that "the essential content...does not lie in the equations; it lies in the ideas that lead to the equations."

[23] Note that I do not include page numbers here. I have searched the book myself for this exact quote. And while much of its Chapter 1 relates to this, it seems that the exact quote comes from Milton Friedman in the field of economics some 10 years or so earlier in a 1953 essay "The methodology of positive economics."

[24] Of course, today, much of this dialog is with oneself since models are easily accessible to all. And we all are using them in one form or another.

[25] Although an alternative hypothesis is that the high runoff may be associated with shallow subsurface stormflow where interfaces become restrictive only during the exceptional rainfall intensities, as we saw in Chapter 4 with the early work of Whipkey.

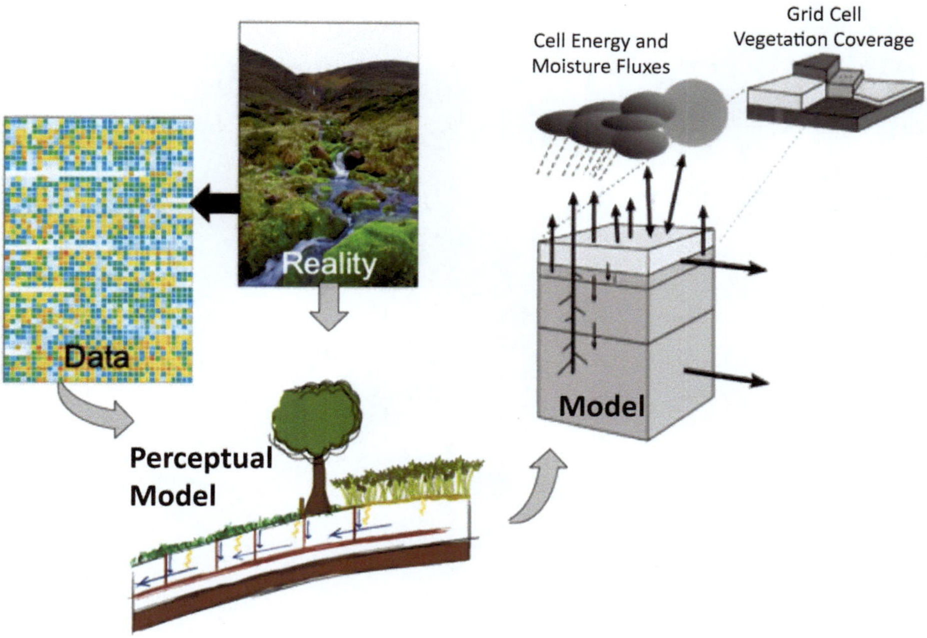

Figure 9.6 The cycle advocated in this book—with perceptual models at the center of data collection and model development and testing.
Source: Thorsten Wagener, 2025, pers. comm.

tracer transport data to inform pore water velocities (McGuire et al., 2024, p. 4). This is the underlying message to modelers in this book: that we need make use of isotope tracers in models and the modeling activity if we are going to make progress with useful and credible model testing.[26]

We now know that tracer-enabled models are a pathway to represent celerity *and* velocity—the combination that is critical to model believability by the experimentalist. And Birkel et al. (2025) now lay out a step-by-step outline for how hydrologists can begin to bring in isotope enabled models into their toolkit. Let's consider the importance of all this going back to the streamwater ages we examined in Chapter 7. If you had a model that perfectly matched a hydrograph but the water in the stream before the storm was some weeks old when the actual mean stream transit time was 50 years, then you might be prompted to say: "Houston, we have a problem." But without such output in the model, you would have blind spots.[27]

We can formalize the dialog between field hydrologist and modeler through fuzzy measures of parameter value acceptability[28] (Figure 9.8). These fuzzy measures, or "soft data" (Seibert and McDonnell, 2002)[29] can enable field hydrologists and modelers to talk the same language as it were, through

[26] But here we still have some ways to go as Birkel et al. (2025) note that there is no mention of isotope-enabled hydrological models even in the latest IPCC report!

[27] Here again I refer the reader to the many recent think pieces by Beven and especially the Beven (2025) paper that outlines the history of philosophies of hydrological model evaluation and hypothesis testing.

[28] Note this is consistent with the "models of everywhere concept" as outlined by Beven (2007) and Beven and Alcock (2012) where, in their examples, local "stakeholders" are shown things like detailed model output visualization and they can provide feedback on whether or not those dynamics fit with their experience of the place. My hope is that for streamflow generation research, we can go beyond this more general stakeholder engagement and get to a specific and detailed dialog between field hydrologist and modeler.

[29] A term introduced by Seibert and McDonnell (2002), soft data are a rather different type of information than traditional hard data measures (a streamflow hydrograph, a well level time series, etc.). Soft data are often "spotty," discontinuous

Figure 9.7 The dialog between field hydrologist and modeler in action. (a) Prof. Yuki Matsushi, field scientist and hydrogeomorphologist on the left discussing with civil engineer and modeler, Prof. Taka Sayama, Kagawa Prefecture, Japan; (b) a sketch of a perceptual model based on the deep cores that will ultimately form the basis for a new numerical flood model.

Source: Photo by McDonnell.

variables and parameters. Soft data is often qualitative knowledge from the field that cannot be used directly for model calibration (or validation). Soft data can sometimes be based on hard data (i.e., direct measurements) but some interpretation of these measurements by a field hydrologist may be needed to make full use of them (e.g., a time series of isotope data for a storm used for hydrograph separation). Figure 9.8 shows how the field person can help define the model evaluation rules and then use something like a fuzzy membership function (the graph part of Figure 9.8) to "punish" the model if it chooses values outside this desired range based on field experience. Examples of soft data could be the percentage of new water at the peak of the storm hydrograph or the mean age of streamflow or perhaps some range of soil depth or saprolite depth as measured in the field. Soft data can then augment the hard data like measured streamflow or water table levels as evaluation measures.

Let's now look at an example of the implementation of this soft data testing of a model as shown in Figure 9.9—one of the first examples of where age information (soil water residence time and catchment mean transit time) was used as a model evaluation criterion. This example is from the Maimai catchment in New Zealand that we examined in previous chapters. It shows how a very simple three-parameter model (Model 1 in Figure 9.9a) can achieve a very good fit for the storm hydrograph. Meaning, the y-axis (represented by a standard efficiency measure) shows values in excess of 0.8, where 1.0 is a perfect fit. Great, right? But

and numerically approximate—but are some of the key signatures for the experimentalist, like % event water in the storm hydrograph or water age in the soil or stream.

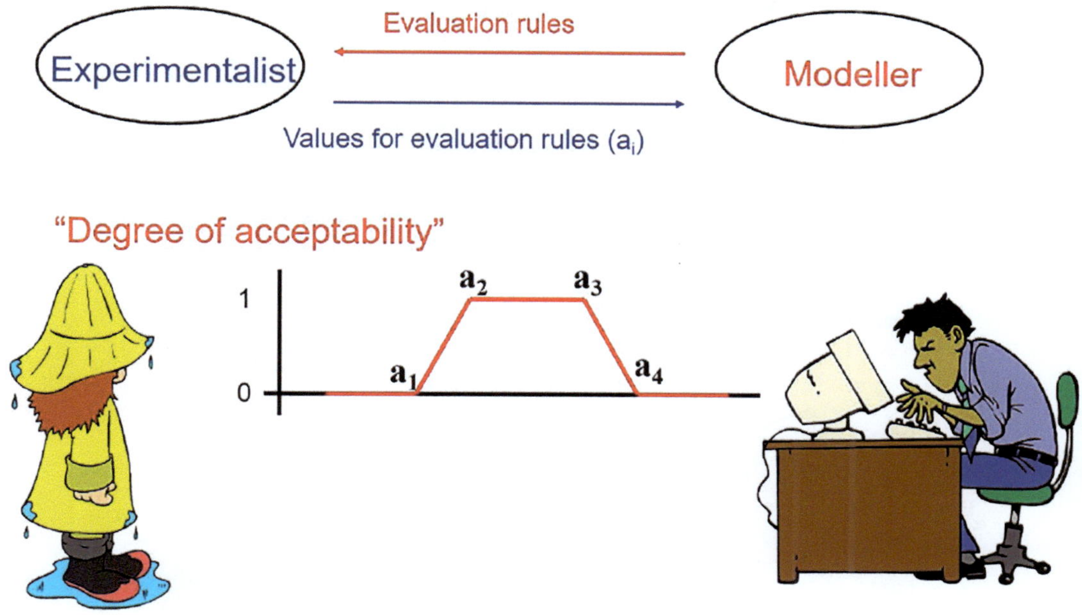

Figure 9.8 A framework for a formalized dialog between experimentalist and modeler using fuzzy membership functions as a way to assign values to the soft data (from Seibert and McDonnell, 2002).

Source: Original figure drafted by Jan Seibert and used with permission.

not so fast. If we look at the two other measures that we know now are essential for believability (the soil water residence time and the catchment mean transit time), Model 1 is hopeless: it gives the right answer (i.e., the hydrograph fit) for the wrong reasons (i.e., fictitious age ranges).

Now look at Model 2 in Figure 9.9b. It shows how including effective porosity as an additional parameter influences the age measures, as does the inclusion of an explicit unsaturated zone volume in Figure 9.9c. Finally, Figure 9.9d, a six-parameter model including each of these parameters, successfully reproduces both the discharge dynamics and the water ages. True, the inclusion of additional parameters increases the overall model uncertainty. But now, the inclusion of residence time and transit time in the evaluation of the accepted model results in a model working for the right reasons[30]; one that matches the streamflow hydrograph well, estimates the soil water age reasonably well and gets the transit time roughly right. But the critical point is that rather than obsessing about the goodness of fit for any one of these measures, the three measures together help align to our understanding of how the catchment stores and releases water.

I admire so much the early TOPMODEL work that really tried to marry model structure with field observations; in the case of TOPMODEL where saturated wedges and topographic convergence from the studies that we examined in Chapter 3 guided a simple, semi-distributed model structure (Beven and Kirkby, 1979). And then with increased evidence that threshold relations existed between slopes and streams and that water tables did not flex up and down in linear way (from studies we reviewed in Chapter 6), how the original

[30] The keen observer might note that with the spread of the green dots, a more accurate statement is that Model 4 is working less right but for the right process reasons. These tradeoffs are discussed in detail in the original soft data paper by Seibert and McDonnell (2002) where they note that such tradeoffs are "worth it" in most cases if the model goal is process realism.

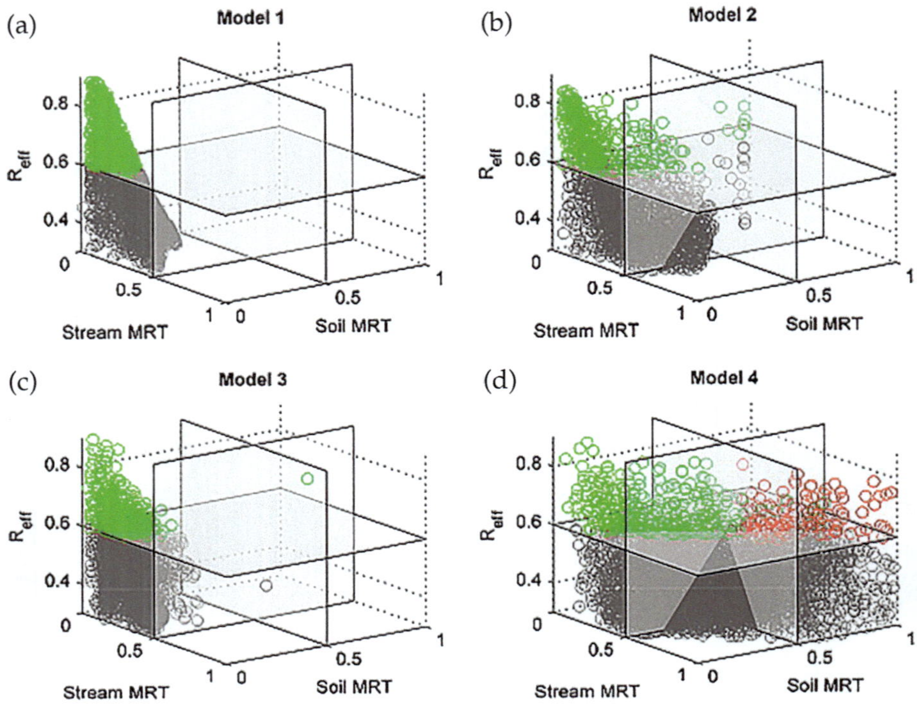

Figure 9.9 Multicriteria plots of the efficiency for runoff (R_{eff}), soil water residence time (soil MRT) and catchment mean transit time (labelled as "Stream MRT") for four different model versions. The upper right octant is the goal: where runoff efficiency is 60% and the residence times and transit times are within 50% of the isotopically derived values.

Source: Vaché and McDonnell (2006).

TOPMODEL was replaced with a "dynamic TOPMODEL" that honored these new research findings (Beven and Freer, 2001). Other models and approaches that have learned from field work to inform model structures include Wrede et al. (2015), Fenecia et al. (2011) and some versions of HBV that are structured to match field data (e.g., Seibert and McDonnell, 2010). Fast forward to today and there are indeed process models that integrate the latest field observations and perceptual models. Perhaps the best example is the EcH$_2$O-iso model (Maneta and Silverman, 2013) and its recent implementation by the team of Doerthe Tetzlaff and Chris Soulsby (Kuppel et al., 2018a, b; Correa et al., 2020) and the STARR Model (Arciniega-Esparza et al., 2023).[31]

Finally, while modelers can learn from field hydrologists, it is important to realize that field hydrologists can learn from modelers to understand mechanisms that are hidden from view or beyond measurement time- or space scales. There are two very personal ones for me. The first is related to the fill-and-spill behavior on the Panola hillslope that we discussed in Chapter 6. We had completed a very detailed mapping of the soil depth on the Panola hillslope and Ilja van Meerveld (then a new PhD student with me), used HYDRUS 2D in a classroom exercise to explore and visualize the 2D flow behavior at the mapped soil–bedrock interface on the Panola hillslope. The model output showed patches of saturation filling and spilling and cascading downslope—in ways that we could not believe.

[31] But likely 99% of catchment modeling studies and applications still use the matching of the stream hydrograph as the *only* evaluation criteria—something we know now from this book that represents only half the story (i.e., the celerity) of the streamflow generation response.

Those early simulations guided the later placement of the many dozens of wells on the slope that would help field test and fully understand the fill-and-spill behavior. The other very personal example where I learned from a model is from the Maimai catchment. This is where a model helped me understand pipeflow generation at the hillslope scale, as shown in Figure 9.10. This is model output from Weiler and McDonnell (2007) where a simple distributed hillslope runoff model (Hill-Vi) explored different realizations of potential pipe networks at the Maimai hillslope. Of course, all this was hidden from view in the field. But the model was an extraordinarily useful learning tool to show us how randomly placed soil pipes could conspire with the subsurface topography to produce emergent network behavior at the hillslope scale.

But beyond my experiences, there have been—and continue to be—numerous great examples where models have been used to test field-based hypotheses. One example I will leave you with is linked to the themes discussed in Chapter 8 regarding ecohydrological separation. And this is where models have really helped explore the reasons for "two water worlds" behavior and where, when and why one might expect it, or not. Figure 9.11 shows the first testing of the ecohydrological separation hypothesis with a reservoir-based model by Cain et al. (2019) where they compared two conceptual rainfall–runoff models that included catchment-scale representations of flow and transport: one that did incorporate ecohydrological separation and one that did not. The models were developed for the site where ecohydrological separation was first found, the H.J. Andrews Experimental Forest, in Oregon, USA. Their model structural variations (as shown in Figure 9.11) resulted in "large differences in catchment storage volumes and fluxes" (Cain et al. 2019, p. 2658). These changes in the model structure influenced their interpreted residence times for the water generating streamflow.

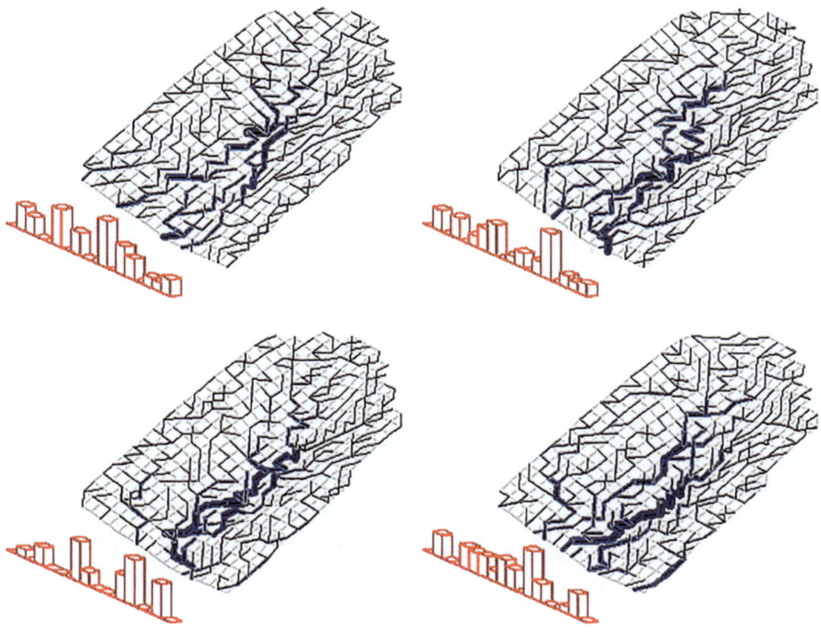

Figure 9.10 Simulations for the Maimai hillslope showing network behavior arising from random soil pipe placement. The thickness of the blue lines represents the relative amount of water flowing in the individual pipes. The red bar graph at the base of the hillslope shows the relative pipe outflow from each grid cell.

Source: Weiler and McDonnell (2007).

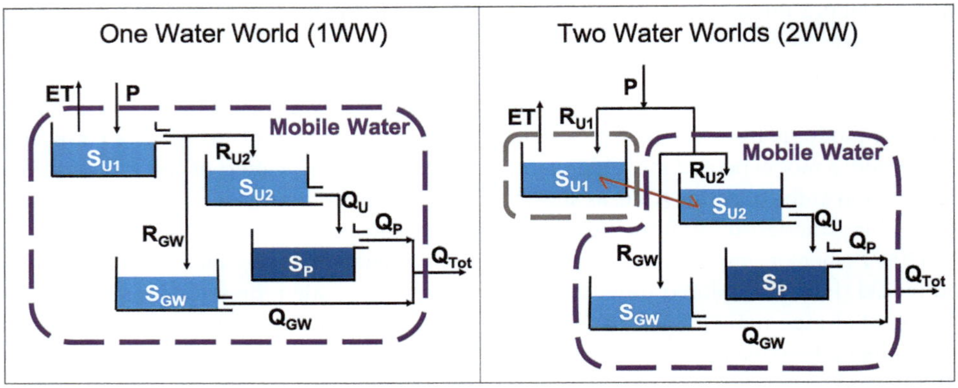

Figure 9.11 Model structures for (a) one water world (1WW) and (b) two water worlds (2WW). All water is mobile for 1WW, but the plant available reservoir SU1 in 2WW is isolated from outflow to the stream. The red line indicates chloride mass transfer between the unsaturated reservoirs SU1 and SU2 in 2WW.

Source: From and quoting directly, Cain et al. (2019, p. 2661).

Others have followed the Cain et al. (2019) analysis to use models as learning tools to explore how the amount of preferential flow influences the two water worlds hypothesis. Radolinski et al. (2021) found that distinct ecohydrological separation may require extreme rainfall intensities. Similarly, Finkenbiner et al. (2022) used a Darcy–Richards based model to test the physical drivers of the ecohydrological separation hypothesis. Unlike anything one could do in the field, they used the model to explore 650 different model configurations of soil, climate and mobile/immobile soil–water domain characteristics. What they learned from the model was that coarser soils and less precipitation led to "reduced separation between pore spaces and drainage" (Finkenbiner et al., 2022, p. 1). Most importantly, despite using the latest and most sophisticated soil physics model, they concluded that conventional "soil physics alone are unlikely to explain large ecohydrological separations observed elsewhere, and further efforts aimed at reducing methodological artifacts, refining understanding of fractionation processes, and investigating new physiochemical mechanisms are needed" (Finkenbiner et al., 2022, p. 1). These examples show great promise to field hydrologists—to use models as learning tools, to test ideas, to help design and inform new field experiments and to explore ranges of behaviors and conditions not possible in the field or on the timescale of natural events.

9.5 Perceptual model urgency in a changing world

Pessimism of the mind, optimism of the will
—Romain Rolland

While I am cautiously optimistic about the future dialog between experimentalist and modeler, I am cautiously pessimistic about our changing world in a hydrological context vis-à-vis streamflow generation. The changes are everywhere and well documented.[32] And just to remind ourselves at the most simple level, a warmer atmosphere can hold more water vapor than a cooler atmosphere. Quoting US Geological Survey projections, at the current mean global surface air temperature of 14°C, the vapor pressure increases about 7% for each one degree increase in temperature. That means that all things being equal (e.g., winds, cloudiness, etc.) we may expect an intensification of the water cycle of 7% per degree of global warming. Of course, this means more intense droughts in some

[32] A very good read is Milly et al. (2008) on non-stationarity and its impact on water resources management.

regions and stronger rainfall in others, as increasingly documented for streams and soil moisture—both changing in unprecedented ways compared to pre-industrial times (Porkka et al., 2024). The last 10 years have been the hottest 10 years in the nearly 200 years of record keeping according to the World Meteorological Organization. And, at the time of this writing, the first paper has been published to show conclusive proof of global-scale terrestrial water-storage changes and an irreversible decline in soil moisture (Seo et al., 2025).

Related to streamflow generation, we are seeing many things linked to climate change, including:

- Increased intensity of individual rain events
- Increased rain event amounts
- Decreasing soil moisture levels (with decreased pre-storm antecedent wetness conditions)
- Increasingly ephemeral and disconnected headwater streams
- Increased glacial melt and disappearance
- Increased and widespread permafrost thaw
- Increase wildfire occurrence and severity of affected surfaces with altered infiltrability
- Wildfire altered evapotranspiration and catchment water cycling dynamics
- Rapid "shrubification" of high-latitude catchments
- Decreasing cloud forest conditions in the humid tropics as cloud bases rise in a warming world.

I could go on. And I haven't even mentioned land use change and land degradation!

There is therefore a special urgency to refine process descriptions and perceptual model development today. This is set against a backdrop of a declining number of field studies and a bias where field work is done and where data are collected (Burt and McDonnell, 2015). Figure 9.12 shows a stunning analysis by Li et al. (2020) for 8089 gauged catchments that showed that most are undergoing change—up or down. And because, as we now know, most of the flow in large rivers comes from headwater catchments, this change is mostly in the headwaters! There is the issue of gauge site bias, as noted by Krabbenhoft et al. (2022, p. 586) as these are located "disproportionally in large, perennial rivers draining more human-occupied watersheds." But like many of the themes of this book, they also note that "gauges are sparsely distributed in protected areas and rivers characterized by non-perennial flow regimes, both of which are critical to freshwater conservation and water security concerns."

Our job now is urgent—to characterize how systems work and to understand how they will change in terms of how they store, mix and release water. And it is not just that the precipitation intensity, duration and amount are changing. The nature of precipitation inputs is changing; many once-snow-dominated catchments are increasingly rain-dominated.

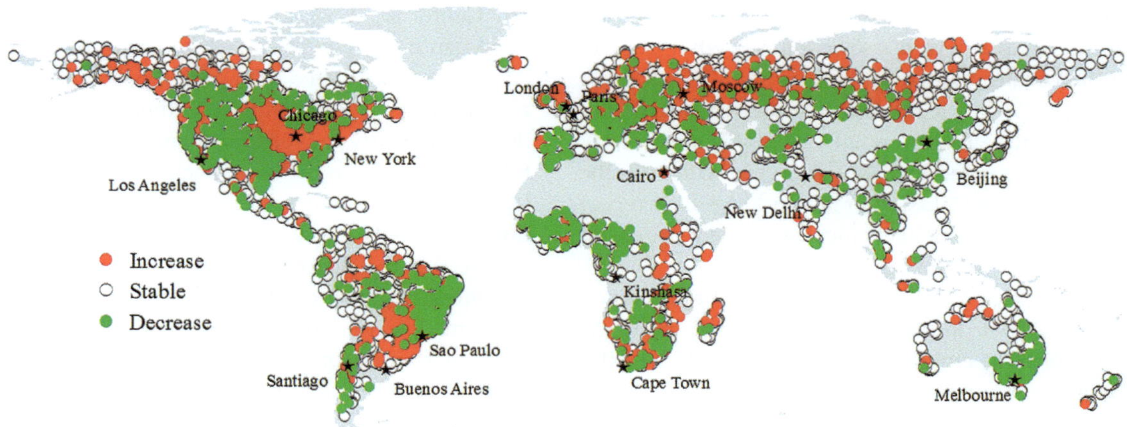

Figure 9.12 Observed changes in streamflow from 8089 gauging stations on 4307 large rivers.
Source: Li et al. (2020).

All this is playing out across the different land uses that comprise our headwaters. This book has been restricted mostly to "natural" landscapes. But such "undisturbed" areas are shrinking fast. Beyond forest removal and replanting, conversion to agriculture and rangeland or suburban areas are affecting soil infiltrability and runoff production across vast swathes of the globe. Much new work is going on in the context of human-modified landscapes (e.g., Tetzlaff et al., 2024). But no global data exists for landuse in the headwater catchments. Recent analysis by Lane et al. (2025) shows, for the USA, the landuse of first- and second-order catchments: forest (25%), shrubland (25%), pasture, hay and cultivated crops (25%) and grassland (13%). These landuses are changing at unprecedented rates.

Globally, streamflow is becoming less seasonal (Wang et al., 2024). Levia et al. (2020) have noted how landuse and land cover changes are accelerating and creating a homogenization of the catchment water cycle. Locally, the all-important boundary conditions—at the surface for infiltration excess overland flow—and in the subsurface for subsurface stormflow are changing; *rapidly*. Be it from altered infiltrabilities due to fire or land cover change, or increased soil water storage due to permafrost thaw. There are also recent suggestions that macroporosity will increase in a warming climate (Hirmas et al., 2018). And linked to these macroporosity changes, rainfall intensity changes combined with higher CO_2 levels could mean significant shifts for plant water selection vs. what goes to groundwater storage and streamflow (Radolinski et al., 2025). While this book has been mostly about catchments with perennial streams, Zimmer et al. (2020) have noted that 20% of gaging stations in the Global Runoff Data Center network regularly report zero flow. This will likely increase in the years ahead as the world warms (Datry et al., 2023).

Think back to all the processes we have learned about. What might occur with such changes? Easy ones to envision might be rainfall-intensity-dependent interfaces where some places that rarely produced infiltration excess overland flow will in the future produce extensive overland flow during extreme events. Or some hillslopes where there was not an interface with a sufficient hydraulic conductivity reduction to generate lateral flow, now, under much higher rainfall intensities, will in the future be sufficient to divert flow laterally.

To pick just one example from the long bulleted list earlier in this section, let's think about fire. Perceptual models of changes to headwaters resulting from wildfires focus on their impacts on surface infiltration and removal of vegetation to protect the soil surface. "Floods follow fires" is an old adage that speaks to a common set of factors where increased rainfall loading (due to removal of canopy interception and transpiration) meets a newly fire-induced hydrophobic (water-repellent) soil surface. This can produce infiltration excess overland flow where there may not have been any previously. But some of the latest field-based discoveries about fire and runoff are showing something different. In a landmark paper, Atwood et al. (2023) show how fire can in fact, *enhance* infiltration and possibly increase subsurface stormflow on steep burned slopes—changing the longstanding narrative that infiltration excess overland flow due to hydrophobicity is the sole change agent for flooding (Figure 9.13). Such findings are wildly at odds with our current models where the best and most comprehensive models of fire alteration of runoff pathways assume a post-fire overland flow conversion. For example, Yu et al. (2023) used the Kineros model[33] to show the simulated impacts of wildfire on flood frequency. For the upper Arroyo Seco catchment in Southern California, USA they noted how a 100-year flood can be "three times larger" than their simulations that considered only peak flows in non-fire-affected years. The open question then is why? Why do we see nearly universal increases in peak flows following fire if, in fact, the mechanism for hydrological alteration may not directly involve overland flow, as shown by Atwood et al. (2023)? While there has been some dialog between field hydrologists and modelers vis-à-vis fire (e.g., Seibert et al., 2010) much more is needed to develop models for burned areas that work for the right reasons. Here stable isotopes could "come to the rescue" where simple

[33] For the latest overall review paper on models and fire, see Ebel et al. (2025).

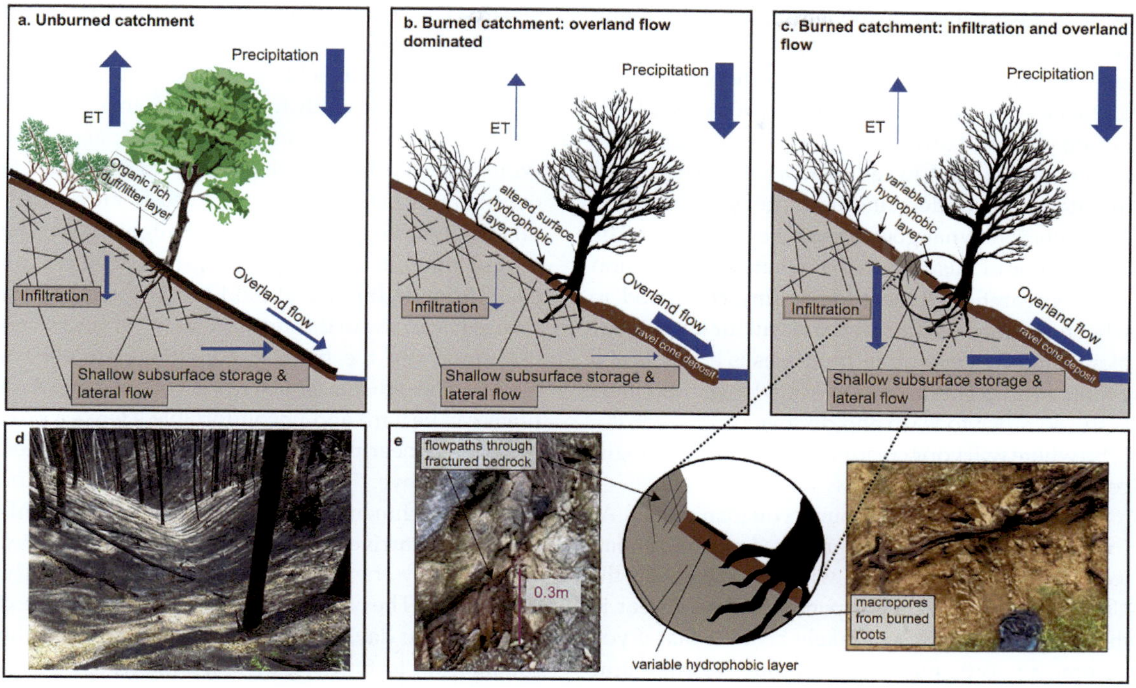

Figure 9.13 Fire, hydrophobicity and production of infiltration excess on slopes. With the new work by Atwood et al. (2023) showing how burned soil may also, unexpectedly, increase funneling of rainfall into the soil to enhance subsurface stormflow contributions to post-burn flooding.
Source: Modified from Atwood et al., (2023) with an image from the 2018 Carr Fire, Northern California (d) from East et al. (2021).

hydrograph separations with event and pre-event water could determine whether or not conversion to overland flow occurs in previously high soil infiltrability environments.[34]

9.6 Launching your career in field hydrology

> *The wise person knows that it is better to sit on the banks of a remote mountain stream than to be emperor of the whole world*
>
> ***—Zhuangzi***

If field work and discovering new things about how water flows in small catchments excites you, then I encourage you to make a career of it! That is another motivation for writing this book—to inspire a next generation of field hydrologists. To let them know that we're not finished with field work just yet![35] I believe that your success in research and your effectiveness as a field hydrologist will be squarely linked to your wonderment of the natural world and extreme curiosity about how things work and operate. To enter the field with a child's mindset is a great way to start. And like others have noted, the creativity quotient plus the passion quotient

[34] This harkens back to early comments by Gremillion et al. (2000) where they noted that "whereas the long-term geochemical, ecological and hydrological effects of changing hydrologic pathways (associated with landuse change) may be unclear, methods to characterize flow components have been thoroughly investigated." These words—aimed at isotope hydrograph separation—remain unacted upon but offer much hope for attacking many of the perceptual model urgencies in disturbed environments.

[35] Nearing the end of my field career, I feel like Isaac Newton when he said: "I do not know what I appear to the world, but to myself I seem to have been only like a boy playing on the seashore, and diverting myself in now and then finding a smoother pebble or a prettier shell than ordinary, whilst the great ocean of truth lay all undiscovered before me." Note that I am not comparing myself to the greatest scientist the world has ever known—merely feeling as he felt.

exceeds the intelligence quotient. So, no matter who you are or what your abilities are, if you can develop enough mastery of the basics and know the literature well enough, you are well on your way. Meaning, on your way to describing what is the status quo understanding and to say then what is *wrong* with this status quo understanding; to get you to discovery science; to what goes *beyond* the status quo. The throughline of this book is that what we understand about streamflow generation is tiny compared to what remains to be discovered. That our knowledge of streamflow generation processes is a tiny beach pebble in a vast ocean of non-knowledge.[36] So there is still so much to learn and discover.

So where will your creativity as a field hydrologist come from?

Let the things you see around you inspire you. As I write these last few paragraphs, I am on sabbatical in Japan and working with a professor at the University of Kyoto who spends 80 days a year in the field. Few of you will likely achieve that if you become an academic researcher. But the more time in the field the better. Being still and thinking on a hillslope can be a wonderful thing; to visualize; to foster your creativity. Nobel laureates like Francis Crick have, over the decades, offered up some useful advice on creativity: "avoid the temptation to work so hard that there is no time left for serious thinking." That advice certainly resonates with my ebbs and flows of creativity (McDonnell, 2024). If we survey other Nobel laureates' career advice, I like the Sydney Brenner quote that "innovation comes from an assault on the unknown." Therefore, in field hydrology, go out and ski the untracked powder of streamflow generation research; to ask the questions no one else has yet asked. To push the bounds of our ignorance and confusion.

For what is field research? It is to see what everyone else has seen *but to think what no one else has yet thought* (paraphrasing Nobel laureate Albert Szent-Gyorgyi).

Field work—the focus of all the examples in this book—is the ultimate challenge and the ultimate reward. The book "A Passion for Field Work" (Burt and Thompson, 2020) gives a sense of why that is and how different field scientists have described their relationship with the joys and challenges of time spent out in the rain and snow. To paraphrase William Beveridge (1950): "a spirit of indomitable perseverance has characterized nearly all successful field hydrologists."[37] Today we might replace the phrase indomitable spirit with "grit." Grittiness can come from endurance of field hardships. The best advice I ever saw in this regard was to "bring a sense of humor into the field."[38] For things are always breaking; weather is almost never cooperating. And times can be hard and uncomfortable. Here's a joke, since we're nearing the end of the book and I need to say goodbye: "A graduate student is sent to the Maimai catchment in New Zealand for their PhD studies. She arrived on a drizzly, dark, overcast day. The next day she woke up to find another cloudy, drizzly day. The next day was the same and the next day too. The day after that she emerged from the Maimai hut to find it gray and drizzly; she saw a small boy passing by on the dirt road outside the research cabin and she asked him "Does the weather here ever change?" "I don't know," he says, "I'm only seven!"[39]

As you can tell by now, I am a big believer in inspirational quotes. I used them heavily in the messaging in my early career mentoring book "Navigating an Academic Career: A brief guide for PhD students, postdocs and early career faculty" (McDonnell, 2020). I find such quotes to be useful tools for mentoring. They can also be useful reminders of the benefits of mentorship. I encourage you to get a mentor in your field hydrology career. In fact, the more mentors the better, as they can be sounding boards for your ideas and help you see

[36] If I had to put what we know about streamflow generation on a scale of 1 to 100, I'd say that we're close to perhaps 10/100—meaning, we've only scratched the surface of process knowledge.

[37] to paraphrase William Beveridge from his classic book from the 1950s "The Art of Scientific Investigation".

[38] I believe that I first read this as an undergraduate in Dunne and Leopold (1978).

[39] Knowing that sometimes humor does not translate if English is not your first language, the joke here is that he is implying that he has only lived 7 years and that for all that time it has rained every day! An example of this is one of my PhD students, Chris Gabrielli, who worked at Maimai where he experienced rainfall on the first 26 days of his one-month drilling campaign and was not ever able to dry out his rain gear.

Figure 9.14 The author on his approach to navigating a career in field hydrology over many decades.
Source: Robert Neubecker, used with permission.

around corners; both in your field science and in your career navigation.

Know that a career as a field hydrologist is a hydrological long game (Figure 9.14). You will be part of a chain of learning with your hydrological ancestors—part of my rationale for the dedication of this book to John Hewlett. So, think about how you will be a good ancestor. Good ancestors embrace the American Geophysical Union motto "unselfish cooperation in research" as their north star. Good ancestors offer ideas from what they observe in the field freely without expectation of co-authorship; knowing that what ideas they put out in the world will come back to them tenfold in terms of people wanting to collaborate with them. They criticize in a positive way; they assume positive intent; they are humble in their estimation of what they know about how catchments work.

Finally, remember the immense complexity of the natural world. And that our job as field hydrologists "is to observe, test, measure and, above all think."[40] To remind ourselves that we are lucky to be part of a timeless tradition of field-based hydrological research. To discover new things about how the hydrological world works and to get paid to do it is a gift and a privilege. In our modeling-obsessed world, it is easy to get caught up in numbers. Remember: "not everything that can be counted counts and not everything that counts can be counted."[41] Field work and field discoveries count. And count most for streamflow generation processes and perceptual models.

[40] to quote one last time our old friend Humboldt from the Preface.

[41] to quote Albert Einstein.

Acknowledgments

The process of writing this book has worked out for me in four ways. The first was a series of commentaries I have written starting in 2003 in the journal *Hydrological Processes*. I was at the time, the editor of the new "HP Today" section. In the spring of 2003, I faced a dilemma of the first-ever issue without a commentary. So, on a flight back to Oregon from an EGU meeting, I quickly penned one to avoid a gap. In the intervening ~20 years, I've written several more commentaries and in different journals: on topics ranging from heterogeneity and how to move beyond it, to catchment transit time conceptualization, or how all runoff processes have certain similarities, to getting to grips with velocities, celerities and residence time distributions, tree water uptake and finally—the last puzzle piece needed for me to imagine this book project—the links between scale and characteristic forms of filling and spilling. So, these commentaries have been my way of thinking through ideas beyond individual research papers. And this book is the synthesis of that thinking; a quilting together of the commentary ideas into what I hope is a coherent narrative.

The second way this book has come about is via the delivery of the annual "Catchment Science Summer School," initiated with Doerthe Tetzlaff and Chris Soulsby and first offered at the University of Aberdeen in 2010. It is now hosted at the University of Birmingham by David Hannah and Stefan Krause. There, I have had the chance to lecture on the opening day to over 400 PhD students over the past 16 years. I've found this to be a very useful endeavor to compress my take on processes and perceptual models in catchment hydrology, their history, current state and future research needs into a compact lecture format. That course has given me a sense of the pacing and tempo needed to tell my story of streamflow generation—how much detail to leave in and most importantly, what to leave out. I thank the other co-instructors—Keith Beven, Josie Geris, Rick Hooper, Nick Chappell, Ilja van Meerveld and Jan Seibert—for discussion and mutual learning along the way.

The third way this book has come about is the result of the 2020–2021 Covid pandemic and the changes it made to daily life. This gave me time at home to get started on the book and time without any travel, propelled by my habit of writing daily. This was also a time I could check in—by Zoom and e-mail and phone—with several senior colleagues on their "take" on some of the historical connections and recollections I was considering to include. I am deeply grateful to the people I interviewed, including Carol Kendall, Dave DeWalle, Wade Nutter, Gordon Grant, Tom Dunne, Mike Kirkby, Pierre Hubert, Breck Bowden, Makoko Tani, Keith Beven, Mike Sklash, Tony Jones, Alan Rodhe, Roy Sidle, Rick Hooper, Renee Brooks, Chris Soulsby, Rhett Jackson and Steve Burgess. Thank you one and all for helping me shape my narrative—but of course, any nonsense in this book is my own.

The fourth way this book came about was a series of field visits to iconic catchments that I had not yet seen (and arid sites I'd had rather little experience with) to fill in my understanding of the literature. Chief among these was a late 2019 field trip with Tim Burt to the famous UK field sites of Bicknoller, East Twin Brook and Slapton and a follow-on field trip to the Plynlimon catchment that same year with Ilja van Meerveld and Jan Seibert. These helped me contextualize much of the early and important UK literature. A late 2023 tour of arid and semi-arid hillslope research sites in Spain near Albacete, Alicante and Valencia really helped cement some of my infiltration excess overland flow thinking. I thank my many field hosts (including Susana Bautista, Manuel Esteban Lucas Borja, Antonio del Campo and María González Sanchis) for their generous gift of field time and helping me "see" some final missing links. An early 2025 field trip to Qatar, Saudi Arabia and Dubai to see truly arid sites helped contextualize such systems even further. And at the other climate extreme, thanks to Taka Sayama for two wonderful early 2025 field trips to steep wet sites on Honshu and Shikoku Islands in Japan where outrageous recorded runoff ratios occur, despite no overland flow. The process diversity on this planet is stunning and humbling.

The general idea for a process hydrology book has gestated over the past 15 years. It began as a collaborative idea with Tim Burt. But we could never imagine an outline that wasn't simply an update of Kirkby's 1978 *Hillslope Hydrology* or Tim's own wonderful book (with Malcolm Anderson), the 1990 *Process Studies in Hillslope Hydrology*. In the end, the perceptual model narrative became a deeply personal one—intertwined with my teaching and early career mentoring ideas—where I wanted the writing to reflect a tone and style very different from a typical textbook. In other words, this book is idiosyncratic to my lived experience as a field hydrologist. And I thank Tim for being so gracious in encouraging me to go it alone.

I have many collaborators to thank—colleagues with me, out in the pouring rain and melting snow. My longest suffering collaborator, Carol Kendall, I met at a Gordon conference in 1989. As a proper isotope geochemist, she had much patience with me as a hydrologist. I learned through her the plight of women scientists at a time when there were very few in field hydrology. Carol made me a better mentor. Others who have influenced my thinking as a catchment hydrologist include other senior mentors: Keith Beven, the most consequential and inspiring hydrologist I have ever known; a friend and a patient teacher. Mike Bonell, an early mentor to me starting in New Zealand and continuing while he led the UNESCO program in water. Tim Burt, who's friendship and mentorship I have enjoyed for many decades now—I came of age scientifically following his field work lead—from deciphering his design of a then-new Scanivalve recording tensiometer system to testing out his UK-based ideas on field measurements at the Maimai catchment at the other end of the world—at a time pre-internet when it really did feel like I was alone at the other end of the world. And Rick Hooper, perhaps the most articulate colleague I've ever had and one who could whip a whiteboard discussion into an intellectual storm of ideas.

There are also four other people I need to thank. The first is the dynamic duo of Chris Soulsby and Doerthe Tetzlaff. They have been comrades in arms for 20+ years including our many memorable field excursions to the Bruntland Burn catchment in Scotland. The other two are Ilja van Meerveld and Jan Seibert; two former members of my team and now leaders in hydrology with their field work and field-driven modeling and working together in Zurich. This whole book could be illustrated with these four people's spectacular contributions. Other colleagues who I want to thank and who I have learned from out in the rain are Laurent Pfister, Dan Penna, Rhett Jackson, Sampurno Bruijnzeel, Jamie Shanley, Jake Peters, Roy Sidle, Yoshio Tsuboyama, Barbara Bond, Renee Brooks, Ying Zhao, Bing Si, Don Siegel, Dave Tarboton and Chris Neale. Thank you one and all. And the many people I have written papers with, who I have learned much from.

I am indebted also to all the colleagues who have hosted me for visiting appointments at their institutions: Siva Sivapalan, Huub Savenije, Troch, Yoshio Tsuboyama, Malcolm Anderson, Stefan Uhlenbrook, Laurent Pfister, Davd Hannah, Andres Iroume, Uwe Morgenstern, the late Bethanna Jackson, Maria Poca, Chris Soulsby, Doerthe Tetzlaff, Esteban Jobbagy, Fuqiang Tian, Ying Zhao, Junguo Liu, Breck Bowden, Jagath Ekanayake, John Herbohn, Jan Seibert, Larry Band and the late Wei-Zu Gu. Thanks especially to Takahiro Sayama for arranging a three-month visiting professorship at the Kyoto University in Winter 2025 that enabled me to complete the book.

But it is to my many PhD students and postdocs that I owe the greatest intellectual debt. Their work has taught me the most. Together these former lab members are too numerous to mention here but some who I must particularly thank for teaching me through their work are (in no particular order): Jan Seibert, Markus Weiler, Kevin McGuire, Ilja van Meerveld, Holly Barnard, Willem van Verseveld, Doug Burns, Magali Nehemy, Luke Pangle, Lyssette Munos-Villers, April James, Anna Coles, Jaivime Evaristo, Yi Nan, Chris Gabrielli, Brian McGlynn, Chris Graham, Mike McHale, the late Chris Cirmo, Jim Freer, Scott Allen, Hannah Green, Takahiro Sayama, Luisa Hopp, Cody Hale, Taro Uchida, Fabrizio Fenecia, Daniele Penna, Richard Keim, Willemijn Appels, Julian Klaus, Kellie Vache, Jay Frentress, Nick Hjerdt, Marcel Gaj, Dyan Pratt, Scott Jasechko, Daryl Janzen, Shreeram Inamdar, Pedro Herve-Fernandes, Ali Ameli, Genevieve Ali, Josie Geris, Leo Peskett and Hongxiu Wang.

Finally, I thank Sonke Adlung and John Smallman at Oxford University Press for their support and patience along the way. Thanks to Harvinder Sidhu and Cody Millar for their painstaking figure work, and Kim Janzen for ferreting out old papers and books. Thanks to Keith Beven and Gordon Grant for early valuable feedback and to Menberu Meles, Jeff Curtis and Jesse Radolinski for late-stage useful discussion. Thanks to Kevin McGuire and Magali Nehemy for detailed feedback on Chapters 7 and 9, respectively. I thank Chris Gabrielli for his ever-pointed feedback on early chapter versions all along the writing journey; and special thanks to Ilja van Meerveld for her final overall, detailed examination of the book. The students have indeed become the professors. Finally, I thank my wife Veva for her love and support and patience with my "one-hour work day"[1] obsession. And my now-adult children, John and Meghan, for gamely accepting a travel-focused childhood that included more "holidays" in remote, wet places than Disneyland excursions. Those many working holidays form much of the basis for this book.

[1] McDonnell, J.J., 2016. The 1-hour workday. *Science*, *353*(6300), 718–718. DOI: 10.1126/science.353.6300.718.

About the author

Jeffrey McDonnell is a field hydrologist. He is Distinguished Professor of Hydrology at the University of Saskatchewan in Canada. He is also Visiting Chair in Water Science at the University of Birmingham and Distinguished Visiting Professor at Tsinghua University, Ludong University and North China University of Water Resources and Electric Power. He is an Officer of the Order of Canada, the winner of the Horton Medal from the American Geophysical Union, the International Hydrology Prize from the International Association of Hydrological Sciences, UNESCO and the World Meteorological Organization and the Dalton Medal from the European Geosciences Union. McDonnell is a fellow of the Royal Society of Canada, the American Association for the Advancement of Science and Academia Europaea. He is a fellow of the American Geophysical Union and the Geological Society of America. He has been the recipient of numerous awards, among them the Outstanding Achievement Award of the New Zealand Hydrological Society, the Ray K. Linsley Award for Surface Water of the American Institute of Hydrology and the Gordon Warwick Award of the British Society for Geomorphology. He is the former President of the AGU Hydrology Section and author of the Wiley / AGU book "Navigating an Academic Career: A Brief Guide for PhD Students, Postdocs and Early Career Faculty."

The author (center) in 2015 with a few of his former graduate students and postdocs (right to left: Markus Weiler, Ilja van Meerveld, Richard Keim, Jeff McDonnell, Natalie Orlowski, Julian Klaus, and Scott Allen)

APPENDIX I

The collected works of John Hewlett

Hewlett[1], J.D., 1957. Coweeta Hydrologic Laboratory. USDA Forest Service, Southeastern Forest Experiment Station, Asheville, NC.

Hewlett, J.D., 1958. Pine and hardwood forest yield. *Journal of Soil and Water Conservation, 13*, 106–109.

Hewlett, J.D. and Metz, L.J., 1960. Watershed management research in the Southeast. *Journal of Forestry, 58*, 269–271.

Hewlett, J.D., 1961. Response of fescue to natural moisture gradient on an artificial slope. *In*: *Research Notes*, Editor (Editor^(Editors). USDA Forest Service, Southeastern Forest Experiment Station, Asheville, NC.

Hewlett, J.D., 1961. Soil moisture as a source of base flow from steep mountain watersheds. *In*: *Station Paper*, Editor (Editor^(Editors). USDA Forest Service, Southeastern Forest Experiment Station, Asheville, NC.

Hewlett, J.D., 1961. Watershed Management. In Editor (Editor^(Editors). University of Georgia, Southeastern Forest Experiment Station, Asheville, N.C., pp. 61–66.

Hewlett, J.D. and Douglass, J.E., 1961. A method for calculating error of soil moisture volumes in gravimetric sampling. *Forest Science, 7*, 265–272.

Hewlett, J.D. and Hibbert, A.R., 1961. Increases in water yield after several types of forest cutting. *International Association Scientific Hydrology Bulletin, 6*, 5–17.

Helvey, J.D. and Hewlett, J.D., 1962. The annual range of soil moisture under high rainfall in the Southern Appalachians. *Journal of Forestry, 60*, 485–486.

Hewlett, John D., 1962. Internal water balance of forest trees on the Coweeta watershed. Unpublished Ph. D. dissertation, Duke University, 122 p.

Hewlett, J.D. and Kramer, P.J., 1963. The measurement of water deficits in broadleaf plants. *Protoplasma, 57*(1), 381–391.

Hewlett, J.D. and Patric, J.H., 1963. A pilot test of multiple use on a small mountain watershed. *In*: *Proceedings, 42nd annual meeting, Society of American Foresters, Appalachian section, 8–9 February 1963, Greenville, SC.* Society of American Foresters, Charlottesville, VA.

Hewlett, J.D. and Hibbert, A.R., 1963. Moisture and energy conditions within a sloping soil mass during drainage. *Journal of Geophysical Research, 68*, 1081–1087.

Hewlett, J.D., 1964. Letter to editor on article Groundwater: Definition by H. E. Thomas and L. B. Leopold. *Science, 144*, 1407–1408.

Hewlett, J.D., 1964. Research in hydrology of forested headwaters of the Coweeta Hydrologic Laboratory. *In*: *Twenty-ninth North American wildlife and natural resources conference, 9–11 March 1964, Las Vegas, NV.* Wildlife Management Institute, Washington, DC.

Hewlett, J.D., 1964. Water or forest - can we have all we need of both? *Frontiers of Plant Science, 17*(2 to 3).

Hewlett, J.D., Douglass, J.E. and Clutter, J.L., 1964. Instrumental and soil moisture variance using the neutron-scattering method. *Soil Science, 97*, 19–24.

Patric, J.H., Douglass, J.E. and Hewlett, J.D., 1965. Soil water absorption by mountain and piedmont forests. *In*: *Proceedings, Soil Society of America.* Soil Society of America, Madison, WI, pp. 303–308.

Waggoner, P.E. and Hewlett, J.D., 1965. Test of a transpiration inhibitor on a forested watershed. *Water Resources Research, 1*, 391–396.

Hewlett, J.D., 1966. Will water demand dominate forest management in the East? *In*: *Proceedings, Society of American Foresters, 1966.* Society Of American Foresters, Seattle, Washington.

Hewlett, J.D. and Hibbert, A.R., 1966. Factors affecting the response of small watersheds to precipitation in humid areas. *In*: *Proceedings of a National Science Foundation advanced science seminar, international symposium of forest hydrology*, 29 August–10 September 1965, University Park, PA. Pergamon Press, New York, NY.

Hewlett, J.D., 1967. A hydrologic response map for the state of Georgia. *Journal of the American Water Resources Association, 3*, 4.

Hewlett, J.D. and Douglass, J.E., 1968. Blending forest uses. *In*: *Research Paper*, Editor (Editor)^(Editors). USDA

[1] downloadable at https://water.usask.ca/hillslope/index.php

Forest Service, Appalachian Forest Experiment Station, Asheville, NC.

Hewlett, J.D., 1969. In Defense of Experimental Watersheds. *Water Resources Research, 5*, 306–316.

Hewlett, J.D., 1969. *Tracing storm and base flow to variable source areas on forested headwaters.* School of Forest Resources, University of Georgia, Athens, GA. 1969.

Hewlett, J.D., 1970. Review of the Catchment Experiment to Determine Water Yield. *In: Proceedings of the Join FAO/U.S.S.R. International Symposium of Forest Influences and Watershed Management*, Food and Agriculture Organization of the United Nations. University of Georgia, Moscow, U.S.S.R.

Hewlett, J.D. and Helvey, J.D., 1970. Effects of forest clear-felling on the storm hydrograph. *Water Resources Research, 6*, 768.

Hewlett, J.D. and Nutter, W.L., 1970. The varying source area of streamflow from upland basins. *In: Proceedings of the symposium, 3–6 August 1970, Bozeman, MT.* American Society of Agricultural Engineers, New York, NY.

Woodruff, J.F. and Hewlett, J.D., 1970. *The hydrologic response of small basins in Georgia*, [n.p.] 1971.

Woodruff, J.F. and Hewlett, J.D., 1970. Predicting and mapping the average hydrologic response for the Eastern United States. *Water Resources Research, 6*, 1312–1326.

Hewlett, J.D., 1971. Comments on the catchment experiment to determine vegetal effects on water yield. *Water Resources Bulletin, 7*, 376–381.

Hewlett, J.D. and Ayers, M., 1971. A study to relate the natural chemistry of stream water to its source on the forested basin. *In*, Editor (Editor^(Editors). Coweeta LTER, University of Georgia, Athens, GA.

Hewlett, J.D., 1972. Reply [to 'Comments on 'Effects of forest clear-felling on the storm hydrograph' by John D. Hewlett and J. D. Helvey']. *Water Resources Research, 8*, 171.

Helvey, J.D., Hewlett, J.D. and Douglass, J.E., 1972. Predicting soil moisture in the Southern Appalachians. *In: Proceedings, Soil Science Society of America 36.* Soil Science Society of America, Madison, WI, pp. 954–959.

Hewlett, J.D. and Pienaar, L., 1973. Design and analysis of the catchment experiment. *In: Proceedings of a Symposium on Use of Small Watersheds in Determining Effects of Forest Land Use on Water Quality, 22–23* May *1973, Lexington, KY*, E. H. Whites. University of Kentucky, Lexington, KY, pp. 88–106.

Hewlett, J.D., 1974. Comments on letters relating to 'Role of subsurface flow in generating surface runoff; 2, upstream source areas' by R. Allan Freeze. *Water Resources Research, 10*, 605–607.

Hewlett, J.D. and Troendle, C.A., 1975. Non-point and diffused water sources: A variable source area problem. *In: Proceedings, Watershed Management Symposium, 11–13 August 1975, Logan, Utah.* American Society of Civil Engineers, New York, NY.

McMinn, J.W. and Hewlett, J.D., 1975. First-year water yield increase after forest cutting: an alternative model. *Journal of Forestry, 73*, 654–655.

Hewlett, J.D., Cunningham, B.B. and Troendle, C.A., 1977. Predicting stormflow and peakflow from small basins in humid areas by the R-index method. *Water Resources Research, 13*, 231–253.

Hewlett, J.D., Fortson, J.C. and Cunningham, B.B., 1977. The effect of rainfall intensity on storm flow and peak discharge from forest land. *Water Resources Research, 13*, 259–265.

Hewlett, J.D., Thompson, W.P. and Brightwell, N., 1979. Erosion control on forest land in Georgia. *In*, Editor (Editor^(Editors). School of Forest Resources, University of Georgia and USDA Soil Conservation Service, Athens, GA.

Hewlett, J.D., 1979. *Forest water quality: an experiment in harvesting and regenerating Piedmont forest.* School of Forest Resources, University of Georgia, Athens, GA.

Hewlett, J.D., 1980. Models in Land Use Hydrology: The need for closing the gap between theory and practice. *Workshop on the Effect of Rural land use and Catchment Management on Water Resources.* Department of Water Affairs, Forestry and Environmental Conservation.

Bosch, J.M. and Hewlett, J.D., 1980. Sediment control in South African forests and mountain catchments. *South African Forestry Journal, 115*, 50–55.

Bernier, P.Y. and Hewlett, J.D., 1982. Test of a revised variable source area simulator (VSAS2) on a forested basin. *Proceedings of the Canadian Hydrology Symposium - Associate Committee on Hydrology*, http://proxy-remote.galib.uga.edu/login?url=http://search.ebscohost.com/login.aspx?direct=true&db=agr&AN=IND83064349&site=eds-live:401-418.

Bosch, J.M. and Hewlett, J.D., 1982. A review of catchment experiments to determine the effect of vegetation changes on water yield and evapotranspiration. *Journal of Hydrology, 55*, 3–23.

Hewlett, J.D. and Fortson, J.C., 1982. Stream temperature under an inadequate buffer strip in the southeast Piedmont. *Water Resources Bulletin, 18*, 983–988.

Burns, R. G. and Hewlett, J.D., 1983. A decision model to predict sediment yield from forest practices. *Water Resources Bulletin, 19*, 9–14.

Hewlett, J.D., 1982. Forests and floods in the light of recent investigation. *Proceedings of the Canadian Hydrology Symposium - Associate Committee on Hydrology*, http://proxy-remote.galib.uga.edu/login?url=http://search.ebscohost.com/login.aspx?direct=true&db=agr&AN=IND83064356&site=eds-live:543-559.

Hewlett, J.D., 1982. Pattern and Roads in the Management of the Forest Property. University of Georgia.

Hewlett, J.D. and Fortson, J.E., 1982. The Paired Catchment Experiment. University of Georgia, Athens, GA.

Hewlett, J.D., Post, H.E., and Doss, R., 1983. Effect of clearcut silviculture on water yield and dissolved mineral export in the Piedmont. *Forest water quality: a collection of papers reporting an experiment in harvesting and regenerating Piedmont forest*, 15–36.

Hewlett, J.D., 1983. Forest cutting and water quality, quantity and timing in the Georgia Piedmont. *Forest water quality: a collection of papers reporting an experiment in harvesting and regenerating Piedmont forest*, 1–10.

Hewlett, J.D., 1983. The use of experimental watersheds in solving land management problems. *In*: *Research Oriented Workshop on Experimental Watersheds*, United States - Uruguay Billateral Cooperation. University of Georgia, Uruguay, South America, p. 19.

Hewlett, J.D. and Doss, R., 1984. Forests, floods, and erosion: a watershed experiment in the southeastern Piedmont. *Forest Science, 30*, 424–434.

Hewlett, J.D., Fortson, J.C. and Cunningham, B.B., 1984. Additional tests on the effect of rainfall intensity on storm flow and peak flow from wild-land basins. *Water Resources Research, 20*, 985–989.

Hewlett, J.D., Post, H.E. and Doss, R., 1984. Effect of clearcut silviculture on dissolved ion export and water yield in the Piedmont. *Water Resources Research*, 20, 1030–1038.

Hewlett, J.D., 1962. Internal water balance of forest trees on the Coweeta watershed. Duke University, Durham, NC.

Hewlett, J.D., 1970. The relation of forests and forestry to water resources. *South African Forestry Journal, 75*, 4.

Hewlett, J.D., 1983. *Forest water quality: a collection of papers reporting an experiment in harvesting and regenerating Piedmont forest*, School of Forest Resources, University of Georgia, Athens, GA.

Hewlett, J.D. and Kramer, P.J., 1963. The measurement of water deficits in broadleaf plants. *Protoplasma, 57*, 381–391.

Hewlett, J.D. and Patric, J.H., 1963. An example of multiple use on a small mountain watershed. *In*: *Proceedings, seventh annual meeting, Soil Conservation Society of America, Georgia Chapter, 24–25* May *1963, Eatonton, GA.* Soil Conservation Society of America, Georgia Chapter, Eatonton, GA.

Hewlett, J.D. and Nutter, W.L., 1968. *The role of forest hydrology in watershed management: a class manual for Forestry 351 (Watershed Management)*.School of Forest Resources, University of Georgia, Atehns, GA.

APPENDIX II

Glossary of Terms

Note[1]

Basin: See catchment or watershed.

Bottom-up analysis: Reductionist science; studying processes or phenomena starting from fundamental units to explain larger behavior.

Bulk water: Water that is extracted from soil using a laboratory technique that gets "all" the water out of the sample (e.g., cryogenic extraction).

Capillary fringe: The tension saturated zone above the water table.

Catchment: A basic unit of the landscape that collects and drains to a stream outlet. See also the terms watershed, basin or drainage basin.

Celerity: Wave speed or speed with which perturbations are transmitted, which control the hydrograph.

Conceptual model: A mathematical model to represent key hydrological processes (e, g. HBV, Topmodel).

Convolution integral: A mathematical technique that allows two functions to be combined to make a third one.

Critical zone: The part of the Earth's surface between the top of the vegetation canopy down through soil and weathered rock down to unweathered rock.

Cryogenic extraction: A laboratory technique for water extraction from soil and plant material.

Deuterium: A heavy isotope of hydrogen (^{2}H) used to trace the water molecule.

Drainage basin: See catchment.

Dual isotope: The approach in hydrology that combines oxygen and hydrogen isotopes to interpret catchment water samples (i.e., streamflow, plant water, soil water).

[1] Note some of these terms follow about ~1/3 of my own, ~1/3 the definitions of terms in *Rainfall-Runoff Modeling: The Primer* by Keith Beven and ~1/3 from ChatGPT5 (Open AI) with suitable edits. This is for consistency between the things new graduate students will be seeing and reading.

Dynamics area: Early term used to denote what would later become the variable source area (concept).

Ecohydrological separation: Plants and streams seemingly drawing from different pools of subsurface water.

Emergent behavior: When a system shows patterns or properties or behaviors that arise from the interaction of its parts.

End-member mixing: A general term for reducing the dimensionality of stream chemistry data; to trace water geographic sources of streamwater, often employing principal component analysis.

Extraction: A means of getting water from soil and plant material. Common types: cryogenic extraction, hydraulic squeezing, centrifugation, etc.

Fill-and-spill: A concept whereby vertical and lateral additions of water to a landscape unit are first placed into storage and only when this storage reaches a critical level (the fill), is a previously infeasible (but subsequently important) outflow pathway activated (the spill).

Fractionation: The selective partitioning of isotopes during physical or chemical processes.

Groundwater ridging: Steepening of hydraulic gradients around the stream where a mound occurs, promoting increased gravity drainage of water to (and away from) the channel.

Headwater catchment: The small, uppermost drainage basins in a river network with maximum second-order streams that drain it.

Headwater stream: Channel draining first- and second-order streams, comprising the vast majority of river lengths globally.

Hydrogen isotopes: ^{1}H (protium), ^{2}H (deuterium) and ^{3}H (tritium).

Hydrogeomorphic concept: A concept that links hydrological processes to geomorphic setting where stormflow sources phase in with increasing wetness, linked to landform characteristics.

Hydrograph separation: To break up the storm hydrograph into its component parts. Graphical hydrograph separation is often done for quickflow and delayed flow; isotope-based hydrograph separation is often done for event water and pre-event water.

Hysteresis: A characteristic form of non-linearity where the value depends on the history.

Idiographic: An approach focused on describing unique, site-specific characteristics or case studies rather than general laws.

Infiltration capacity: The maximum rate at which water can enter the soil surface.

Infiltration excess overland flow: When the rain or snowmelt intensity exceeds the soil infiltration capacity and ponding (and resulting lateral downslope flow) occurs.

Interflow: See subsurface stormflow.

Isotope: Atoms of the same element that differ in the number of neutrons and therefore in atomic mass; commonly used in hydrology (e.g., ^{18}O and ^{2}H) as tracers of the time sources of flow and residence time and transit time.

Macropore: Non-capillary void space within the soil that enables preferential flow.

Meteoric water line: A global or regional linear relationship between the stable isotope ratios of hydrogen and oxygen in precipitation.

Mixing model: A quantitative method to estimate the proportion of different water sources contributing to a mixture (e.g., streamflow) using tracers such as isotopes or solutes.

Mobile water: Soil water in pores of an equivalent cylindrical diameter greater than approximately 3.7 μm that is equilibrated and sampled by suction cup lysimeters.[2]

Nomothetic: A law-based approach that seeks universal principles or rules governing hydrological processes, as opposed to site-specific descriptions.

Oxygen isotopes: the light (^{16}O) and heavy (^{18}O) isotopes of oxygen used to trace the water molecule.

Partial area: A fixed area in a catchment where infiltration excess overland flow is produced

Perceptual model: Mental model of a key hydrological process or processes.

Perched water table: Temporary saturation above an interface.

Phloem: The zone between the bark and the xylem.

[2] And getting really technical: corresponding to an applied pressure of −400 cm H_2O in the Young–Laplace equation; from Dusek and Vogel (2024)

Plant water deficit: The difference between the water required by plants for optimal growth and the water actually available in the soil.

Preferential flow: Water flow in non-capillary flow in void space within the soil. See macropore.

Preferred states hypothesis: The idea that hydrological systems exhibit dominant, recurring states or patterns (e.g., catchment wetness levels) that govern the threshold generation of runoff.

Quickflow: The portion of the storm hydrograph above a line that is projected from the first part of the rise at 0.0055 l/s/ha/hr.

Representative elementary area: The smallest spatial unit beyond which hydrological processes can be considered statistically homogeneous and representative of the larger catchment.

Residence time: The time a water molecule spends in storage (e.g., in soil, groundwater, etc.).

Rhizosphere: The soil region directly influenced by plant roots, where water uptake and root–soil interactions strongly affect moisture dynamics.

Runoff: An often catch-all term for stormflow in the channel. It is also used in the literature to mean the shedding of water from the sideslopes of the catchment. It can also be used to describe "discharge divided by area" to yield a depth of flow in mm.

Runoff ratio: Runoff for an event divided by the precipitation amount for that event.

Sapflow: The upward ascent of water in the xylem via capillary action, cohesion and tension.

Saturated area: The area around the stream channel margins where groundwater rises to intersect the soil surface.

Saturation overland flow: See the more precise term "Saturation excess overland flow."

Saturation excess overland flow: The combination of rainfall (or snowmelt) onto near-stream saturated area and groundwater seepage that promotes rapid runoff to the stream.

Scaling (of flow, event water percentage, transit time): The study of how hydrological processes and parameters change with spatial or temporal scale, and how small-scale measurements can be extrapolated to larger areas or timescales.

Soft data: Qualitative knowledge from the field that cannot be used directly for model simulation or parameter assessment but nonetheless can be made useful through fuzzy measures and fuzzy membership functions for model and parameter assessment.

Soil water: Water in and from the unsaturated zone within the soil profile.

Stable isotope: An isotope that does not undergo radioactive decay (e.g., ^{18}O, ^{2}H), commonly used in hydrology as a conservative tracer of water sources and pathways.

Stemflow: Water moving vertically down the bark tree during a rainfall event.

Stormflow: Event runoff in the stream channel. Also known as storm runoff.

Storm runoff: See stormflow.

Stream number: The classification system for determining channel hierarchy where channel tips are first-order channels, second-order channels are defined when two first-order channels come together, etc, etc.

Storage: The volume of water retained in the critical zone. Can be passive or active.

Storage–discharge relation: The functional relationship between the amount of water stored in a catchment and the rate of water released as streamflow.

Substitution hypothesis: The idea that snowmelt or rainfall displaces water in the ground where that groundwater appears in substitution for the input in the channel.

Subsurface stormflow: Lateral subsurface flow above a restricting layer of lower hydraulic conductivity that occurs during and following storm events.

Suction lysimeter: A porous cup device that is inserted into the soil, where vacuum is created, to draw in water from the unsaturated zone for later chemical or isotopic analysis. See mobile water.

Tensiometer: A porous cup device connected to a pressure gauge to record the suction (sometimes called soil water tension) in soil. Some tensiometer systems can also record pressure potential (i.e., positive pore pressures) when saturation occurs.

Threshold: A critical point or condition in a hydrological system beyond which behavior changes abruptly.

Throughfall: Rainfall or snowmelt that makes its way through the forest or plant canopy and reaches the soil surface.

Throughflow: See subsurface stormflow.

Top-down analysis: A hydrological modeling approach that begins with observed catchment-scale responses (e.g., streamflow data) and works backward to infer dominant processes.

Transit time: The time water spends travelling subsurface through a catchment to the stream network.

Translatory flow: Lateral piston-like flow displacement through the soil profile.

Transmissivity feedback: Occurs when the water table (permanent or perched) rises into a zone of higher saturated hydraulic conductivity and lateral flow increases rapidly.

Travel time: See transit time.

Tritium: The radioactive isotope of hydrogen (^{3}H) with a half life of 12.3 years.

Two- or three-component mixing model: A method to estimate the proportions of two or three distinct water sources in a mixture (e.g., event and pre-event water; or event water, soil water and groundwater, respectively).

Two water worlds hypothesis: See ecohydrological separation.

Variable source area (concept): A generalized concept whereby subsurface flow of water from the upslope exceeds the capacity of the soil profile to transmit it, the water will come to the surface and the channel length will grow.[3]

Velocity: The speed at which water particles move through a porous medium. Can apply also to particle transport in overland flow and in the stream channel.

Virtual experiment: Numerical experiments with a model driven by a field-based perceptual model where the experimentalist and modeler work together to analyze and interpret the results.

Water age: See residence time or transit time.

Watershed: See catchment.

Xylem: The vascular tissue (i.e., the dead, interconnected channels) in a tree or plant where water is transported upwards from the roots to the leaves or needles.

[3] Note that this is the original definition by Hewlett and Hibbert (1967, p. 279).

APPENDIX III

Chapter Quizzes

Chapter 1 Quiz[1]

Q1.1. According to the author, what is the primary danger of the modern over-reliance on hydrological models, a state described as "model land"?

A. Students lose touch with the complexity of natural hydrological processes by focusing only on model calibration and validation.
B. It leads to the development of too many different model types, causing confusion in the field.
C. It requires expensive computing power that is not accessible to all researchers.
D. It makes forecasting and predicting future "what if" scenarios impossible.

Hint: Consider the author's concern about the cycle where students no longer go into the field to develop their own understanding.

Q1.2. What is the definition of a "perceptual model" as presented in the book chapter?

A. A universally applicable theory of hydrology that is independent of place and time.
B. A computer program that uses a set of equations to simulate streamflow generation.
C. A detailed, yet qualitative, mental model of how hydrological processes operate and interact within a specific catchment.
D. A statistical model that finds correlations between rainfall and runoff without considering physical processes.

Hint: Think about the distinction the author makes between a "mind-drawing" and a set of mathematical equations.

Q1.3. Why does the book focus primarily on "headwater catchments"?

A. They are where most streamflow is generated and are at a scale suitable for detailed field measurements.
B. They are the only locations where government stream-gauging networks provide reliable data.
C. They represent the simplest and most uniform hydrological systems, making them easy to model.
D. They are dominated by alluvial geomorphic processes, which are the main subject of modern hydrology.

Hint: Consider both the hydrological importance and the practical constraints of doing fieldwork.

Q1.4. How does the author distinguish a perceptual model from a work of art, like Edvard Munch's "The Scream"?

A. A perceptual model uses complex equations, while a work of art uses paint and canvas.
B. Given the same data, different hydrologists should arrive at the same perceptual model, whereas art is unique to the artist.
C. A perceptual model is purely objective, while a work of art is purely subjective.
D. A perceptual model can be created by anyone, whereas a work of art requires a famous artist.

Hint: Think about the role of data and reproducibility in science versus personal vision in art.

Q1.5. What is the author's critique of the common undergraduate textbook diagrams (like Figure 1.6) showing runoff generation processes?

A. They are outdated, representing understanding from the 1970s, and fail to capture the complexity of real catchments.
B. They focus too much on subsurface flow and neglect the importance of overland flow.
C. They are too complex for an undergraduate audience and should be reserved for graduate-level study.

[1] Note—all questions generated by AI in ChatGPT in October 2025

D. They are fundamentally incorrect and describe processes that do not occur in nature.

Hint: The author compares the state of these diagrams to their taste in music.

Q1.6. The author describes a "creative tension" in teaching hydrology. This tension exists between which two approaches?

A. An engineering approach, assuming general laws, and an earth science approach emphasizing that place matters.
B. Using historical data versus collecting new, real-time data.
C. Field-based hydrology versus laboratory-based experiments.
D. Qualitative descriptions versus quantitative measurements.

Hint: Consider the difference in perspective between a discipline that seeks universal formulas and one that studies unique locations.

Q1.7. What reason does the author give for encouraging students to read "old papers"?

A. To see what has stood the test of time and to understand the historical development of key ideas and discoveries.
B. To find simple equations and models that have been forgotten in the modern era of complex computing.
C. Because older papers are generally better written and more impactful than modern research articles.
D. To understand that modern hydrological questions are entirely different from those asked in the past.

Hint: The author compares scientific discovery to summiting a mountain for the first time.

Q1.8. The author uses the term "epistemic arrogance" to describe a key problem in modern hydrology. What does this term refer to?

A. The tendency for senior scientists to dismiss the ideas of beginning graduate students.
B. The idea that hydrology is the most important of all the environmental sciences.
C. Overestimating what is known and underestimating the uncertainty and complexity of catchment processes.
D. The belief that field hydrologists are superior to modelers.

Hint: This concept has a "double effect" related to what we know and the uncertainty we perceive.

Q1.9. Why does the author claim the path of learning in process hydrology is "more squiggle than line"?

A. Because hydrological processes themselves follow meandering, non-linear flowpaths.
B. Because the field lacks clear direction and has made little real progress over the decades.
C. Because research funding is unpredictable, forcing scientists to change topics frequently.
D. Because progress is messy and non-linear, making it difficult for new students to trace the development of ideas through the literature.

Hint: Consider the challenge a new graduate student faces when trying to make sense of decades of journal articles.

Q1.10. What is the author's concern regarding the success of machine learning (AI) in hydrological modeling?

A. Machine learning algorithms are too difficult for the average hydrologist to understand and implement.
B. The predictive success of AI could lead to process knowledge becoming fully decoupled from modeling.
C. AI models require vast amounts of data that are only available for large, well-gauged river basins.
D. AI models will become so accurate that they will make field hydrologists obsolete.

Hint: Think about the relationship between being able to predict an outcome and understanding why that outcome occurs.

Chapter 2 Quiz

Q2.1. According to the "reverse hydrological cycle" theory held by early thinkers like Aristotle and Seneca, what was the ultimate source of water for streams and springs?

A. A subsurface connection to the ocean.
B. Meltwater from distant, high-elevation snow and ice.
C. Deep infiltration of rainwater that was stored for long periods.
D. Condensation of air within underground caverns.

Hint: Consider the explanation these early thinkers proposed for a seemingly permanent water supply that was independent of local precipitation.

Q2.2. What was Pierre Perrault's primary contribution in 1674 that marked the beginning of modern catchment hydrology?

A. He was the first to propose that streamflow was directly proportional to the catchment area.
B. He accurately mapped a catchment area and measured streamflow to create the first complete water balance.

C. He used field measurements of rainfall and streamflow to argue that precipitation was sufficient to sustain rivers.
D. He discovered that rain infiltrated deep into the soil to feed springs and streams.

Hint: His work represented a fundamental shift from philosophical speculation to data-driven investigation regarding the origin of streamflow.

Q2.3. Robert Horton described the process of infiltration excess overland flow. Under what key condition does this process occur?

A. When the ground is already saturated from previous rainfall.
B. When the rate of rainfall exceeds the soil's infiltration capacity.
C. When water flows rapidly through large macropores like root channels and cracks.
D. When subsurface stormflow emerges at the base of a slope.

Hint: Think of the soil surface acting like a gate with a limited capacity to let water through.

Q2.4. Charles Hursh, working at the Coweeta Hydrological Laboratory, provided the first formal definition of subsurface stormflow. How did he characterize this process?

A. As the primary mechanism for runoff generation in arid regions with intense rainfall.
B. As overland flow that is hidden by a thick mat of grass or forest litter.
C. As a type of turbulent groundwater flow that only occurs during major floods.
D. As water that enters the soil and moves laterally through upper soil horizons much faster than normal groundwater seepage.

Hint: Consider how this process is distinct from both surface runoff and deep groundwater flow.

Q2.5. What is the central idea of the "partial area concept" as developed by Betson (1964)?

A. Runoff is primarily generated from fixed, limited zones within a catchment with low permeability.
B. Subsurface stormflow from different parts of the catchment contributes water at different times.
C. Streamflow is generated from the entire catchment area, but at varying rates.
D. The area contributing to runoff expands and contracts depending on rainfall intensity.

Hint: This concept challenges the idea that infiltration excess overland flow occurs uniformly across a watershed.

Q2.6. Why did the "era of infiltration" dominate hydrological thinking from approximately 1930 to 1960?

A. New technologies made it easy to measure soil permeability across entire catchments.
B. Overland flow was a visible process that was conceptually and mathematically simpler to model than unseen subsurface flow.
C. Foresters like Charles Hursh demonstrated that subsurface flow was only important in mountainous regions.
D. Engineers like Robert Horton had disproven the existence of subsurface stormflow.

Hint: Consider the practical advantages of studying a process that can be directly observed versus one that occurs underground.

Q2.7. What limitation in Pierre Perrault's work was corrected over a century later by John Dalton?

A. Dalton showed that one-sixth of precipitation becoming runoff was a universal constant.
B. Dalton clearly articulated the relationship between a river's flow and the area of the catchment draining to it.
C. Dalton proved that rainfall, not snowmelt, was the primary source of streamflow.
D. Dalton developed a method to directly measure evapotranspiration losses.

Hint: This correction involved a fundamental principle of how watersheds are delineated and quantified.

Q2.8. The text describes the catchment water balance as a "black box" model. What does this metaphor imply?

A. It is an outdated model that has been completely replaced by physical process descriptions.
B. It is a complex model that requires powerful computers to solve.
C. It accurately describes the flowpaths and age of water within the catchment.
D. It accounts for annual inputs and outputs but does not explain the internal processes of streamflow generation.

Hint: Think about what an accounting ledger tells you about a company's day-to-day operations.

Q2.9. In his 1919 text, how did Adolph Meyer advance the understanding of runoff beyond a simple percentage of rainfall?

A. He developed the first mathematical formulas to predict infiltration rates.
B. He proved that snowmelt contributes less to streamflow than rainfall.

C. He stated runoff is a residual found by deducting losses like evaporation and seepage from precipitation.
D. He argued that runoff was mainly composed of deep groundwater.

Hint: He described runoff not as a fraction of the input, but as what remains after other demands are met.

Q2.10. What was a key element of Arnold Engler's 1919 perceptual model for subsurface stormflow in the Emmental catchment?
A. The conclusion that subsurface flow was negligible compared to overland flow in forested areas.
B. A uniform, slow seepage of water through the entire soil matrix.
C. Flow was concentrated in preferential pathways, which he imagined as "uncountable veins."
D. The idea that all stormflow was generated by water flowing over an impermeable bedrock layer.

Hint: His model emphasized that water does not move uniformly through the soil, but follows specific routes.

Q2.11. How did Robert Horton view the phenomenon now known as subsurface stormflow?
A. He called it "concealed-surface runoff" and found the term "subsurface-runoff" confusing.
B. He denied its existence, arguing all quick runoff was infiltration excess overland flow.
C. He believed it was simply a form of laminar groundwater flow.
D. He embraced it as the primary runoff mechanism in forested landscapes.

Hint: He acknowledged the process but objected to the common terminology for it.

Q2.12. What new discovery about the global water cycle is changing modern perceptual models, as mentioned at the end of the chapter?
A. Subsurface flow has been proven to be the dominant runoff process in all catchments globally.
B. Human activities have now become the largest component of the water cycle, exceeding precipitation.
C. The total volume of groundwater is much smaller than previously thought.
D. A vast amount of deep groundwater is "fossil water" that does not actively participate in the surface water cycle.

Hint: This new understanding relates to the age of water and the connectivity between deep and shallow parts of the Earth's crust.

Chapter 3 Quiz

Q3.1. What is the core idea of the variable source area concept as defined by Hewlett and Hibbert (1967)?
A. Subsurface water flow from upslope exceeds the soil's transmission capacity, causing it to surface and extend the channel network.
B. Streamflow is generated solely by direct precipitation onto a fixed, saturated area immediately adjacent to the stream channel.
C. The entire catchment contributes equally to streamflow as soon as rainfall begins, a process known as translatory flow.
D. Runoff is primarily generated when rainfall intensity exceeds the soil's infiltration capacity across the entire catchment.

Hint: Consider the mechanism that causes the source area to expand.

Q3.2. What was the primary conclusion of the "sloping block" experiment conducted by Hewlett and Hibbert (1963)**?**
A. It quantified the expansion and contraction of saturated areas due to a rising water table.
B. It showed that a log–log plot of outflow recession had a distinct break, indicating a shift from fast to slow drainage.
C. It proved that infiltration excess overland flow was the dominant runoff process on steep slopes.
D. It demonstrated piston flow displacement using tritium tracers in a controlled hillslope environment.

Hint: Focus on the graphical representation of the outflow data from the artificial hillslope.

Q3.3. How did Tom Dunne's work at Sleepers River, Vermont, expand upon Hewlett's original variable source area concept?
A. It introduced the idea of "translatory flow," where new rain displaces old soil water.
B. It confirmed that subsurface stormflow was the only significant contributor to the storm hydrograph.
C. It provided the first field-based evidence for infiltration excess overland flow in forested catchments.
D. It documented and quantified saturation excess overland flow, where return flow and direct precipitation on saturated areas dominate runoff.

Hint: Think about the specific runoff process Dunne observed and measured in the low-relief, glaciated landscape of Vermont.

Q3.4. What was a key critique raised by Beven (2006) regarding Hewlett and Hibbert's use of the Horton and Hawkins (1965) **tracer study?**

A. The displacement timescale in the experiments (weeks) was much longer than that of a storm runoff hydrograph (hours).
B. The tracer used was not tritium, leading to inaccurate conclusions about water displacement.
C. The tracer experiment was conducted in clay soils, which were not representative of the Coweeta site.
D. The experiment failed to show any displacement effect, contradicting the concept of translatory flow.

Hint: Consider the difference in the duration of the experimental process versus the natural event it was used to explain.

Q3.5. According to the ranking by Hewlett and Hibbert (1967), which factor was considered the MOST important in explaining variations in quick flow among the catchments they studied?
A. Average size and number of large storms.
B. Land use.
C. Average land slope and its effect on slope length.
D. Average soil mantle depth to an impermeable layer.

Hint: Review the four factors Hewlett and Hibbert listed on page 288 of their paper; the text identifies which was "foremost."

Q3.6. How did Ragan's (1968) findings in a Vermont catchment with deep sandy deposits differ from Dunne's observations at Sleepers River?
A. Ragan found that the contributing saturated area expanded to cover up to 50% of the catchment during storms.
B. Ragan identified a small, relatively fixed runoff-contributing area near the stream, driven by a rapidly responding groundwater ridge.
C. Ragan concluded that deep groundwater was the sole source of all streamflow, with no contribution from recent rainfall.
D. Ragan's work was the first to show that subsurface stormflow from distant hillslopes was the dominant runoff mechanism.

Hint: Think about the size and stability of the runoff-generating area that Ragan observed.

Q3.7. What term did Hewlett and Hibbert (1967) use for the process where new rainfall displaces stored soil water, contributing it to streamflow?
A. Saturation excess overland flow.
B. Translatory flow.
C. Infiltration excess overland flow.
D. Return flow.

Hint: This term describes a process analogous to pushing a piston through a tube filled with water.

Q3.8. In his letter to Leopold, how did Hewlett characterize Dunne's work on "partial source areas"?
A. As a revolutionary new concept that completely replaced the variable source area idea.
B. As fundamentally flawed due to incorrect measurements of subsurface flow.
C. As a special case of the variable source area concept, that would become variable with more rainfall.
D. As interesting data that ultimately refuted the importance of any source area concepts.

Hint: Consider how a proponent of a general theory might view a specific example that seems different.

Q3.9. According to the recent global study by McMillan et al. (2025), how does the prevalence of saturation excess overland flow compare to infiltration excess overland flow?
A. Infiltration excess is more common in all climate types.
B. Saturation excess is only dominant in humid, temperate regions.
C. They are equally common across all climate types.
D. Saturation excess is more frequently described than infiltration excess across all aridity classes.

Hint: Focus on the study's findings regarding the frequency of the two main types of overland flow in research watersheds.

Q3.10. What was a significant, though perhaps unintended, consequence of the design of the Hewlett and Hibbert (1963) **sloping soil block experiment?**
A. It destroyed the natural soil structure and placed an artificial bottom on the slope.
B. It allowed for the precise measurement of natural preferential flowpaths in the soil.
C. It perfectly simulated deep groundwater conditions found in natural catchments.
D. It demonstrated that surface vegetation had no impact on subsurface drainage rates.

Hint: Consider the limitations mentioned in the text regarding the construction of the experimental setup.

Chapter 4 Quiz

Q4.1. What was the primary argument in R. Allan Freeze's (1974) influential review paper regarding subsurface stormflow?
A. He doubted that it could deliver sufficient water quickly enough to be a significant component of storm runoff.
B. It could not exist without the presence of extensive soil pipe networks.
C. It was simply a different name for saturation overland flow, and the terms were interchangeable.
D. It was the dominant storm runoff mechanism in all catchments, regardless of size or geology.

Hint: Consider the controversy this review paper caused and how it contrasted with findings from catchments like HJ Andrews.

Q4.2. The work of Anderson and Burt (1978) **at the Bicknoller site was transformative for understanding subsurface stormflow generation. What was their key finding?**

A. The length of the hillslope was the single most important factor controlling the volume of stormflow.
B. Topographic convergence in hollows was a critical control, concentrating flow and generating the most significant subsurface stormflow.
C. Soil pipes were responsible for over 90% of the subsurface flow, regardless of topography.
D. Subsurface stormflow was generated uniformly across all parts of the hillslope, regardless of shape.

Hint: Think about how the three-dimensional shape of a hillslope might focus or disperse water moving through the soil.

Q4.3. According to the study by Harr (1977) in the HJ Andrews catchment, what was the role of subsurface stormflow during large rainstorms?

A. It was completely overshadowed by saturation overland flow, which generated nearly all the quick flow.
B. It was only significant after the storm had ended, contributing primarily to the hydrograph recession.
C. It was "too small, too late and too insensitive" to contribute significantly to the storm hydrograph.
D. It accounted for more than 96% of the quickflow for each of the large events studied.

Hint: Recall the evidence presented in the text that directly challenged the conclusions of Freeze's 1974 review.

Q4.4. The work of Weyman (1973) at East Twin Brook was pivotal in developing which specific concept related to subsurface stormflow?

A. The mapping of regional soil pipe networks and their connection to channel extension.
B. The idea of "percolines" where moisture accumulates in areas of topographic convergence.
C. The dominance of infiltration excess overland flow on steep, forested hillslopes.
D. The development of a "saturated wedge" that grows upward from the slope base during a storm.

Hint: This concept describes the dynamic, expanding source area for subsurface flow at the base of a hillslope.

Q4.5. What is the primary mechanism by which subsurface interfaces, as described by Whipkey (1965), generate lateral subsurface stormflow?

A. They increase the vertical permeability, allowing water to drain faster to the bedrock.
B. They exclusively channel water into large soil pipes and macropores.
C. They are composed of highly porous material that actively pulls water sideways through capillary action.
D. They impede downward percolation, causing a temporary zone of saturation to form above them, which then flows downslope.

Hint: Think about what happens when flowing water encounters a layer that is much harder to pass through.

Q4.6. Based on Figure 4.1 from Pearce et al. (1982), what is the general relationship between storm size and the dominant runoff generation process in the depicted catchments?

A. Saturation overland flow is the dominant process for small events, but subsurface stormflow becomes dominant as storm size increases.
B. Subsurface stormflow dominates runoff generation for all storm sizes, large and small.
C. Antecedent wetness is the only factor that controls the runoff process, with storm size being irrelevant.
D. The dominant runoff process is random and shows no correlation with the size of the rainfall event.

Hint: Examine the labels in the perceptual model linked to the plot and see how they change as you move from left to right along the x-axis.

Q4.7. Mosley's (1979) work in the Maimai catchment revealed what important characteristic of hydrographs generated by soil pipe flow?

A. The hydrographs showed a long, slow recession, indicating slow drainage from the soil matrix.
B. The hydrograph peaks were significantly delayed and attenuated as water moved downslope.
C. The hydrographs were identical in volume, regardless of whether they were measured near the ridge or near the stream.
D. The hydrograph peaks from soil pits at different slope positions and the stream were all synchronous.

Hint: Consider how a network of open channels within the soil would affect the timing of water delivery compared to flow through the soil matrix.

Q4.8. What was a major terminological issue during the International Hydrological Decade (1965–1974) concerning subsurface stormflow?

A. There was only one accepted term, "subsurface stormflow," which was applied incorrectly to different processes.
B. The term "infiltration excess overland flow" was frequently and incorrectly used to describe subsurface flow.
C. All terms related to subsurface flow were abandoned in favor of Horton's "concealed runoff."
D. Researchers often used multiple, alternative terms like "throughflow" and "interflow" to describe the same process, causing confusion.

Hint: The author quotes Winston Churchill at the beginning of the chapter to allude to this specific problem.

Q4.9. According to Weyman (1973), what is the condition generally required for the initiation of lateral subsurface stormflow on a hillslope?
A. The presence of deep, sandy soils with uniform permeability from the surface to the bedrock.
B. Some form of break in the vertical permeability of the soil profile, such as an impeding layer or bedrock.
C. A period of at least 42 days without rainfall to allow the soil to drain completely.
D. A significant contribution from lateral unsaturated flow, which directs water toward the slope base.

Hint: Think about what physical barrier would be needed to divert vertically percolating water into a horizontal path.

Q4.10. The text highlights four main process advancements regarding subsurface stormflow during the International Hydrological Decade. Which of the following is NOT one of those four advancements?
A. The development of stable isotope tracing to determine water age.
B. The recognition of the importance of topographic convergence.
C. The recognition of saturated wedges developing from the slope base.
D. The recognition of how soil pipes could aid high rates of subsurface stormflow.

Hint: Review the bulleted list and Figure 4.2, which summarize the key conceptual developments of the era.

Chapter 5 Quiz

Q5.1. According to the source material, what was the primary reason that, prior to the 1970s, most hydrologists were skeptical that groundwater could be a major source of storm runoff?
A. The belief that groundwater moved too slowly through porous media to contribute significantly to rapid storm hydrographs.
B. The observation that stream baseflow, which is groundwater, decreased during storm events.
C. The prevailing theory that all storm runoff was generated by infiltration excess overland flow.
D. The lack of instrumentation to measure deep water table levels accurately during storm events.

Hint: Consider the discrepancy between observed runoff rates and the calculated speed of water moving through soil.

Q5.2. The "substitution hypothesis" proposed by Martinec (1975) explains the rapid response of streamflow to snowmelt. Which statement best describes this mechanism?
A. Meltwater causes the water table to rise, which then saturates the surface and generates widespread overland flow.
B. Infiltrating meltwater exerts pressure on the existing groundwater, causing a rapid "piston-like" displacement of stored "old" water into the stream.
C. Infiltrating meltwater mixes completely with existing groundwater, and this mixture is then discharged to the stream.
D. Meltwater rapidly travels through large soil macropores and pipes directly to the stream, bypassing the groundwater system.

Hint: Think of the garden hose analogy used in the text, which distinguishes between the speed of a pressure pulse and the speed of the water molecules themselves.

Q5.3. The development of the "groundwater ridging" hypothesis was based on the idea that a specific near-stream soil property allows for a very rapid water table response. What is this property?
A. Low soil porosity that prevents infiltration and promotes surface runoff.
B. High saturated hydraulic conductivity in deep soil layers.
C. A large network of soil macropores and pipes connected to the stream.
D. A substantial tension-saturated capillary fringe that extends close to the ground surface.

Hint: This mechanism is particularly effective in fine-grained materials where water is held under tension just above the water table.

Q5.4. How does the "transmissivity feedback" mechanism explain the rapid increase in groundwater discharge to a stream?

A. It suggests that transient perched water tables on hillslopes are activated and drain rapidly through a network of soil pipes.
B. It proposes that infiltrating water pushes out old groundwater in a piston-flow manner throughout the entire catchment.
C. It relies on the conversion of a tension-saturated capillary fringe into phreatic water near the stream bank.
D. It posits that the water table rises into overlying soil layers with much higher hydraulic conductivity, drastically increasing lateral flow.

Hint: Consider how the ability of a hillslope to transmit water changes as the saturated zone gets thicker.

Q5.5. What was the major conundrum that isotope studies by Pearce et al. (1986) and Sklash et al. (1986) created for the understanding of runoff generation at the Maimai catchment?

A. They found no evidence of groundwater contributing to streamflow, suggesting all runoff was from surface flow.
B. Their results indicated that the "substitution hypothesis" was the only viable mechanism, despite the catchment's steep terrain.
C. They showed that most of the streamflow was "new" rainwater, contradicting the groundwater ridging theory.
D. They demonstrated that soil pipes, previously thought to rapidly transport "new" rainwater, were actually delivering a hydrograph dominated by "old" water.

Hint: Think about the conflict between what was observed about the physical pathways (pipes) and what the water chemistry revealed about the water's age.

Q5.6. The work by Torres et al. (1998) at Metman Ridge provided a more mechanistic explanation for pressure wave dynamics. What was their key finding?

A. Pressure waves are primarily generated by flow interactions between shallow colluvium and deep, fractured bedrock.
B. In certain soils, a small increase in pressure head near saturation can cause a massive (orders of magnitude) increase in hydraulic conductivity, allowing for rapid pressure propagation.
C. Pressure waves can only be generated when the soil is fully saturated before a rainfall event begins.
D. The speed of the pressure wave is always equal to the velocity of the water molecules.

Hint: Focus on the relationship between soil water pressure and the ease with which water can move through the soil, as shown in Figure 5.8.

Q5.7. Early isotopic studies, such as Martinec's, relied heavily on measuring bomb-produced tritium (^3H). Why was this isotope particularly useful as a tracer in the 1960s and 1970s?

A. Atmospheric thermonuclear testing created a massive, globally distributed spike in its concentration around 1963, providing a distinct timestamp in precipitation.
B. It is a stable isotope that does not undergo radioactive decay, making measurements simple.
C. It has a very short half-life, allowing for the precise dating of water that infiltrated within the last year.
D. It is naturally produced in large quantities in the subsurface, allowing hydrologists to trace deep groundwater flow paths.

Hint: Consider the major global events of the mid-twentieth century that dramatically altered the composition of the atmosphere.

Q5.8. What is the key difference between a "top-down" research approach, like isotope tracing, and a "bottom-up" approach in hydrology?

A. Top-down approaches measure surface processes like rainfall, while bottom-up approaches measure groundwater.
B. A top-down approach makes measurements at the catchment outlet to understand the integrated system, while a bottom-up approach scales up measurements from small points or hillslopes.
C. Top-down methods focus on water velocity, while bottom-up methods focus on pressure propagation (celerity).
D. Top-down refers to computer modeling based on theory, while bottom-up refers to direct field measurements.

Hint: Think about the scale at which the primary measurement is made and what that measurement represents.

Q5.9. The final perceptual model for Maimai, proposed by McDonnell (1990), explained how soil pipes could deliver "old" water. Which sequence of events describes this model?

A. Rainwater preferentially flows to the soil–bedrock interface, perches, mixes with stored "old" soil water and is then rapidly drained laterally by the pipe network.

B. Rainfall creates a groundwater ridge in the valley bottom, which pushes old water up into the hillslope soil pipes.
C. Rainfall is collected by surface depressions and funneled directly into pipe inlets as "new" water, which pushes a small amount of old water out ahead of it.
D. Rainfall saturates the soil from the top down, creating a perched water table that slowly displaces old water into the pipes.

Hint: The mechanism involves new water taking a shortcut to the bottom of the soil profile to activate the drainage of stored water.

Q5.10. The case study from the HJ Andrews catchment (Figure 5.21) showed a large streamflow response to rainfall, but the isotopic signature of the streamwater remained virtually unchanged. What does this extreme example demonstrate?
A. The rainfall had the exact same isotopic composition as the streamwater, so no change could be detected.
B. In this catchment, all storm runoff is generated by direct precipitation onto the stream channel.
C. Isotope tracing methods are ineffective in steep, wet catchments like the HJ Andrews.
D. The concept of celerity (a rapid pressure response) can be completely decoupled from velocity (the actual movement of new water molecules to the stream).

Hint: Revisit the distinction between a pressure pulse and the movement of the water itself, as described by the garden hose analogy.

Chapter 6 Quiz

Q6.1. What is the primary implication of "non-linearity" in catchment hydrology, as introduced in Chapter 6?
A. The runoff response is not directly proportional to the precipitation input across all storm sizes.
B. All precipitation events contribute equally to streamflow, differing only by the volume of rainfall.
C. Runoff generation is best understood by averaging the contributions from all parts of the catchment.
D. Catchments wet up and dry down in a predictable, linear sequence based on the growth of saturated wedges.

Hint: Recall the quote from John Henry Holland and the dictionary definition of non-linearity provided in the text.

Q6.2. In the Western and Grayson (1998) study at the Tarawarra catchment, what was the key observation regarding soil moisture patterns between dry and wet periods?
A. During wet periods, soil moisture patterns became random due to complex subsurface flow, while they were organized during dry periods.
B. The average soil moisture content was the only factor that mattered for runoff, with spatial patterns being irrelevant.
C. Soil moisture was consistently highest in hillslope hollows during both dry and wet periods.
D. During dry periods, soil moisture patterns were random, but during wet periods, they became topographically organized.

Hint: Think about how the dominant direction of water movement (vertical vs. lateral) changes as the catchment wets up.

Q6.3. The research at the Panola hillslope by Tromp-van Meerveld and McDonnell was crucial for understanding subsurface stormflow. What was identified as the primary control on the location and connection of saturated patches?
A. The distribution of preferential flow paths like root channels.
B. The topography of the ground surface.
C. The uniform depth of the loamy sand soil.
D. The topography of the soil–bedrock interface.

Hint: The text emphasizes that the "key surface" for subsurface stormflow was not the one visible from above.

Q6.4. The rainfall threshold for initiating significant subsurface stormflow was ~15 mm at the Maimai hillslope but ~60 mm at Panola. According to the text, what factors explain this difference?
A. Maimai receives more intense rainfall events than Panola.
B. Panola has much deeper soils, allowing more water storage before runoff occurs.
C. Maimai is steeper, has wetter antecedent conditions and less permeable underlying bedrock.
D. Panola's surface is much rougher, leading to more surface storage before subsurface flow begins.

Hint: Consider the key differences in climate, topography and geology between the New Zealand and Georgia sites.

Q6.5. How does the chapter describe the concept of "connectivity" as a form of emergent behavior?
A. It is a property that transcends the individual runoff processes and arises when different parts of the system become linked, leading to a threshold response.

B. It is the simple sum of all individual runoff processes like infiltration excess and saturation excess overland flow.
C. It is a property that depends solely on the static, structural patterns of the catchment, not on the hydrologic processes.
D. It is a linear process where the expansion of saturated areas is directly proportional to rainfall input.

Hint: Recall the Aristotle quote mentioned in the text: "the whole is greater than the sum of its parts."

Q6.6. What is the primary reason that continuous, hillslope-to-stream subsurface stormflow connectivity is considered the "exception rather than the rule" for most hillslopes?
A. Significant "loss" of lateral flow to deeper percolation along the flowpath.
B. The permeability contrast at the soil–bedrock interface is usually too low.
C. Riparian zones effectively block and reabsorb all incoming hillslope flow.
D. Most hillslopes are not steep enough to generate lateral flow.

Hint: Consider the findings of the Klaus and Jackson (2018) synthesis on the downslope travel distance of subsurface stormflow.

Q6.7. At the Maimai site, the gauged hillslope exhibits a distinct threshold response to rainfall, but the larger catchment hydrograph does not. What is the reason for this difference?
A. Measurement errors at the catchment scale are too large to detect the threshold response seen at the hillslope.
B. The hillslope was not representative of the catchment, generating much more runoff than other areas.
C. The catchment response is an average of many hillslopes, which smooths out the threshold behavior of any single slope.
D. The riparian zone is always connected to the stream and contributes flow during small events that do not activate the hillslopes.

Hint: Think about which part of the catchment is "always on" and which part "turns on" only after a certain amount of rain.

Q6.8. In arid regions like the Walnut Gulch catchment, how does storm runoff typically change with increasing catchment area?
A. Runoff per unit area increases as larger areas collect more overland flow.
B. Runoff per unit area remains constant, indicating uniform runoff generation.
C. Runoff per unit area becomes zero beyond a certain catchment size as all water evaporates.
D. Runoff per unit area decreases because of large infiltration losses in dry stream channels.

Hint: Consider what happens to the water once it leaves the hillslope and enters the typically dry, sandy stream bed.

Q6.9. The chapter describes a modern view where infiltration excess overland flow and subsurface stormflow are governed by similar processes. What is the unifying concept?
A. Both processes require the entire soil profile to be fully saturated before they can occur.
B. Both processes only occur when antecedent soil moisture is above a critical threshold of 80%.
C. Both processes are primarily controlled by the steepness of the hillslope.
D. Both processes can be described as "fill-and-spill" behavior occurring at a flow-restricting interface.

Hint: Look at the summary diagram (Figure 6.28) that shows emergent behavior from a "partitioning surface."

Q6.10. At the Oregon Coast Range site (Figure 6.12), researchers failed to measure any shallow subsurface stormflow despite seemingly ideal conditions. What was the ultimate explanation for the catchment's rapid streamflow response?
A. Widespread saturation excess overland flow occurring in near-stream areas.
B. A measurement error in the tipping bucket gauges which failed to record the flow.
C. Deeper forms of connectivity, where water infiltrated past the shallow interface and flowed through deeper weathered rock.
D. Infiltration excess overland flow from the highly transmissive forest soils.

Hint: If the water isn't flowing out sideways in the shallow soil, where else could it go to reach the stream quickly?

Q6.11. What is the role of soil pipes and macropores in relation to subsurface stormflow connectivity?
A. They create a linear and predictable runoff response, damping the threshold behavior of the soil matrix.
B. They enhance lateral flow connectivity and can lead to a threshold-like response.
C. They are only effective at draining water at the hillslope scale if they are continuous from the ridge to the stream.

D. They primarily increase vertical infiltration, reducing the potential for lateral subsurface stormflow.

Hint: Think about whether these features would make it easier or harder for saturated patches to connect downslope.

Q6.12. The tracer study at the Panola catchment (Figure 6.19) revealed the sequence of contributing water sources to the stream during a storm. Which sequence best describes the findings?

A. All three sources contribute in equal proportions throughout the entire storm event.
B. Hillslope subsurface stormflow, then riparian groundwater, then bedrock outcrop overland flow.
C. Bedrock outcrop overland flow is the only significant contributor during the storm peak.
D. Riparian groundwater, then bedrock outcrop overland flow, then hillslope subsurface stormflow.

Hint: Examine the mixing plot in Figure 6.19 and trace the path of the streamwater samples from pre-storm conditions through the event.

Chapter 7 Quiz

Q7.1. According to the text, what is the primary distinction between "transit time" and "residence time" in the context of catchment hydrology?

A. Transit time is measured using stable isotopes like ^{18}O, while residence time is measured using radioactive isotopes like tritium.
B. Transit time is the age of water in passive storage, while residence time is the age of water in active storage.
C. Transit time refers to the time it takes for water to leave storage and become streamflow, while residence time is the time water spends in storage.
D. They are interchangeable terms that became more distinct after 2010 to avoid philosophical problems.

Hint: Consider the journey of a water molecule from entering the catchment to either staying put or leaving via the stream.

Q7.2. The text presents a paradox where groundwater age is often far greater than streamwater transit time in the same catchment. What is the primary explanation for this phenomenon?

A. Tracer-dating methods for groundwater, like carbon-14, are inherently less accurate than the stable isotope methods used for streamwater.
B. The existence of large volumes of "passive storage" holds very old water that contributes minimally or very slowly to streamflow.
C. Groundwater follows much longer, deeper flowpaths, but it travels at a velocity high enough that it should match streamwater age.
D. Streamwater is primarily composed of recent precipitation that flows over the surface, bypassing the older groundwater.

Hint: Think about the different components of subsurface water and how actively they participate in generating runoff.

Q7.3. In comparing the HJ Andrews and Alsea catchments in Oregon, what was the key finding regarding the factors that control catchment mean transit time?

A. The steeper slopes at Alsea led to younger water, while the gentler slopes at HJ Andrews resulted in older water.
B. Annual rainfall amounts were the dominant control, with the wetter Alsea catchments having much shorter transit times than HJ Andrews.
C. Despite similar climates and topographies, geology was the critical factor, leading to different scaling relationships for transit time.
D. In both catchments, mean transit time scaled directly with catchment area, showing a universal principle.

Hint: The chapter emphasizes that outward appearances, like topography, can hide very different internal "plumbing."

Q7.4. What is the "inverse storage effect" as described in the chapter?

A. The finding that tracer-based storage estimates are inversely proportional to water-balance storage estimates.
B. The observation that larger catchments paradoxically have smaller dynamic storage volumes.
C. A situation where passive storage becomes active during extreme drought, releasing very old water.
D. The principle that as catchment storage increases (wetter conditions), the transit time of streamwater tends to decrease.

Hint: Consider how the dominant flowpaths might change between a dry summer and a wet winter.

Q7.5. The chapter highlights the comparison of soil water aging at the Maimai (New Zealand) and Fudoji (Japan) catchments. What was the main difference in their aging patterns?

A. Both catchments showed that soil water aged primarily with depth, with the oldest water at the soil–bedrock interface.
B. Maimai showed no clear aging pattern, while Fudoji showed water getting older with proximity to the stream.

C. Fudoji's soil water was much older (years) than Maimai's (days) due to thicker soils and deeper bedrock.
D. Maimai exhibited clear downslope aging of soil water due to lateral flow over impermeable bedrock, while Fudoji showed aging primarily with depth.

Hint: Think about what happens to water at the soil–bedrock interface. Does it move sideways or continue downward?

Q7.6. What technological advance around 2006 was credited with enabling more widespread analysis of catchment transit times?
A. New drilling techniques that allowed for easy access to deep groundwater samples.
B. The invention of automated weather stations for accurate precipitation data.
C. The commercial availability of laser spectrometers for stable isotope analysis.
D. The development of high-precision GPS for mapping catchment flowpaths.

Hint: Calculating transit time requires comparing tracer concentrations in rainfall and streamflow. What tool would make analyzing those concentrations easier?

Q7.7. At the Panola catchment, researchers found a strong nonlinear relationship between monthly runoff and streamwater transit time (Figure 7.20). What did this relationship show?
A. Streamwater was oldest during low-flow conditions and became progressively younger as flow increased.
B. There was no clear relationship, indicating that transit time at Panola is controlled by factors other than discharge.
C. Streamwater age remained consistently old (5–8 years) regardless of the flow rate.
D. Streamwater was youngest during low-flow conditions, as it represented recent, shallow groundwater.

Hint: Consider which water sources sustain the stream during a long dry period versus a major rainstorm.

Q7.8. The study of the Loess Plateau in China provided a dramatic example of water aging. What unique tracer signal was identified at approximately 8 meters depth in the unsaturated zone?
A. The 1963 peak of tritium (^3H) from atmospheric thermonuclear testing.
B. A layer of fossil groundwater from the Pleistocene epoch.
C. The seasonal variation of stable isotopes (^{18}O and ^2H), which had been completely damped out.
D. The peak of acid rain deposition from the 1980s.

Hint: Think about global events in the mid-twentieth century that left a distinct isotopic signature in the Earth's atmosphere and precipitation.

Q7.9. What is meant by the "young water fraction" (F_{yw}), a concept developed by Kirchner (2016)?
A. The total volume of water stored in the active, or dynamic, storage zone of a catchment.
B. The proportion of streamflow with a transit time younger than a threshold of roughly 2–3 months.
C. The percentage of rainfall from a single storm that appears in the stream as "quickflow."
D. The fraction of groundwater that is less than one year old.

Hint: This concept uses the seasonal cycles of isotopes in rain and streamwater to determine the contribution of recent water.

Q7.10. Based on the text, what is the fundamental difference between "active storage" (or dynamic storage) and "passive storage"?
A. Active storage is soil water, while passive storage is bedrock groundwater.
B. Active storage can be measured by water balance (P–Q–ET), whereas passive storage can only be measured by drilling wells.
C. Active storage contains young water (months old), and passive storage contains fossil water (thousands of years old).
D. Active storage is the water that actively contributes to streamflow under typical conditions, while passive storage does not.

Hint: Consider which portion of the total subsurface water is responsible for the rises and falls of the stream hydrograph.

Q7.11. The study by Pfister et al. (2017) in Luxembourg examined 16 nested catchments and found a strong relationship between mean transit time and which catchment characteristic?
A. The percentage of agricultural land use.
B. The drainage density of the stream network.
C. The permeability of the underlying bedrock.
D. The average slope of the catchment.

Hint: This study reinforces the "geology is destiny" theme by looking at how easily water can move through the parent material.

Q7.12. In the comparison of Maimai, HJ Andrews and Panola catchments, which one was characterized as being always wet, with thin soils over poorly permeable bedrock, resulting in very short transit times of a few months?

A. HJ Andrews
B. Elk River
C. Panola
D. Maimai

Hint: Consider which catchment's physical description implies that water has little space to be stored for long and is quickly flushed out.

Chapter 8 Quiz

Q8.1. According to reviews by Bosch & Hewlett (1982) **and others, what is the primary reason that the paired catchment approach has low predictive power regarding water yield changes after forest removal?**

A. The approach largely ignores critical differences in subsurface geology, storage and structure between catchments.
B. The replanting of different tree species post-treatment invalidates the comparison with the original forest cover.
C. Most studies did not continue for a long enough duration to account for long-term climate cycles.
D. The statistical methods used in before–after tests are not robust enough for hydrological data analysis.

Hint: Consider what crucial aspect of a catchment system, emphasized throughout the book, is overlooked when it is treated as a simple "black box."

Q8.2. What is the key difference in the type of soil water collected by a suction lysimeter compared to the cryogenic vacuum distillation method?

A. Suction lysimeters collect mobile water held at low tension, whereas cryogenic extraction collects the bulk soil water, including tightly bound water.
B. Suction lysimeters can only be used in wet soils, while cryogenic extraction is designed exclusively for sampling very dry soils.
C. Suction lysimeters preserve the water's isotopic signature perfectly, while cryogenic extraction can cause fractionation during the heating process.
D. Suction lysimeters sample only water from large macropores, while cryogenic extraction samples only water adsorbed to clay particles.

Hint: Think about the physical force each method uses to remove water from the soil matrix.

Q8.3. In the influential study by Brooks et al. (2010), why did the xylem water from Douglas fir trees plot below the local meteoric water line in dual isotope space (^{2}H vs. ^{18}O)?

A. The trees were using soil water from a source that had undergone evaporation, which preferentially removes lighter isotopes.
B. The trees were accessing ancient groundwater that had a fundamentally different isotopic signature from modern precipitation.
C. The process of water transport through the xylem causes isotopic fractionation, enriching the water in heavier isotopes.
D. The annual average precipitation for the site was isotopically lighter than the water from the specific storms the trees were using.

Hint: Consider the physical process that changes the phase of water and systematically sorts light and heavy isotopes.

Q8.4. What is the central concept of the "ecohydrological separation" hypothesis?

A. A spatial separation where streamside trees use streamwater and upslope trees use soil water.
B. A temporal separation where trees use summer rainfall and streams are fed by winter rainfall.
C. A separation between plant-available water in the soil and inaccessible water stored in the underlying bedrock.
D. Two poorly mixed water pools exist in the soil: a mobile pool for streamflow and a tightly bound pool for transpiration.

Hint: This hypothesis is sometimes referred to by the phrase "two water worlds."

Q8.5. The case study of the Jarrah forest in southwestern Australia illustrates a major ecohydrological consequence of land-use change. What happened after the deep-rooted forest was replaced with shallow-rooted crops?

A. The regional climate became drier, leading to the desertification of former croplands.
B. Stream baseflow remained the same, but the frequency of large floods increased.
C. Stream baseflow increased, and stream salinity rose to extremely high levels.
D. Stream baseflow decreased, and soil erosion increased dramatically.

Hint: Consider the dual role of the original deep-rooted trees in both transpiring water and preventing the mobilization of substances stored deep in the soil profile.

Q8.6. What is "rock moisture" as a source for plant water uptake, according to the work of Rempe and Dietrich (2018)**?**

A. Water held in the unsaturated, weathered bedrock layer that sits between the soil and deeper groundwater.

B. Groundwater that is stored in large caverns within limestone bedrock.
C. Precipitation that is directly intercepted by and channeled through exposed rock outcrops on the surface.
D. The thin film of hygroscopic water that clings to rock fragments within the soil profile.

Hint: This water is found in a zone below the soil but above the water table, in a material that is transitioning from soil to unweathered rock.

Q8.7. The study of Aleppo pine and holm oak in Spain during a drought by del Castillo et al. (2016) is an example of what ecological concept?

A. Mutualism, where both species benefit from growing together by modifying the soil environment.
B. Competitive exclusion, where one species completely outcompetes the other for a single water source.
C. Niche segregation, where species reduce competition by using different resources, especially under stress.
D. A hydraulic lift, where one species shares deep water with another in the shallow soil.

Hint: Think about how competing species might coexist by evolving to use the environment in slightly different ways.

Q8.8. What did the research by Allen et al. (2019) across 182 sites in Switzerland reveal about the seasonal origin of water used by trees for midsummer transpiration?

A. Trees predominantly used the recent summer rainfall, showing a tight coupling with current weather.
B. Most sampled tree species primarily used water that had recharged the soil from the previous winter's precipitation.
C. The water used was a thoroughly mixed combination of precipitation from all seasons of the previous year.
D. All tree species switched to using deep groundwater of an indeterminate seasonal origin during the summer.

Hint: This finding supports the idea that the "grow" season for plants can be disconnected from the "flow" season for water input.

Q8.9. The chapter distinguishes between the age of water transpired by a plant and the transit time of water through the plant. What is the key difference?

A. There is no difference; the terms are used interchangeably to mean the time from rainfall to transpiration.
B. Age refers to stable isotope dating, while transit time is measured using radioactive tracers like tritium.
C. Transit time is the time spent in the soil, while age is the time spent moving through the xylem.
D. Age includes the residence time in the soil before uptake, while transit time is only the period of movement within the plant's vascular system.

Hint: Consider the total journey of a water molecule from when it falls as rain to when it exits a leaf, and break that journey into two main parts.

Q8.10. What major methodological issue is highlighted in the chapter summary as a key challenge for advancing the understanding of plant–water relations?

A. The difficulty of measuring precipitation accurately under a dense forest canopy.
B. The lack of standardized methods for water extraction from soil and plant tissues, with different techniques sampling different water pools.
C. The high cost and complexity of mass spectrometers, which limits the number of samples that can be analyzed.
D. The inability to track water movement in real time, as all isotope methods rely on destructive sampling.

Hint: The text discusses how different lab procedures might be getting water held under different tensions, leading to a need for "standard operating procedures."

Chapter 9 Quiz

Q9.1. According to the author in Section 9.1, why are the 1970s-era process diagrams of infiltration excess, saturation excess and subsurface stormflow considered insufficient as perceptual models of streamflow generation?

A. They only apply to large river basins and are not relevant to the headwater catchments discussed in the book.
B. They focus too heavily on the role of vegetation, which is now understood to be of minimal importance.
C. They are fundamentally incorrect and have been completely replaced by modern theories.
D. They fail to capture the aggregate behavior of interacting components, such as nonlinearities and storage dynamics.

Hint: The author describes these diagrams as a necessary but not sufficient part of a larger, more complex picture.

Q9.2. In Section 9.4, the author challenges the mindset that accurately predicting a streamflow hydrograph

is the ultimate test of a model. What is the primary reason for this critique?

A. Field measurements of streamflow are often too inaccurate to provide a reliable basis for rigorous model evaluation.
B. Modern hydrological models are too complex to be evaluated by a single performance metric.
C. Multiple models with different internal structures can produce the same hydrograph, while failing to correctly represent the water's origin, age or pathways.
D. The hydrograph only represents the celerity (pressure wave) and does not reveal the velocity (particle movement) of water.

Hint: Consider what the hydrograph shape represents versus what it does *not* represent about the water itself.

Q9.3. What are the two primary factors identified in Section 9.3 that thwart the understanding of streamflow generation in headwater catchments?

A. The dominance of simple overland flow and the uniformity of subsurface pathways.
B. The excessive cost of hydrological models and a lack of trained modelers.
C. The failure of remote sensing technology and the political difficulty of establishing research sites.
D. A lack of gauging data and extraordinary levels of process heterogeneity.

Hint: The section title offers a clue about one of the factors, and the text explicitly mentions a "dual problem."

Q9.4. The author strongly rejects the notion that "every catchment is a law unto itself." What alternative perspective is advocated for in the book?

A. There is a firm belief in grouping dominant behaviors to show that similar perceptual models can explain streamflow in multiple places.
B. Catchment behavior is entirely random and unpredictable, making any attempt at grouping or classification futile.
C. Each catchment's uniqueness is so profound that only site-specific, idiographic descriptions are scientifically valid.
D. All catchments can be described by a single, universal hydrological model if enough data is provided.

Hint: This question is about finding similarities and general principles versus focusing only on unique differences.

Q9.5. According to the text, how did modeling contribute to the understanding of the "fill-and-spill" behavior on the Panola hillslope?

A. The model disproved the "fill-and-spill" hypothesis that was based on initial field observations.
B. The model demonstrated that "fill-and-spill" was a minor process only possible under extreme, unrealistic rainfall conditions.
C. The model was parameterized with detailed well data to confirm that "fill-and-spill" was the dominant process.
D. An early 2D model simulation visualized patches of saturation filling and spilling, which guided the subsequent placement of wells for field testing.

Hint: Think about the sequence of events and the dialogue between modeling and fieldwork described in this personal example.

Q9.6. What is the author's primary concern regarding the recent large-scale efforts to group and classify catchment behaviors using datasets like CAMELS?

A. The classification methods are statistically flawed and produce groupings that are not meaningful.
B. These efforts might give students the impression that more is known than is actually true, disincentivizing new fieldwork.
C. The groupings focus too much on subsurface stormflow and largely ignore overland flow processes.
D. The datasets used are too small and geographically biased to be scientifically useful.

Hint: The author describes these developments as "exciting and useful" but then expresses a specific "fear" about their impact on future research.

Q9.7. What is the critical distinction between "residence time" and "transit time" as defined in the provided glossary?

A. Transit time is the instantaneous speed of a pressure wave, while residence time is the speed of a water molecule.
B. Residence time is the duration a water molecule spends in storage, while transit time is the duration it spends traveling to the stream.
C. Residence time applies to groundwater, while transit time applies only to surface runoff.
D. They are interchangeable terms for the amount of time water is present in a catchment.

Hint: One term relates to being held within a system, while the other relates to the journey to exit that system.

Q9.8. As defined in the glossary and discussed in Section 9.4, what distinguishes "soft data" from traditional "hard data" in hydrological modeling?

A. Soft data is any data collected manually, while hard data is collected by automated sensors.

B. Soft data only includes social or economic information from stakeholders.
C. Soft data is often qualitative, spotty or approximate, and is used to assess a model's process realism.
D. Soft data is derived from model outputs, whereas hard data consists of direct field measurements.

Hint: Think about the nature of the information (precise time series vs. conceptual or approximate values) and its purpose in model evaluation.

Q9.9. What is the author's proposed "top-down" approach for starting research at a new site, as described in Section 9.2?

A. Using three largely orthogonal measures: the runoff response signature, the percentage of event water and stream water age.
B. First calculating the total annual precipitation, average air temperature and evapotranspiration rate for the water balance.
C. Measuring soil permeability, topographic slope and vegetation density to classify the catchment first.
D. Determining the catchment area, stream order and channel gradient from topographic maps.

Hint: The text describes a set of three "non-redundant" measures that help an investigator understand "what sort of beast you might be dealing with."

Q9.10. What is the primary reason the author highlights the urgency of refining perceptual models and conducting new fieldwork today?

A. Global changes, such as shifts in precipitation patterns and land use, are altering how headwater catchments function.
B. The bias in gauging station locations means that our existing understanding is fundamentally flawed and must be replaced.
C. A new generation of complex hydrological models requires more precise input parameters from the field.
D. The number of academic positions for field hydrologists is declining, creating a knowledge gap.

Hint: Think about the external, large-scale pressures on hydrological systems that are mentioned toward the end of the chapter.

Q9.11. The author uses a personal example from the Maimai catchment to illustrate a key benefit of modeling. What did the Hill-Vi model help the author understand?

A. The exact depth and geographic location of every major soil pipe on the Maimai hillslope.
B. That subsurface topography was the sole factor controlling runoff, with soil pipes having a negligible effect.
C. That soil pipes were not a significant contributor to streamflow, contrary to what was believed from field observations.
D. How randomly placed soil pipes could interact with subsurface topography to produce emergent, organized network behavior.

Hint: The example is about how unseen components can "conspire" to create a larger, observable pattern at the hillslope scale.

Q9.12. Which of the following best describes the concept of "preferential flow" as it is used throughout the book?

A. Water movement through non-capillary void spaces like macropores or soil pipes, bypassing the main soil matrix.
B. The tendency for water to flow toward areas with the lowest topographic elevation in a catchment.
C. The movement of water exclusively through the weathered bedrock layer, completely avoiding the soil.
D. The uniform, piston-like displacement of old water by new water as it moves through the soil profile.

Hint: The glossary provides a concise definition and links the term to "macropore flow."

References

Chapter 1

Beven, K.J., 1987. Towards a new paradigm in hydrology. In *Water for the Future: Hydrology in Perspective*, IAHS Press Publication, pp. 393–403.

Beven, K., 2021. The era of infiltration. *Hydrology and Earth System Sciences Discussions*, *25*(2), 851–866.

Beven, K.J. and Chappell, N.A., 2021. Perceptual perplexity and parameter parsimony. *Wiley Interdisciplinary Reviews: Water*, *8*(4), e1530.

Beven, K., Archfield, S., Batelaan, O., Chen, C., Fenicia, F., Gascuel-Odoux, C., Marshall, L., McDonnell, J., Mendiondo, E.M., Prieto, C. and Vimal, S., 2025. On the value of a history of hydrology and the establishment of a History of Hydrology Working Group. *Hydrological Sciences Journal*, *70*(5), 717–729.

Brutsaert, W., 2005. *Hydrology: An Introduction*. Cambridge University Press.

Burt, T.P. and McDonnell, J.J., 2015. Whither field hydrology? The need for discovery science and outrageous hydrological hypotheses. *Water Resources Research*, *51*(8), 5919–5928.

Cook, H.L., 1946. The infiltration approach to the calculation of surface runoff. *Eos, Transactions American Geophysical Union*, *27*(5), 726–747.

Deng, N., Gu, X., Fan, Y., Havlin, S. and Zeng, A., 2025. The critical role of persistent disruption in advancing science. *Nature Computational Science*, *5*(6), 492–501.

Gleason, C.J. and Brown, C.M., 2025. The once and future hydrology is whole Earth hydrology. *Perspectives of Earth and Space Scientists*, *6*(1), e2024CN000267.

Golden, H.E., Christensen, J.R., McMillan, H.K., Kelleher, C.A., Lane, C.R., Husic, A., Li, L., Ward, A.S., Hammond, J., Seybold, E.C., Jaeger, K.L., Zimmer, M., Sando, R., Jones, C.N., Segura, C., Mahoney, D.T., Price, A.N. and Cheng, F., 2025. Advancing the science of headwater streamflow for global water protection. *Nature Water*, *3*(1), 16–26.

Grant, G.E. and Dietrich, W.E., 2017. The frontier beneath our feet. *Water Resources Research*, *53*(4), 2605–2609.

Gregory, K.J. and Walling, D.E., 1973. *Drainage basin form and process: A geomorphological approach*. Arnold.

Kirchner, J.W., 2006., Getting the right answers for the right reasons: Linking measurements, analyses, and models to advance the science of hydrology. *Water Resources Research*, *42*(3), W03S04.

Kirchner, J.W., Benettin, P. and van Meerveld, I., 2023. Instructive surprises in the hydrological functioning of landscapes. *Annual Review of Earth and Planetary Sciences*, *51*(1), 277–299.

Kirkby, M.J., 2004. *Hillslope geomorphology*. In *Geomorphology: Critical Concepts in Geography*, *2*. Routledge, pp. 562–594.

Kuhn, T.S., 1997. *The structure of scientific revolutions* (Vol. 962). Chicago: University of Chicago Press.

McGuire, K.J., Klaus, J., and Jackson, C.R., 2024. Interflow, subsurface stormflow and throughflow: A synthesis of field work and modelling. *Hydrological Processes*, *38*, e15263.

McMillan, H., 2022. A taxonomy of hydrological processes and watershed function. *Hydrological Processes*, *36*(3), e14537.

McMillan, H., Araki, R., Bolotin, L., Kim, D.H., Coxon, G., Clark, M. and Seibert, J., 2025. Global patterns in observed hydrologic processes. *Nature Water*, *3*(4), pp. 497–506.

Montgomery, D.R., and Foufoula-Georgiou, E., 1993. Channel network source representation using digital elevation models. *Water Resources Research*, *29*(12), 3925–3934.

Nearing, G.S., Kratzert, F., Sampson, A.K., Pelissier, C.S., Klotz, D., Frame, J.M., Prieto, C. and Gupta, H.V., 2021. What role does hydrological science play in the age of machine learning? *Water Resources Research*, *57*(3), e2020WR028091.

Park, M., Leahey, E. and Funk, R.J., 2023. Papers and patents are becoming less disruptive over time. *Nature*, *613*(7942), 138–144.

Pfister, L. and Kirchner, J.W., 2017. Debates—Hypothesis testing in hydrology: Theory and practice. *Water Resources Research.*, *53*(3), 1792–1798.

Pilkey, O.H., and Pilkey-Jarvis, L., 2007. *Useless arithmetic: Why environmental scientists can't predict the future.* Columbia University Press.

Poff, N.L., Bledsoe, B.P. and Cuhaciyan, C.O., 2006. Hydrologic variation with land use across the contiguous United States: geomorphic and ecological consequences for stream ecosystems. *Geomorphology*, *79*(3–4), 264–285.

Sayama, T. and McDonnell, J.J., 2009. A new time–space accounting scheme to predict stream water residence time and hydrograph source components at the watershed scale. *Water Resources Research*, *45*(7), W07401.

Smith, C.T., 1969. *The drainage basin as an historical basis for human activity*. In Introduction to Geographical Hydrology. Routledge, pp. 20–29.

Strahler, A.N., 1952. Hypsometric (area–altitude) analysis of erosional topography. *Geological Society of America Bulletin*, *63*(11), 1117–1142.

Strahler, A.N., 1957. Quantitative analysis of watershed geomorphology. *Eos, Transactions American Geophysical Union*, *38*(6), 913–920.

Taleb, N.N., 2007. Black swans and the domains of statistics. *The American Statistician*, *61*(3), pp. 198–200.

Thompson, E., 2022. *Escape from Model Land: How Mathematical Models can Lead us Astray and What We Can Do About It*. Basic Books.

Van Stan, J.T., Allen, S.T., Aubrey, D.P., Berry, Z.C., Biddick, M., Coenders-Gerrits, M.A., Giordani, P., Gotsch, S.G., Gutmann, E.D., Kuzyakov, Y. and Magyar, D., 2023. Shower thoughts: why scientists should spend more time in the rain. *Bioscience*, *73*(6), 441–452.

Wagener, T., Gleeson, T., Coxon, G., Hartmann, A., Howden, N., Pianosi, F., Rahman, M., Rosolem, R., Stein, L. and Woods, R., 2021. On doing hydrology with dragons: Realizing the value of perceptual models and knowledge accumulation. *Wiley Interdisciplinary Reviews: Water*, *8*(6), e1550.

Chapter 2

Adams F., 1938. *The Birth and Development of the Geological Sciences*. Williams & Wilkins Co.

Anderson, M.G. and Burt, T.P., 1977. Automatic monitoring of soil moisture conditions in a hillslope spur and hollow. *Journal of Hydrology*, *33*(1–2), 27–36.

Beardmore, N., 1862. *Manual of Hydrology*. Waterlow and Sons.

Berg, E.V., 1945. Some results of storm runoff investigation in small catchments of the Valdaj Highland. *Papers of the Research Institute of the Central Board of the Hydrometeorological Service*, *4*(17), 47–63

Betson, R.P., 1964. What is watershed runoff? *Journal of Geophysical Research*, *69*(8), 1541–1552.

Beven, K.J., 2001. *Rainfall-runoff Modelling: The Primer*, 1st Edition. John Wiley and Sons.

Beven, K., 2004a. Robert E. Horton's perceptual model of infiltration processes. *Hydrological Processes*, *18*(17), 3447–3460.

Beven, K., 2004b. Robert E. Horton and abrupt rises of ground water. *Hydrological Processes*, *18*(18), 3687–3696.

Beven, K.J., 2006. *Streamflow generation processes*. International Association of Hydrological Sciences.

Beven, K., 2021. The era of infiltration. *Hydrology and Earth System Sciences Discussions*, *25*, 851–866.

Beven, K., Archfield, S., Batelaan, O., Chen, C., Fenicia, F., Gascuel-Odoux, C., Marshall, L., McDonnell, J., Mendiondo, E.M., Prieto, C. and Vimal, S., 2025. On the value of a history of hydrology and the establishment of a History of Hydrology Working Group. *Hydrological Sciences Journal*, *70*(5), 717–729.

Biswas, A.K., 1970. *History of Hydrology*. North Holland Publishing Company.

Brutsaert, W., 2005. *Hydrology: An Introduction*. Cambridge University Press.

Burt, T.P. and Swank, W.T., 2010. Hursh CR and Brater EF (1941) Separating storm-hydrographs from small drainage-areas into surface-and subsurface-flow. Transactions, American Geophysical Union 22: 863–871. *Progress in Physical Geography*, *34*(5), 719–726.

Burykin, A.M., 1957. On subsurface flow in mountain conditions of humid subtropics. *Soil Science*, *12*, 90–97.

Cappus, P., 1960. Etude des lois de l'écoulement-Application au calcul et à la prévision des débits. *La houille blanche*, *46*(1), 493–520.

Chorley, R.J., 1978. *The hillslope hydrological cycle*. In Hillslope Hydrology. John Wiley and Sons, pp. 1–42.

Cook, H.L., 1946. The infiltration approach to the calculation of surface runoff. *Eos, Transactions American Geophysical Union*, *27*(5), 726–747.

Dalton, J., 1798. *Memoirs of the Literary and Philosophical Society of Manchester, Volume 5*. Manchester Literary and Philosophical Society.

Dalton, J., 1802. Experiments and observations to determine whether the quantity of rain and dew is equal to the quantity of water carried off by the rivers and raised by evaporation; with an enquiry into the origin of springs. *Memoirs of the Literary and Philosophical Society of Manchester*, *5*(2), pp. 346–372.

Daraselija, K.M. 1949. *Red and Podzol soils of Georgia and their Use for Subtropical Crops*. Agriculture, *445*.

Duffy, C.J., 2017. The terrestrial hydrologic cycle: an historical sense of balance. *Wiley Interdisciplinary Reviews: Water*, *4*(4), e1216.

Engler, A., 1919. *Untersuchungen über den Einfluss des Waldes auf den Stand der Gewässer, 12*. Kommissionsverlag von Beer & Cie.

Ferguson, G., McIntosh, J.C., Jasechko, S., Kim, J.H., Famiglietti, J.S. and McDonnell, J.J., 2023. Groundwater deeper than 500 m contributes less than 0.1% of global river discharge. *Communications Earth & Environment, 4*(1), 48.

Fletcher, J.E., Harris, K., Peterson, H.G. and Chandler, V.N., 1954. Piping. *Transactions of the American Geophysical Union, 35*, 258–263.

Freeze, R.A., 1974. Streamflow generation. *Reviews of Geophysics, 12*(4), 627–647.

Hall, S., 1918. Stream-flow and percolation-water. *Transactions of the Institution of Water Engineers, 23*, 92–127.

Holko, L., Pekárová, P., Szolgay, J., Miklánek, P. and Babiaková, G., 2025. History of hydrological research in Slovakia and its links to water management. *Hydrological Sciences Journal, 70*(1), 10–26.

Horton, R.E., 1931. The field, scope, and status of the science of hydrology. *Eos, Transactions American Geophysical Union, 12*(1), 189–202.

Horton, R.E., 1933. The role of infiltration in the hydrologic cycle. *Eos, Transactions American Geophysical Union, 14*(1), 446–460.

Horton, R.E., 1935. Surface Runoff Phenomena, Part I – Analysis of the hydrograph, Publication 101. *Horton Hydrological Laboratory*.

Horton, R.E., 1936. Maximum ground-water levels. *Eos, Transactions American Geophysical Union, 17*(2), 344–357.

Horton R.E., 1937. Determination of infiltration-capacity for large drainage basins. *Transactions, American Geophysical Union, 18*, 371–385

Horton, R.E., 1942. Remarks on hydrologic terminology. *Eos, Transactions American Geophysical Union, 23*(2), 479–482.

Horton. R. E., 1943. Discussion of "Infiltration capacities of some plant· soil complexes on Utah range watershed lands" by Lowell Woodward. *American Geophysical Union Transactions, 24*(2), 473–475.

Hoyt, W.G., 1936. *Studies of Relations of Rainfall and Run-Off in the United States* (Vol. 772). US Government Printing Office.

Hoyt, W.G. and Langbein, W.B., 1939. Some general observations of physiographic and climatic influences on floods. *Eos, Transactions American Geophysical Union, 20*(2), 166–174.

Hursh, C.R., 1936. Storm water and absorption. *Transactions, American Geophysical Union, 17*, 301–302.

Hursh, C.R., 1938. Mulching for road bank fixation. Technical Note 31. *USDA Forest Service.*

Hursh, C.R., 1944. Appendix B—Report of sub-committee on subsurface-flow. *Eos, Transactions American Geophysical Union, 25*(5), 743–746.

Hursh, C.R. and Brater, E.F., 1941. Separating storm-hydrographs from small drainage-areas into surface-and subsurface-flow. *Eos, Transactions American Geophysical Union, 22*(3), 863–871.

Hursh, C.R. and Fletcher, P.W., 1942. The soil profile as a natural reservoir. *Proceedings of Soil Science Society of America, 6*, 414–422.

Jasechko, S., Wada, Y., Birks, J., Welker, J., Gleeson, T., Fawcett, P., McDonnell, J.J and Sharp, Z., 2014. The pronounced seasonality of global groundwater recharge, *Water Resources Research, 50*, 8845–8867.

Jasechko, S., Perrone, D., Befus, K.M., Bayani Cardenas, M., Ferguson, G., Gleeson, T., Luijendijk, E., McDonnell, J.J., Taylor, R.G., Wada, Y. and Kirchner, J.W., 2017. Global aquifers dominated by fossil groundwaters but wells vulnerable to modern contamination. *Nature Geoscience, 10*(6), 425–429.

Kirkby, M.J. and Chorley, R.J., 1967. Throughflow, overland flow and erosion. *Hydrological Sciences Journal, 12*(3), 5–21.

Kuchment, L. and Gelfan, A., 2024. Development of runoff generation models in the former USSR and Russia: a historical overview. *Hydrological Sciences Journal, 69*(16), 2323–2336.

La Rocque, A., 1967. *On the origin of springs by Pierre Perrault*. Hafner Publishing Company.

Lowdermilk, W.C., 1934. The role of vegetation in erosion control and water conservation. *Journal of Forestry, 32*(5), 529–536.

McDonnell, J.J., 2023. Wei-Zu Gu and the remarkable rise of hydrological process research in China. *Hydrological Processes, 37*(5), e14896.

McDonnell, J.J., Beven, K., Morgenstern, U. and Pfister, L., 2025. The first catchment water balance: new insights into Pierre Perrault, his perceptual model and his peculiar catchment. *Hydrological Sciences Journal, 70*(1), 27–36.

McGuire, K.J., Klaus, J., and Jackson, C.R., 2024. Interflow, subsurface stormflow and throughflow: A synthesis of field work and modelling. *Hydrological Processes, 38*, e15263.

Meyer, A.F., 1917. *Elements of Hydrology*. John Wiley and Sons.

Nakamura, S., Tanaka, T., Iizumi, Y., Kubota, T., Kimura, M., Yoshida, T., Katsuyama, M. and Teramura, J., 2025. Emergence of, and developments in, hydrology (suimongaku) in Japan from the late 19th century to 1970. *Hydrological Sciences Journal, 70*(8), 1–11.

Nan, Y., Chen, C., Zhao, Y., Tian, F. and McDonnell, J., 2024. A historical overview of experimental hydrology in China. *Hydrological Processes, 38*(7), e15233.

Pavlić, K., Oskoruš, D., Ožanić, N. and Tadić, L., 2025. The history of hydrology in Croatia. *Hydrological Sciences Journal, 70*(6), 860–875.

Perrault, P., 1674. *De L'Origine des Fontaines*. Paris: Chez Pierre le Petit, Printer and Bookseller in ordinary to the King, Rue Saint Jacques à la Croix d'or.

Ramser, C.E., 1927. Run-off from small agricultural areas. *Journal of Agricultural Research, 34*, 797–823.

Reshetnikov, A.I., 1945. Snowmelt and rainfall runoff from the elementary plots in Valdaj region. *Transactions of the Research Institute of the Central Board of the Hydrometeorological Service, 4*(17), 38–46.

Richards, L.A., 1931. Capillary conduction of liquids through porous mediums. *Physics, 1*(5), 318–333.

Rode, A.A., 1954. About soil water and soil flow. *Soil Science, 9*, 52–63.

Rodnikov N.V., 1940. About Water Regime of the Podzol Soils under Forest and Meadow in Connection with Water Conservation Role of a Forest (O vodnom rezhime podzolistykh pochv pod lesom I lugom v svyazi s vodookhrannymi svoistvami lesa). Goslestekhizdat: Moscow; 40 pp.

Rutkovskij, V.I., 1940. A hydrological role of the forest and forestry industry. *Transaction of the All-Union Research Institute of Forestry, 18*, 65–112.

Schumacher, W., 1864. *Die physik des bodens*. Wiegandt & Hempel.

Smathktin, V.U., 2002. Some early Russian studies of subsurface storm-flow processes. *Hydrological Processes, 16*(13), 2613–2620.

Snyder, F.F., 1939. A conception of runoff-phenomena. *Eos, Transactions American Geophysical Union, 20*(4), 725–738.

Sozykin, N.F., 1940. Forest effects on soil moisture characteristics. *Transaction of the All-Union Research Institute of Forestry, 18*, 197–231.

Tóth, J., 1963. A theoretical analysis of groundwater flow in small drainage basins. *Journal of Geophysical Research, 68*(16), 4795–4812.

Tuan, Y-F., 1968. *The Hydrologic Cycle and the Wisdom of God: A theme in geoteleology*. University of Toronto Press.

Vasiliev I.S., 1948. The experience of studying surface and subsurface flow in a forest podzol soil. *Soil Science (Opyt izuchenija poverhnostnogo i vnutripochvennogo stoka v lesnoj podzolistoj pochve. Pochvovedenije) 5*, 312–324.

Vasiliev, I.S., 1950. Water regime of the podzol soils. *Papers of the Soil Science Institute of the Academy of Sciences of the USSR, 32*, 74–296.

Vimal, S., et al. 2025. Curating Robert Elmer Horton's bibliography. *Hydrological Sciences Journal*, in press.

Volftsun, I.B., 1954. Investigations of stormflow generation. *Papers of the State Hydrological Institute, 46*(100), 5–47.

Voskresenskij, K.P., 1948. A scheme for stormflow prediction accounting for catchment infiltration capacity and variable graph of runoff distribution. *Papers of the State Hydrological Institute, 6*(60), 5–41.

Weiler, M., McDonnell, J.J., Tromp-van Meerveld, I. and Uchida, T., 2005. *Subsurface stormflow*. In Encyclopedia of Hydrological Sciences, *20*(2), 377–394.

Werritty, A. and Black, A., 2025. WN McClean–pioneer of Scottish hydrometry. *Hydrological Sciences Journal, 70*(3), 422–435.

Chapter 3

Balek, J. and Perry, J., 1973. Hydrology of seasonally inundated African headwater swamps. *Journal of Hydrology, 19*(3), 227–249.

Beven, K.J., 2006. *Streamflow generation processes*. International Association of Hydrological Sciences.

Boelter, D.H. and Verry, E.S., 1977. *Peatland and water in the northern lake states* (No. 31). Department of Agriculture, Forest Service, North Central Forest Experiment Station.

Bonell, M., 1993. Progress in the understanding of runoff generation dynamics in forests. *Journal of Hydrology, 150*, 217–275.

Bonell, M. and Gilmour, D.A., 1978. The development of overland flow in a tropical rainforest catchment. *Journal of Hydrology, 39*(3–4), 365–382.

Bonell, M., Gilmour, D.A. and Sinclair, D.F., 1981. Soil hydraulic properties and their effect on surface and subsurface water transfer in a tropical rainforest catchment. *Hydrological Sciences Journal, 26*(1), 1–18.

Buttle, J., 2020. Clarifying misconceptions regarding the relationship between Hewlett and Hibbert's translatory flow process and ecohydrological separation. *Hydrological Processes, 34*(26), 5686–5689.

Cappus, P., 1960. Etude des lois de l'écoulement-Application au calcul et à la prévision des débits. *La houille blanche, 46*(1), 493–520.

Dewalle, D.R., Swistock, B.R. and Sharpe, W.E., 1988. Three-component tracer model for stormflow on a small Appalachian forested catchment. *Journal of Hydrology, 104*(1–4), 301–310.

Dickinson, W.T. and H. Whiteley., 1970. Watershed areas contributing to runoff. *International Association of Scientific Hydrology Bulletin, 96*(1), 12–26.

Dietrich, W.E., 1982. Geology, climate and hydrology of Barro Colorado Island. In *The ecology of tropical forest: seasonal rhythms and long-term changes*, Smithsonian Institution Press, 21–46.

Dunne, T., 1978. Field studies of hillslope flow processes. In, M.J. Kirkby (ed) *Hillslope hydrology*, John Wiley and Sons, pp. 227–293.

Dunne, T., 1983. Relation of field studies and modeling in the prediction of storm runoff. *Journal of Hydrology, 65*(1–3), 25–48.

Dunne, T. and Black, R.D., 1970a. An experimental investigation of runoff production in permeable soils. *Water Resources Research, 6*(2), 478–490.

Dunne, T. and Black, R.D., 1970b. Partial area contributions to storm runoff in a small New England watershed. *Water Resources Research, 6*(5), 1296–1311.

Dunne, T. and Leopold, L.B., 1978. *Water in Environmental Planning*. Freeman.

Dunne, T., Moore, T. and Taylor, C.H., 1975. Recognition and prediction of runoff-producing zones in humid regions. *Hydrological Sciences Bulletin, 20*(3), 305–327.

Elsenbeer, H., 2001. Hydrologic flowpaths in tropical rainforest soilscapes—a review. *Hydrological Processes, 15*(10), 1751–1759.

Elsenbeer, H. and Cassel, D.K., 1990. Surficial processes in the rainforest of western Amazonia. *Research Needs and Applications to Reduce Erosion and Sedimentation in Tropical Steeplands, 192*, 289–297.

Harr, R.D., 1977. Water flux in soil and subsoil on a steep forested slope. *Journal of Hydrology, 33*, 37–58.

Hewlett, J.D., 1961. *Watershed Management*. USDA Forest Service.

Hewlett, J.D. and Hibbert, A.R., 1963. Moisture and energy conditions within a sloping soil mass during drainage. *Journal of Geophysical Research, 68*(4), 1081–1087.

Hewlett, J.D. and Hibbert, A.R., 1967. Factors affecting the response of small watersheds to precipitation in humid areas. In Sopper, W.E. and Lull, H.W., editors, *Forest Hydrology*, Pergamon Press, pp. 275–290.

Hewlett, J.D. and Nutter, W.L., 1970. The varying source area of streamflow from upland basins. In *Proceedings of the Symposium on Interdisciplinary Aspects of Watershed Management*. American Society of Civil Engineers, New York, pp. 65–83.

Horton, J.H. and Hawkins, R.H., 1965. Flow path of rain from the soil surface to the water table. *Soil Science, 100*(6), 377–383.

Hursh, C.R. and Brater, E.F., 1941. Separating storm-hydrographs from small drainage-areas into surface-and subsurface-flow. *Eos, Transactions American Geophysical Union, 22*(3), 863–871.

Jenny, H., 1941. *Factors of soil formation: a system of quantitative pedology*. Republished 1994, Courier Corporation.

Kendall, K.A., Shanley, J.B. and McDonnell, J.J., 1999. A hydrometric and geochemical approach to test the transmissivity feedback hypothesis during snowmelt. *Journal of Hydrology, 219*(3–4), 188–205.

Kirchner, J.W., 2003. A double paradox in catchment hydrology and geochemistry. *Hydrological Processes, 17*(4), 871–874.

Kirkby, M.J. and Chorley, R.J., 1967. Throughflow, overland flow and erosion. *Hydrological Sciences Journal, 12*(3), 5–21.

McDonnell, J.J., 2009. Classics in Physical Geography Revisited: Hewlett JD, Hibbert AR. 1967. Factors affecting the response of small watersheds to precipitation in humid areas. *Progress in Physical Geography, 33*(2), 1–6.

McDonnell, J.J., 2009. Hewlett, J.D. and Hibbert, A.R., 1967: Factors affecting the response of small watersheds to precipitation in humid areas. In Sopper, W.E. and Lull, H.W., editors, *Forest Hydrology*, Pergamon Press, New York, pp. 275–290. *Progress in Physical Geography, 33*(2), 288–293.

McDonnell, J.J. and Beven, K., 2014. Debates-The future of hydrological sciences: A (common) path forward? A call to action aimed at understanding velocities, celerities and residence time distributions of the headwater hydrograph. *Water Resources Research, 50*(6), 5342–5350.

McDonnell, J.J. and Taylor, C.H., 1987. Surface and subsurface water contributions during snowmelt in a small Precambrian Shield watershed, Muskoka, Ontario. *Atmosphere-Ocean, 25*(3), 251–266.

McGlynn, B.L., McDonnell, J.J., Seibert, J. and Kendall, C., 2004. Scale effects on headwater catchment runoff timing, flow sources, and groundwater-streamflow relations. *Water Resources Research, 40*(7).

McMillan, H., Araki, R., Bolotin, L., Kim, D. H., Coxon, G., Clark, M. and Seibert, J., 2025. Global patterns in observed hydrologic processes. *Nature Water, 3*(4), 497–506.

Mosley, M.P., 1979. Streamflow generation in a forested watershed, New Zealand. *Water Resources Research, 15*(4), 795–806.

Musgrave, G.W. and Holtan, H.N., 1964. Infiltration. In *Handbook of Applied Hydrology*. McGraw-Hill.

Nan, Y., Chen, C., Zhao, Y., Tian, F. and McDonnell, J., 2024. A historical overview of experimental hydrology in China. *Hydrological Processes, 38*(7), e15233.

O'Brien, A.L., 1977. Hydrology of two small wetland basins in eastern Massachusetts. *Journal of the American Water Resources Association, 13*(2), 325–340.

Ragan, R.M., 1968. An experimental investigation of partial-area contributions. *International Association of Scientific Hydrology, 76*, 241–249.

Taylor, C.H., 1982. The effect on storm runoff response of seasonal variations in contributing zones in small watersheds. *Hydrology Research, 13*(3), 165–182.

Verry, E.S. and Boelter, D.H., 1975. The influence of bogs on the distribution of streamflow from small bog-upland catchments. *Forest Service*, United States Department of Agriculture, 469–478.

Zhao, R. and Zhuang, Y., 1963. Regional patterns of rainfall–runof f relationship. *Journal of Hohai University (Natural Sciences), S2*, pp. 53–68.

Chapter 4

Anderson, M.G. and Burt, T.P., 1978. The role of topography in controlling throughflow generation. *Earth Surface Processes, 3*(4), 331–344.

Arnett, R.R., 1974. Environmental factors affecting the speed and volume of topsoil interflow. In *Fluvial Processes in Instrumented Watersheds: Studies of small watersheds in the British Isles*. Institute of British Geographers, pp. 7–22.

Atkinson, T.C., 1978. Techniques for measuring subsurface flow on hillslopes. In Kirkby, M.J. editor, *Hillslope Hydrology*, John Wiley and Sons, pp. 73–120.

Betson, R.P. and Marius, J.B., 1969. Source areas of storm runoff. *Water Resources Research, 5*(3), 574–582.

Betson, R.P., Marius, J.B. and Joyce, R.T., 1968. Detection of saturated interflow in soils with piezometers. *Soil Science Society of America Journal, 32*(4), 602–604.

Beven, K., 2018. A century of denial: Preferential and nonequilibrium water flow in soils, 1864–1984. *Vadose Zone Journal, 17*(1), 1–17.

Beven, K.J. and Kirkby, M.J., 1979. A physically based, variable contributing area model of basin hydrology. *Hydrological Sciences Journal, 24*(1), 43–69.

Beven, K.J., Lamb, R., Kirkby, M.J. and Freer, J.E., 2021. A history of TOPMODEL. *Hydrology and Earth System Sciences Discussions, 25*, 527–549.

Bonell, M., 1993. Progress in the understanding of runoff generation dynamics in forests. *Journal of Hydrology, 150*, 217–275.

Bunting, B.T., 1961. The role of seepage moisture in soil formation, slope development and stream initiation. *American Journal of Science, 259*, 503–518.

Bunting, B.T., 1964. Slope development and soil formation on some British sandstones. *The Geographical Journal, 130*(1), 73–79.

Burt, T.P., 1978. An automatic fluid-scanning switch tensiometer system. British Geomorphological Research Group, Technical Bulletins, p. 30

Chappell, N.A., Bonell, M., Barnes, C.J. and Tych, W., 2012. Tropical cyclone effects on rapid runoff responses: Quantifying with new continuous-time transfer function models. *International Association of Hydrological Sciences Publication, 353*, 82–92.

Chifflard, P., Blume, T., Maerker, K., Hopp, L., Meerveld, I.V., Graef, T., Gronz, O., Hartmann, A., Kohl, B., Martini, E. and Reinhardt-Imjela, C., 2019. How can we model subsurface stormflow at the catchment scale if we cannot measure it? *Hydrological Processes, 33*(9), 1378–1385.

Dunne, T. and Black, R.D., 1970. An experimental investigation of runoff production in permeable soils. *Water Resources Research, 6*(2), 478–490.

Elsenbeer, H., 2001. Hydrologic flowpaths in tropical rainforest soilscapes—a review. *Hydrological Processes, 15*(10), 1751–1759.

Engler, A., 1919. *Untersuchungen über den Einfluss des Waldes auf den Stand der Gewässer, 12*. Kommissionsverlag von Beer & Cie.

Freeze, R.A., 1974. Streamflow generation. *Reviews of Geophysics, 12*(4), 627–647.

Gilman, K. and Newson, M.D., 1980. Soil pipes and pipeflow: A hydrological study in upland Wales. GeoBooks, p. 114.

Harr, R.D., 1977. Water flux in soil and subsoil on a steep forested slope. *Journal of Hydrology, 33*, 37–58.

Hewlett, J.D. and Hibbert, A.R., 1963. Moisture and energy conditions within a sloping soil mass during drainage. *Journal of Geophysical Research, 68*(4), 1081–1087.

Hursh, C.R., 1936. Storm water and absorption. *Transactions, American Geophysical Union, 17*, 301–302.

Jones, J.A.A., 1971. Soil piping and stream channel initiation. *Water Resources Research, 7*(3), 602–610.

Jones, J.A.A., 1979. Extending the Hewlett model of stream runoff generation. *Area*, 110–114.

Jones, J.A.A., 1994. Soil piping and its hydrogeomorphic function. *Cuaternario y Geomorfología, 8*(3–4), 77–102.

Jones, J.A.A., 2025. The view from the Welsh mountains: on the rise and proliferation of small basin studies in the UK and their legacy. *Hydrological Sciences Journal, 70*(8), 1260–1290.

Kearey, P., 2001. *The New Penguin Dictionary of Geology*, 2nd edition, Penguin Books Ltd.

Kendall, C., McDonnell, J.J. and Gu, W., 2001. A look inside 'black box' hydrograph separation models: a study at the Hydrohill catchment. *Hydrological Processes, 15*(10), 1877–1902.

Kirkby, M.J., 1978. *Hillslope Hydrology*. John Wiley and Sons.

Kirkby, M.J. and Chorley, R.J., 1967. Throughflow, overland flow and erosion. *Hydrological Sciences Journal, 12*(3), 5–21.

McDonnell, J.J., 2023. Wei-Zu Gu and the remarkable rise of hydrological process research in China. *Hydrological Processes, 37*(5), e14896.

McGuire, K.J., Klaus, J. and Jackson, C.R., 2024. Interflow, subsurface stormflow and throughflow: A synthesis of field work and modelling. *Hydrological Processes*, *38*, e15263.

McMillan, H., 2022. A taxonomy of hydrological processes and watershed function. *Hydrological Processes*, *36*(3), e14537.

Mosley, M.P., 1979. Streamflow generation in a forested watershed, New Zealand. *Water Resources Research*, *15*(4), 795–806.

Pearce, A.J. and McKerchar, A.I., 1979. Upstream generation of storm runoff. In *Physical Hydrology — New Zealand Experience.* New Zealand Hydrological Society, pp. 165–192.

Pearce, A.J., Rowe, L.K. and O'Loughlin, C.L., 1982. Hydrologic regime of undisturbed mixed evergreen forests, South Nelson, New Zealand. *Journal of Hydrology*, *21*(2), 98–116.

Ragan, R.M., 1968. An experimental investigation of partial-area contributions. *International Association of Scientific Hydrology*, *76*, 241–249.

Tsukamoto, Y., 1963. Storm discharge from an experimental watershed. *Journal of the Japanese Forestry Society*, *45*(6), 186–190.

Weyman, D.R., 1970. Throughflow on hillslopes and its relation to the stream hydrograph. *Hydrological Sciences Journal*, *15*(3), 25–33.

Weyman, D.R., 1973. Measurements of the downslope flow of water in a soil. *Journal of Hydrology*, *20*(3), 267–288.

Whipkey, R.Z., 1965. Subsurface stormflow from forested slopes. *Hydrological Sciences Journal*, *10*(2), 74–85.

Whipkey, R.Z. and Kirkby, M.J., 1978. Flow within soil. In Kirkby, M.J., editor, Hillslope Hydrology, John Wiley & Sons, pp. 121–144

Chapter 5

Abdul, A. S. and Gillham, R.W., 1984. Laboratory studies of the effects of the capillary fringe on streamflow generation. *Water Resources Research*, *20*(6), 691–698.

Abdul, A. and Gillham, R., 1989. Field studies of the effects of the capillary fringe on streamflow generation. *Journal of Hydrology*, *112*(1–2), 1–18.

Anderson, A.E., Weiler, M., Alila, Y. and Hudson, R.O., 2010. Piezometric response in zones of a watershed with lateral preferential flow as a first-order control on subsurface flow. *Hydrological Processes*, *24*(16), 2237–2247.

Anderson, S.P., Dietrich, W.E., Montgomery, D.R., Torres, R., Conrad, M.E. and Loague, K., 1997. Subsurface flow paths in a steep, unchanneled catchment. *Water Resources Research*, *33*(12), 2637–2653.

Begemann, F. and Libby, W.F., 1957. Continental water balance, ground water inventory and storage times, surface ocean mixing rates and world-wide water circulation patterns from cosmic-ray and bomb tritium. *Geochimica et Cosmochimica Acta*, *12*(4), 277–296.

Berman, E.S.F., Gupta, M., Gabrielli, C., Garland, T. and McDonnell, J.J., 2009. High-frequency field-deployable isotope analyzer for hydrological applications. *Water Resources Research*, *45*(10), W10201.

Biggin, D.S., 1971. *The use of natural tritium in hydrograph analysis*. Subsurface Hydrology Report No. 26, NERC Institute of Hydrology, 16.

Bishop K.H., 1991. Episodic increase in stream acidity, catchment flow pathways and hydrograph separation. Doctoral Dissertation, University of Cambridge.

Bottomley, D., Craig, D. and Johnston, L., 1984. Neutralization of acid runoff by groundwater discharge to streams in Canadian Precambrian Shield watersheds. *Journal of Hydrology*, *75*(1–4), 1–26.

Bottomley, D., Craig, D. and Johnston, L., 1986. Oxygen-18 studies of snowmelt runoff in a small Precambrian shield watershed: Implications for streamwater acidification in acid-sensitive terrain. *Journal of Hydrology*, *88*(3–4), 213–234.

Brown, R.M., 1970. Distribution of hydrogen isotopes in Canadian waters. In *Isotope Hydrology 1970. Proceedings of a Symposium on Use of Isotopes in Hydrology*. IAEA, p. 935.

Carsel, R.F. and Parrish, R.S., 1988. Developing joint probability distributions of soil water retention characteristics. *Water Resources Research*, *24*(5), 755–769.

Christophersen, N. and Hooper, R.P., 1992. Multivariate analysis of stream water chemical data: The use of principal components analysis for the end-member mixing problem. *Water Resources Research*, *28*(1), 99–107.

Christophersen, N., Kjærnsrød, S., Rodhe, A. and Johansson, I., 1985. Preliminary evaluation of flow patterns in the Birkenes catchment using natural 18O as a tracer. In *Hydrologic and Hydrogeochemical Mechanisms and Model Approaches to the Acidification of Ecological Systems*, Nordic Hydrological Program Report 10, 29–40.

Cloke, H.L., Anderson, M.G., McDonnell, J.J. and Renaud, J.P., 2006. Using numerical modelling to evaluate the capillary fringe groundwater ridging hypothesis of streamflow generation. *Journal of Hydrology*, *316*(1–4), 141–162.

Crouzet, E., Hubert, P., Olive, P., Siwertz, E. and Marce, A., 1970. Le tritium dans les mesures d'hydrologie de surface. Determination experimentale du coefficient de ruissellement. *Journal of Hydrology*, *11*(3), 217–229.

Detty, J.M. and McGuire, K.J., 2010. Threshold changes in storm runoff generation at a till-mantled headwater catchment. *Water Resources Research*, *46*(7).

Dewalle, D.R., Swistock, B.R. and Sharpe, W.E., 1988. Three-component tracer model for stormflow on a small Appalachian forested catchment. *Journal of Hydrology, 104*(1–4), 301–310.

Dinçer, T., Payne, B.R., Florkowski, T., Martinec, J. and Tongiorgi, E.G.E.I., 1970. Snowmelt runoff from measurements of tritium and oxygen-18. *Water Resources Research, 6*(1), 110–124.

Dracos, T., 1984. Reaction of groundwater bodies to infiltration. *Eos Transactions American Geophysical Union, 65*(16), p. 215.

Eriksson, E., 1958. The possible use of tritium for estimating groundwater storage. *Tellus, 10*(4), 472–478.

Fritz, P., Cherry, J.A., Sklash, M. and Weyer, K.U., 1976. Storm runoff analysis using environmental isotopes and major ions. In *Interpretation of Environmental Isotope and Hydrochemical Data in Groundwater.* International Atomic Energy Agency, 111–130.

Gillham, R.W., 1984. The capillary fringe and its effect on water-table response. *Journal of Hydrology, 67*(1–4), 307–324.

Grip, H. and Rodhe, A., 1985. *Water's way from rain to stream.* Forskningsrådens förlagstjänst, p. 156.

Harris, D.M., McDonnell, J.J. and Rodhe, A., 1995. Hydrograph separation using continuous open system isotope mixing. *Water Resources Research, 31*(1), 157–171.

Herrmann, A. and Stichler, W., 1980. Groundwater–runoff relationships. *Catena, 7*(2–3), 251–263.

Hooper, R.P. and Shoemaker, C.A., 1986. A comparison of chemical and isotopic hydrograph separation. *Water Resources Research, 22*(10), 1444–1454.

Hursh, C.R., 1944. Appendix B—Report of sub-committee on subsurface-flow. *Eos, Transactions American Geophysical Union, 25*(5), 743–746.

Jacks, G., Olofsson, E. and Werme, G., 1986. An acid surge in a well-buffered stream. *Ambio, 15*(5), 282–285.

Jasechko, S., 2019. Global isotope hydrogeology—review. *Reviews of Geophysics, 57*(3), 835–965.

Jones, J.A.A., 2025. The view from the Welsh mountains: on the rise and proliferation of small basin studies in the UK and their legacy. *Hydrological Sciences Journal, 70*(8), 1260–1290.

Kendall, K.A., Shanley, J.B. and McDonnell, J.J., 1999. A hydrometric and geochemical approach to test the transmissivity feedback hypothesis during snowmelt. *Journal of Hydrology, 219*(3–4), 188–205.

Kennedy, V.C., Kendall, C., Zellweger, G.W., Wyerman, T.A. and Avanzino, R.J., 1986. Determination of the components of stormflow using water chemistry and environmental isotopes, Mattole River basin, California. *Journal of Hydrology, 84*(1–2), 107–140.

Kirchner, J.W., 2016. Aggregation in environmental systems–Part 1: Seasonal tracer cycles quantify young water fractions, but not mean transit times, in spatially heterogeneous catchments. *Hydrology and Earth System Sciences, 20*(1), 279–297.

Klaus, J. and McDonnell, J.J., 2013. Hydrograph separation using stable isotopes: Review and evaluation. *Journal of Hydrology, 505*, 47–64.

Klaus, J., Zehe, E., Elsner, M., Külls, C. and McDonnell, J.J., 2013. Macropore flow of old water revisited: experimental insights from a tile-drained hillslope. *Hydrology and Earth System Sciences, 17*(1), 103–118.

Krouse, H., Holecek, G. and Steppuhn, H., 1978. Detecting snow water in streamflow by D/H and $^{18}O/^{16}O$ abundances in two western Canadian basins. In *Stable Isotopes in the Earth Sciences*. DSIR Science Information Division, 87–95.

Laudon, H., Taberman, I., Ågren, A., Futter, M., Ottosson-Löfvenius, M. and Bishop, K., 2013. The Krycklan Catchment Study—A flagship infrastructure for hydrology, biogeochemistry, and climate research in the boreal landscape. *Water Resources Research, 49*(10), 7154–7158.

Lundin, L., 1982. Soil moisture and ground water in till soil and the significance of soil type for runoff. Doctoral dissertation, Uppsala University, UNGI Report Nr. 56. 216 pp.

Martinec, J., Siegenthaler, U., Oeschger, H., and Tongiorgi, E., 1974. New insights into the run-off mechanism by environmental isotopes. In *Proceedings Symposium on Isotope Techniques in Groundwater Hydrology*. International Atomic Energy Agency, pp. 129–143.

Martinec, J., 1975. Subsurface flow from snowmelt traced by tritium. *Water Resources Research., 11*(3), 496–498.

McDonnell, J.J., 1990. A rationale for old water discharge through macropores in a steep, humid catchment. *Water Resources Research, 26*(11), 2821–2832.

McDonnell, J.J. and Buttle, J.M., 1998. Comment on "A deterministic—empirical model of the effect of the capillary-fringe on near-stream area runoff. 1. Description of the model" By Jayatilaka, CJ and Gillham, RW (*Journal of Hydrology* vol. 184 (1996) 299–315). *Journal of Hydrology, 207*(3–4), 280–285.

McDonnell, J.J., Gabrielli, C., Ameli, A., Ekanayake, J., Fenicia, F., Freer, J., Graham, C., McGlynn, B., Morgenstern, U., Pietroniro, A., Sayama, T., Seibert, J., Stewart, M., Vaché, K., Weiler, M. and Woods, R., 2021. The Maimai M8 experimental catchment database: Forty years of process-based research on steep, wet hillslopes. *Hydrological Processes, 35*(5), e14112.

McGlynn, B.L., McDonnel, J.J. and Brammer, D.D., 2002. A review of the evolving perceptual model of hillslope flowpaths at the Maimai catchments, New Zealand. *Journal of Hydrology, 257*(1–4), 1–26.

Montgomery, D.R., Dietrich, W.E., Torres, R., Anderson, S.P., Heffner, J.T. and Loague, K., 1997. Hydrologic response of a steep, unchanneled valley to natural and applied rainfall. *Water Resources Research, 33*(1), 91–109.

Mook, W.G., Groeneveld, D.J., Brouwn, A.E., and Van Ganswijk, A.J., 1974. Analysis of a run-off hydrograph by means of natural 18O. In *Isotope Techniques in Groundwater Hydrology*. International Atomic Energy Agency, pp. 145–156.

Mosley, M.P., 1979. Streamflow generation in a forested watershed, New Zealand. *Water Resources Research, 15*(4), 795–806.

Obradovic, M.M. and Sklash, M.G., 1986. An isotopic and geochemical study of snowmelt runoff in a small arctic watershed. *Hydrological Processes, 1*(1), 15–30.

Pearce, A.J., Stewart, M.K. and Sklash, M.G., 1986. Storm runoff generation in humid headwater catchments: 1. Where does the water come from? *Water Resources Research, 22*(8), 1263–1272.

Ragan, R.M., 1968. An experimental investigation of partial-area contributions. *International Association of Scientific Hydrology, 76*, 241–249.

Rasmussen Jr, T.C., Baldwin, R.H., Dowd, J.F. and Williams, A.G., 2000. Tracer vs. pressure wave velocities through unsaturated saprolite. *Soil Science Society of America Journal, 64*(1), 75–85.

Rodhe, A., 1981. Spring flood meltwater or groundwater? *Hydrology Research, 12*(1), 21–30.

Rodhe, A., 1984. Groundwater contribution in Swedish forested till soil as estimated by oxygen-18. In *Isotope Hydrology*, International Atomic Energy Agency, pp. 56–66.

Rodhe, A., 1987. The origin of streamwater traced by Oxygen-18. University of Uppsala.

Schwartz, F.W., 1980. Hydrogeological investigation of Muskeg River basin, Alberta. AOSERP Report 87.

Sklash, M.G., 1990. Environmental isotope studies of storm and snowmelt runoff. In *Process Studies in Hillslope Hydrology*, John Wiley and Sons, pp. 401–435.

Sklash, M.G., Beven, K.J., Gilman, K. and Darling, W.G., 1996. Isotope studies of pipeflow at Plynlimon, Wales, UK. Hydrological Processes, *10*(7), 921–944.

Sklash, M.G. and Farvolden, R.N., 1979. The role of groundwater in storm runoff. *Journal of Hydrology, 43*(1–4), 45–65.

Sklash, M.G. and Farvolden, R.N., 1980. Groundwater response during runoff events. In *Proceedings of the Symposium on Watershed Management*, American Society of Civil Engineers, pp. 743–754.

Sklash, M.G., Farvolden, R.N. and Fritz, P., 1976. A conceptual model of watershed response to rainfall, developed through the use of oxygen-18 as a natural tracer. *Canadian Journal of Earth Sciences, 13*(2), 271–283.

Sklash, M.G., Stewart, M.K. and Pearce, A.J., 1986. Storm runoff generation in humid headwater catchments: 2. A case study of hillslope and low-order stream response. *Water Resources Research, 22*(8), 1273–1282.

Torres, R., Dietrich, W.E., Montgomery, D.R., Anderson, S.P. and Loague, K., 1998. Unsaturated zone processes and the hydrologic response of a steep, unchanneled catchment. *Water Resources Research, 34*(8), 1865–1879.

van Genuchten, M.T., 1980. A closed-form equation for predicting the hydraulic conductivity of unsaturated soils. *Soil Science Society of America Journal, 44*(5), 892–898.

Vischer, D., 1970. Einfluss eines schwankenden Vorfluterspiegels auf den Grundwasserspiegel. *Schweizerische Bauzeitung, 88*(12), 243–246.

Waddington, J.M., Morris, P.J., Kettridge, N., Granath, G., Thompson, D.K. and Moore, P.A., 2015. Hydrological feedbacks in northern peatlands. *Ecohydrology, 8*(1), pp.113–127.

Waswa, G.W. and Lorentz, S.A., 2015. Energy considerations in groundwater-ridging mechanism of streamflow generation. *Hydrological Processes, 29*(23), 4932–4946.

Weiler, M., McGlynn, B.L., McGuire, K.J. and McDonnell, J.J., 2003. How does rainfall become runoff? A combined tracer and runoff transfer function approach. *Water Resources Research, 39*(11).

Zaltsberg, E., 1986. Comment on "Laboratory studies of the effects of the capillary fringe on streamflow generation" by AS Abdul and RW Gillham. *Water Resources Research, 22*(5), 837–838.

Zimmermann, U., Munnich, K.O., Roether, W., Kruetz, W., Schubach, K. and Siegel, O., 1966. Tracers determine movement of soil moisture and evapotranspiration. *Science, 152*(3720), 346–347.

Chapter 6

Ameli, A.A., Gabrielli, C., Morgenstern, U. and McDonnell, J.J., 2018. Groundwater subsidy from headwaters to their parent water watershed: A combined field-modeling approach. *Water Resources Research, 54*(7), 5110–5125.

Anderson, M.G. and Burt, T.P., 1978. The role of topography in controlling throughflow generation. *Earth Surface Processes, 3*(4), 331–344.

Anderson, A.E., Weiler, M., Alila, Y. and Hudson, R.O., 2009. Dye staining and excavation of a lateral preferential flow network. *Hydrology and Earth System Sciences, 13*(6), 935–944.

Antoine, M., Javaux, M. and Bielders, C., 2009. What indicators can capture runoff-relevant connectivity properties of the micro-topography at the plot scale? *Advances in Water Resources, 32*(8), 1297–1310.

Appels, W.M., Graham, C.B., Freer, J.E. and McDonnell, J.J., 2015. Factors affecting the spatial pattern of bedrock groundwater recharge at the hillslope scale. *Hydrological Processes*, *29*(21), 4594–4610.

Asano, Y., Kawasaki, M., Saito, T., Haraguchi, R., Takatoku, K., Saiki, M. and Kimura, K., 2020. An increase in specific discharge with catchment area implies that bedrock infiltration feeds large rather than small mountain headwater streams. *Water Resources Research*, *56*(9), e2019WR025658.

Asano, Y., Kawasaki, M., Takatoku, K., Uchida, T. and Hiroki, S., 2023. Lateral flow in bedrock dominates baseflow and flood flow in a catchment underlain with permeable sedimentary rock in the high-relief Chichibu Mountains. *Hydrological Processes*, *37*(11), e15011.

Betson, R.P. and Marius, J.B., 1969. Source areas of storm runoff. *Water Resources Research*, *5*(3), 574–582.

Beven, K. and Germann, P., 2013. Macropores and water flow in soils revisited. *Water Resources Research*, *49*(6), 3071–3092.

Beven, K.J. and Kirkby, M.J., 1979. A physically based, variable contributing area model of basin hydrology. *Hydrological Sciences Journal*, *24*(1), 43–69.

Blume, T. and van Meerveld, H.J., 2015. From hillslope to stream: methods to investigate subsurface connectivity. *Wiley Interdisciplinary Reviews: Water*, *2*(3), 177–198.

Bracken, L.J. and Croke, J., 2007. The concept of hydrological connectivity and its contribution to understanding runoff-dominated geomorphic systems. *Hydrological Processes: An International Journal*, *21*(13), 1749–1763.

Bracken, L.J., Wainwright, J., Ali, G.A., Tetzlaff, D., Smith, M.W., Reaney, S.M. and Roy, A.G., 2013. Concepts of hydrological connectivity: Research approaches, pathways and future agendas. *Earth Science Reviews*, *119*, 17–34.

Burns, D.A., McDonnell, J.J., Hooper, R.P., Peters, N.E., Freer, J.E., Kendall, C. and Beven, K., 2001. Quantifying contributions to storm runoff through end-member mixing analysis and hydrologic measurements at the Panola Mountain Research Watershed (Georgia, USA). *Hydrological Processes*, *15*(10), 1903–1924.

Chen, X. and Duan, Z., 2015. Impacts of soil crusts on soil physicochemical characteristics in different rainfall zones of the arid and semi-arid desert regions of northern China. *Environmental Earth Sciences*, *73*, 3335–3347.

Chu, X., Yang, J., Chi, Y. and Zhang, J., 2013. Dynamic puddle delineation and modeling of puddle-to-puddle filling–spilling–merging–splitting overland flow processes. *Water Resources Research*, *49*(6), 3825–3829.

Clothier, B.E., Green, S.R. and Deurer, M., 2008. Preferential flow and transport in soil: progress and prognosis. *European Journal of Soil Science*, *59*(1), 2–13.

Coles, A.E. and McDonnell, J.J., 2018. Fill and spill drives runoff connectivity over frozen ground. *Journal of Hydrology*, *558*, 115–128.

Darboux, F., Davy, P. and Gascuel-Odoux, C., 2002a. Effect of depression storage capacity on overland-flow generation for rough horizontal surfaces: water transfer distance and scaling. *Earth Surface Processes and Landforms: The Journal of the British Geomorphological Research Group*, *27*(2), 177–191.

Darboux, F., Davy, P., Gascuel-Odoux, C. and Huang, C.H., 2002b. Evolution of soil surface roughness and flowpath connectivity in overland flow experiments. *Catena*, *46*(2–3), 125–139.

de Haan, J., 2006. How emergence arises. *Ecological Complexity*, *3*(4), 293–301.

Detty, J.M. and McGuire, K.J., 2010a. Topographic controls on shallow groundwater dynamics: implications of hydrologic connectivity between hillslopes and riparian zones in a till-mantled catchment. *Hydrological Processes*, *24*(16), 2222–2236.

Detty, J.M. and McGuire, K.J., 2010b. Threshold changes in storm runoff generation at a till-mantled headwater catchment. *Water Resources Research*, *46*(7), W07525.

Du, E., Jackson, C.R., Klaus, J., McDonnell, J.J., Griffiths, N.A., Williamson, M.F., Margaret, F., Greco, J.L. and Bitew, M., 2016. Interflow dynamics on a low relief forested hillslope: Lots of fill, little spill. *Journal of Hydrology*, *534*, 648–658.

Dunne, T., Zhang, W. and Aubry, B.F., 1991. Effects of rainfall, vegetation, and microtopography on infiltration and runoff. *Water Resources Research*, *27*(9), 2271–2285.

Freer, J., McDonnell, J.J., Beven, K.J., Peters, N.E., Burns, D.A., Hooper, R.P., Aulenbach, B. and Kendall, C., 2002. The role of bedrock topography on subsurface storm flow. *Water Resources Research*, *38*(12), 5–16.

Gabrielli, C.P., Morgenstern, U., Stewart, M.K. and McDonnell, J.J., 2018. Contrasting groundwater and streamflow ages at the Maimai watershed. *Water Resources Research*, *54*(6), 3937–3957.

Graham, C.B., Woods, R.A. and McDonnell, J.J., 2010. Hillslope threshold response to rainfall:(1) A field based forensic approach. *Journal of Hydrology*, *393*(1–2), 65–76.

Grayson, R.B. and Western, A.W., 1998. Towards areal estimation of soil water content from point measurements: time and space stability of mean response. *Journal of Hydrology*, *207*(1-2), 68–82.

Grayson, R.B., Western, A.W., Chiew, F.H. and Blöschl, G., 1997. Preferred states in spatial soil moisture patterns: Local and nonlocal controls. *Water Resources Research*, *33*(12), 2897–2908.

Guo, L., Lin, H., Fan, B., Nyquist, J., Toran, L. and Mount, G.J., 2019. Preferential flow through shallow fractured bedrock and a 3D fill-and-spill model of hillslope subsurface hydrology. *Journal of Hydrology, 576*, 430–442.

Hoogmoed, W.B. and Stroosnijder, L., 1984. Crust formation on sandy soils in the Sahel I. Rainfall and infiltration. *Soil and Tillage Research, 4*(1), 5–23.

Hopp, L. and McDonnell, J.J., 2009. Connectivity at the hillslope scale: Identifying interactions between storm size, bedrock permeability, slope angle and soil depth. *Journal of Hydrology, 376*(3–4), 378–391.

Jarvis, N.J., 2007. A review of non-equilibrium water flow and solute transport in soil macropores: Principles, controlling factors and consequences for water quality. *European Journal of Soil Science, 58*(3), 523–546.

Jencso, K.G., McGlynn, B.L., Gooseff, M.N., Wondzell, S.M., Bencala, K.E. and Marshall, L.A., 2009. Hydrologic connectivity between landscapes and streams: Transferring reach-and plot-scale understanding to the catchment scale. *Water Resources Research, 45*(4), W04428.

Kirkby, M. and Cerdà, A., 2021. Following the curve? Reviewing the physical basis of the SCS curve number method for estimating storm runoff. *Hydrological Processes, 35*(11), e14404.

Klaus, J. and Jackson, C.R., 2018. Interflow is not binary: A continuous shallow perched layer does not imply continuous connectivity. *Water Resources Research, 54*(9), 5921–5932.

Kosugi, K.I., Katsura, S.Y., Katsuyama, M. and Mizuyama, T., 2006. Water flow processes in weathered granitic bedrock and their effects on runoff generation in a small headwater catchment. *Water Resources Research, 42*(2), W02414.

Maestre, F.T., Bowker, M.A., Cantón, Y., Castillo-Monroy, A P., Cortina, J., Escolar, C., Escudero, A., Lázaro, R. and Martínez, I., 2011. Ecology and functional roles of biological soil crusts in semi-arid ecosystems of Spain. *Journal of Arid Environments, 75*(12), 1282–1291.

McDonnell, J.J., 1997. Comment on "The changing spatial variability of subsurface flow across a hillside" by Ross Woods and Lindsay Rowe. *Journal of Hydrology, New Zealand, 36*(1), 97–100.

McDonnell, J.J., 2013. Are all runoff processes the same? *Hydrological Processes, 27*(26), 4103–4111.

McDonnell, J.J., Freer, J., Hooper, R., Kendall, C., Burns, D., Beven, K. and Peters, J., 1996. New method developed for studying flow on hillslopes. *Eos, Transactions American Geophysical Union, 77*(47), 465–472.

McGlynn, B.L., McDonnell, J.J., Seibert, J. and Kendall, C., 2004. Scale effects on headwater catchment runoff timing, flow sources, and groundwater-streamflow relations. *Water Resources Research, 40*(7), W07504.

McGuire, K.J., Klaus, J. and Jackson, C.R., 2024. Interflow, subsurface stormflow and throughflow: A synthesis of field work and modelling. *Hydrological Processes, 38*, e15263.

McGuire, K.J. and McDonnell, J.J., 2010. Hydrological connectivity of hillslopes and streams: Characteristic time scales and nonlinearities. *Water Resources Research, 46*(10), W10543.

Messager, M.L., Lehner, B., Cockburn, C., Lamouroux, N., Pella, H., Snelder, T., Tockner, K., Trautmann, T., Watt, C. and Datry, T., 2021. Global prevalence of non-perennial rivers and streams. *Nature, 594*(7863), 391–397.

Montgomery, D.R., Dietrich, W.E., Torres, R., Anderson, S.P., Heffner, J.T. and Loague, K., 1997. Hydrologic response of a steep, unchanneled valley to natural and applied rainfall. *Water Resources Research, 33*(1), 91–109.

Morin, J. and Benyamini, Y., 1977. Rainfall infiltration into bare soils. *Water Resources Research, 13*(5), 813–817.

Onstad, C.A, 1984. Depressional storage on tilled soil surfaces. *Transactions of the ASAE, 27*(3), 729–732.

Penna, D., Borga, M., Norbiato, D. and Dalla Fontana, G., 2009. Hillslope scale soil moisture variability in a steep alpine terrain. *Journal of Hydrology, 364*(3-4), 311–327.

Penna, D., Tromp-van Meerveld, H.J., Gobbi, A., Borga, M. and Dalla Fontana, G., 2011. The influence of soil moisture on threshold runoff generation processes in an alpine headwater catchment. *Hydrology and Earth System Sciences, 15*(3), 689–702.

Penna, D., Van Meerveld, H.J., Oliviero, O., Zuecco, G., Assendelft, R.S., Dalla Fontana, G. and Borga, M., 2015. Seasonal changes in runoff generation in a small forested mountain catchment. *Hydrological Processes, 29*(8), 2027–2042.

Peñuela, A., Darboux, F., Javaux, M. and Bielders, C.L., 2016. Evolution of overland flow connectivity in bare agricultural plots. *Earth Surface Processes and Landforms, 41*(11), 1595–1613.

Peñuela, A., Javaux, M. and Bielders, C.L., 2013. Scale effect on overland flow connectivity at the plot scale. *Hydrology and Earth System Sciences, 17*(1), 87–101.

Renard, K.G., Nichols, M.H., Woolhiser, D.A. and Osborn, H.B., 2008. A brief background on the US Department of Agriculture Agricultural Research Service Walnut Gulch experimental watershed. *Water Resources Research, 44*(5), W05S02.

Rinderer, M., Ali, G. and Larsen, L.G., 2018. Assessing structural, functional and effective hydrologic connectivity with brain neuroscience methods: State-of-the-art and research directions. *Earth Science Reviews, 178*, 29–47.

Rinderer, M., McGlynn, B.L. and van Meerveld, H.J., 2017. Groundwater similarity across a watershed derived from time-warped and flow-corrected time series. *Water Resources Research, 53*(5), 3921–3940.

Rinderer, M., van Meerveld, I., Stähli, M. and Seibert, J., 2016. Is groundwater response timing in a pre-alpine catchment controlled more by topography or by rainfall? *Hydrological Processes*, *30*(7), 1036–1051.

Rodhe, A., 1987. The origin of streamwater traced by Oxygen-18. University of Uppsala.

Sidle, R.C., Tsuboyama, Y., Noguchi, S., Hosoda, I., Fujieda, M. and Shimizu, T., 2000. Stormflow generation in steep forested headwaters: a linked hydrogeomorphic paradigm. *Hydrological Processes*, *14*(3), 369–385.

Soulsby, C., Tetzlaff, D., Rodgers, P., Dunn, S. and Waldron, S., 2006. Runoff processes, stream water residence times and controlling landscape characteristics in a mesoscale catchment: An initial evaluation. *Journal of Hydrology*, *325*(1–4), 197–221.

Tani, M., 1997. Runoff generation processes estimated from hydrological observations on a steep forested hillslope with a thin soil layer. *Journal of Hydrology*, *200*(1–4), 84–109.

Tetzlaff, D., Birkel, C., Dick, J., Geris, J. and Soulsby, C., 2014. Storage dynamics in hydropedological units control hillslope connectivity, runoff generation, and the evolution of catchment transit time distributions. *Water Resources Research*, *50*(2), 969–985.

Tromp-van Meerveld, H.J. and McDonnell, J.J., 2006a. Threshold relations in subsurface stormflow: 1. A 147-storm analysis of the Panola hillslope. *Water Resources Research*, *42*(2), W02410.

Tromp-van Meerveld, H.J. and McDonnell, J.J., 2006b. Threshold relations in subsurface stormflow: 2. The fill and spill hypothesis. *Water Resources Research*, *42*(2), W02411.

Tromp-van Meerveld, I. and Weiler, M., 2008. Hillslope dynamics modeled with increasing complexity. *Journal of Hydrology*, *361*(1-2), 24–40.

Tromp-van Meerveld, H.J., Peters, N.E. and McDonnell, J.J., 2007. Effect of bedrock permeability on subsurface stormflow and the water balance of a trenched hillslope at the Panola Mountain Research Watershed, Georgia, USA. *Hydrological Processes: An International Journal*, *21*(6), 750–769.

Tsuboyama, Y., Sidle, R.C., Noguchi, S. and Hosoda, I., 1994. Flow and solute transport through the soil matrix and macropores of a hillslope segment. *Water Resources Research*, *30*(4), 879–890.

Uchida, T., Kosugi, K.I. and Mizuyama, T., 2001. Effects of pipeflow on hydrological process and its relation to landslide: a review of pipeflow studies in forested headwater catchments. *Hydrological Processes*, *15*(11), 2151–2174.

Uchida, T., Tromp-van Meerveld, I. and McDonnell, J.J., 2005. The role of lateral pipe flow in hillslope runoff response: an intercomparison of non-linear hillslope response. *Journal of Hydrology*, *311*(1–4), 117–133.

Unnikrishna, P.V., McDonnell, J.J., Tarboton, D.G. and Kendall, C., 1995. Isotopic analysis of hydrologic processes in a small semiarid catchment. *IAHS Publications-Series of Proceedings and Reports-Intern Assoc Hydrological Sciences*, *229*, 295–304.

Van Meerveld, H.J., Seibert, J. and Peters, N.E., 2015. Hillslope–riparian-stream connectivity and flow directions at the Panola Mountain Research Watershed. *Hydrological Processes*, *29*(16), 3556–3574.

Wainwright, J., Turnbull, L., Ibrahim, T G., Lexartza-Artza, I., Thornton, S.F. and Brazier, R.E., 2011. Linking environmental régimes, space and time: Interpretations of structural and functional connectivity. *Geomorphology*, *126*(3–4), 387–404.

Western, A.W. and Grayson, R.B., 1998. The Tarrawarra data set: Soil moisture patterns, soil characteristics, and hydrological flux measurements. *Water Resources Research*, *34*(10), 2765–2768.

Western, A. and Grayson, R., 2000. Soil moisture and runoff processes at Tarrawarra. In R. Grayson and G. Bloeschl (eds), *Spatial Patterns in Catchment Hydrology: Observations and Modelling*.Cambridge University Press, pp. 209–246.

Western, A.W., Grayson, R.B., Blöschl, G., Willgoose, G.R. and McMahon, T.A., 1999. Observed spatial organization of soil moisture and its relation to terrain indices. *Water Resources Research*, *35*(3), 797–810.

Wheater, H.S., 2025. Why is arid zone hydrology a scientific desert - and how does the hydrological community move forward? Environmental Modelling and Software Journal, *192*, 106561.

Wohl, E., Brierley, G., Cadol, D., Coulthard, T.J., Covino, T., Fryirs, K.A., Grant, G., Hilton, R.G., Lane, S.N., Magilligan, F.J., Meitzen, K.M., Passalacqua, P., Poeppl, R.E., Rathburn, S.L. and Sklar, L.S., 2019. Connectivity as an emergent property of geomorphic systems. *Earth Surface Processes and Landforms*, *44*(1), 4–26.

Woods, R. and Rowe, L., 1996. The changing spatial variability of subsurface flow across a hillside. *Journal of Hydrology (New Zealand)*, 35(1), 49–84.

Woods, R. and Rowe, L, 1997. Reply to "Comment on 'The changing spatial variability of subsurface flow across a hillside' By Ross Woods and Lindsay Rowe", By Jeffrey J. McDonnell. *Journal of Hydrology (New Zealand)*, *36*(1), 101–103.

Yair, A., 1990. Runoff generation in a sandy area—the Nizzana sands, Western Negev, Israel. *Earth Surface Processes and Landforms*, *15*(7), 597–609.

Yair, A.,1992. The control of headwater area on channel runoff in a small arid watershed. In *Overland Flow*, CRC Press, pp. 48–62.

Yair, A., Sharon, D. and Lavee, H., 1980. Trends in runoff and erosion processes over an arid limestone hillside, northern Negev, Israel. *Hydrological Sciences Journal*, *25*(3), 243–255.

Zhang, J., Wang, S., Fu, Z., Wang, F., Wang, K. and Chen, H., 2024. Soil thickness influences the control effect of micro-topography on subsurface runoff generation in the karst hillslope critical zone. *Catena*, *239*, 107957.

Chapter 7

Asano, Y., Uchida, T. and Ohte, N., 2002. Residence times and flow paths of water in steep unchannelled catchments, Tanakami, Japan. *Journal of Hydrology*, *261*(1–4), 173–192.

Benettin, P., Bailey, S.W., Campbell, J.L., Green, M.B., Rinaldo, A., Likens, G.E., McGuire, K.J. and Botter, G., 2015. Linking water age and solute dynamics in streamflow at the Hubbard Brook Experimental Forest, NH, USA. *Water Resources Research*, *51*(11), 9256–9272.

Benettin, P., Rodriguez, N.B., Sprenger, M., Kim, M., Klaus, J., Harman, C. J., van der Velde, Y., Hrachowitz, M., Botter, G., McGuire, K.J., Kirchner, J.W., Rinaldo, A. and McDonnell, J.J., 2022. Transit time estimation in catchments: recent developments and future directions. *Water Resources Research.*, *58*(11), e2022WR033096.

Benettin, P., Soulsby, C., Birkel, C., Tetzlaff, D., Botter, G. and Rinaldo, A., 2017. Using SAS functions and high-resolution isotope data to unravel travel time distributions in headwater catchments. *Water Resources Research*, *53*(3), 1864–1878.

Berman, E.S.F., Gupta, M., Gabrielli, C., Garland, T. and McDonnell, J.J., 2009. High-frequency field-deployable isotope analyzer for hydrological applications. *Water Resources Research*, *45*(10), W10201.

Berghuijs, W.R. and Kirchner, J.W., 2017. The relationship between contrasting ages of groundwater and streamflow. *Geophysical Research Letters*, *44*(17), 8925–8935.

Birkel, C., Tetzlaff, D., Dunn, S.M. and Soulsby, C., 2010. Towards a simple dynamic process conceptualization in rainfall–runoff models using multi-criteria calibration and tracers in temperate, upland catchments. *Hydrological Processes: An International Journal*, *24*(3), 260–275.

Birkel, C., Tetzlaff, D., Dunn, S.M. and Soulsby, C., 2011. Using time domain and geographic source tracers to conceptualize streamflow generation processes in lumped rainfall–runoff models. *Water Resources Research*, *47*(2), W02515.

Botter, G., Bertuzzo, E. and Rinaldo, A., 2011. Catchment residence and travel time distributions: The master equation. *Geophysical Research Letters*, *38*(11).

Broxton, P.D., Troch, P.A. and Lyon, S.W., 2009. On the role of aspect to quantify water transit times in small mountainous catchments. *Water Resources Research*, *45*(8), W08427.

Burgman, J.O., Calles, B., and Westman, F., 1987. Conclusions from a ten-year study of oxygen-18 in precipitation and runoff in Sweden. In *Isotope Techniques in Water Resources Development*, IAEA Division of Publications, pp. 579–590.

Burns, D.A., 2002. Stormflow-hydrograph separation based on isotopes: the thrill is gone—what's next? *Hydrological Processes*, *16*(7), 1515–1517.

Burns, D.A., Plummer, L.N., McDonnell, J.J., Busenberg, E., Casile, G.C., Kendall, C., Hooper, R. P., Freer, J.E., Peters, N.E., Beven, K. and Schlosser, P., 2003. The geochemical evolution of riparian ground water in a forested Piedmont catchment. *Ground Water*, *41*(7), 913–925.

Calabrese, S. and Porporato, A., 2017. Multiple outflows, spatial components, and nonlinearities in age theory. *Water Resources Research*, *53*(1), 110–126.

Chouard, T., 2008. Beneath the surface: you might think that once evolution has found one way to get something done, it will stick with it. But similar physical forms can hide radically different wiring, finds Tanguy Chouard. *Nature*, *456*(7220), 300–304.

Condon, L.E., Markovich, K.H., Kelleher, C.A., McDonnell, J.J., Ferguson, G. and McIntosh, J.C., 2020. Where is the bottom of a watershed? *Water Resources Research*, *56*(3), e2019WR026010.

Dunn, S.M., McDonnell, J.J and Vaché, K.B., 2007. Factors influencing the residence time of catchment waters: A virtual experiment approach. *Water Resources Research*, *43*(6).

Fan, Y., 2019. Are catchments leaky? *Wiley Interdisciplinary Reviews, Water*, *6*(6), e1386.

Fenicia, F., Wrede, S., Kavetski, D., Pfister, L., Hoffmann, L., Savenije, H.H.G. and McDonnell, J.J., 2010. Assessing the impact of mixing assumptions on the estimation of streamwater mean residence time, *Hydrological Processes*, *24*, 1730–1741.

Ferguson, G. and Grasby, S.E., 2011. Thermal springs and heat flow in North America. *Geofluids*, *11*(3), 294–301.

Ferguson, G., McIntosh, J.C., Jasechko, S., Kim, J.H., Famiglietti, J.S. and McDonnell, J.J., 2023. Groundwater deeper than 500 m contributes less than 0.1% of global river discharge. *Communications Earth & Environment*, *4*(1), 48.

Ferguson, G., McIntosh, J.C., Perrone, D. and Jasechko, S., 2018. Competition for shrinking window of low salinity groundwater. *Environmental Research Letters*, *13*(11), 114013.

Gabrielli, C.P., Morgenstern, U., Stewart, M.K. and McDonnell, J.J., 2018. Contrasting groundwater and streamflow ages at the Maimai watershed. *Water Resources Research*, *54*(6), 3937–3957.

Gleeson, T. and Manning, A.H., 2008. Regional groundwater flow in mountainous terrain: Three-dimensional simulations of topographic and hydrogeologic controls. *Water Resources Research*, *44*(10), W10403.

Hale, V.C. and McDonnell, J.J., 2016. Effect of bedrock permeability on stream base flow mean transit time scaling relations: 1. A multiscale catchment intercomparison. *Water Resources Research*, *52*(2), 1358–1374.

Hale, V.C., McDonnell, J.J., Stewart, M.K., Solomon, D.K., Doolittle, J., Ice, G.G. and Pack, R.T., 2016. Effect of bedrock permeability on stream base flow mean transit time scaling relationships: 2. Process study of storage and release. *Water Resources Research*, *52*(2), 1375–1397.

Harman, C.J., 2015. Time-variable transit time distributions and transport: Theory and application to storage-dependent transport of chloride in a watershed. *Water Resources Research*, *51*(1), 1–30.

Heidbüchel, I., Troch, P.A., Lyon, S.W. and Weiler, M., 2012. The master transit time distribution of variable flow systems. *Water Resources Research*, *48*(6), W06520.

Hewlett, J.D. and Hibbert, A.R., 1963. Moisture and energy conditions within a sloping soil mass during drainage. *Journal of Geophysical Research*, *68*(4), 1081–1087.

Horton, J.H. and Hawkins, R.H., 1965. Flow path of rain from the soil surface to the water table. *Soil Science*, *100*(6), 377–383.

Hrachowitz, M., Soulsby, C., Tetzlaff, D., Dawson, J.J.C. and Malcolm, I.A., 2009. Regionalization of transit time estimates in montane catchments by integrating landscape controls. *Water Resources Research*, *45*(5), W05421.

Hrachowitz, M., Savenije, H., Bogaard, T.A., Tetzlaff, D. and Soulsby, C., 2013. What can flux tracking teach us about water age distribution patterns and their temporal dynamics? *Hydrology and Earth System Sciences*, *17*, 533–564.

Jasechko, S., 2019. Global isotope hydrogeology—review. *Reviews of Geophysics*, *57*(3), 835–965.

Jasechko, S., Kirchner, J.W., Welker, J.M. and McDonnell, J.J., 2016. Substantial proportion of global streamflow less than three months old. *Nature Geoscience*, *9*(2), 126–129.

Jasechko, S., Perrone, D., Befus, K.M., Bayani Cardenas, M., Ferguson, G., Gleeson, T., Luijendijk, E., McDonnell, J.J., Taylor, R.G., Wada, Y. and Kirchner, J.W., 2017. Global aquifers dominated by fossil groundwaters but wells vulnerable to modern contamination. *Nature Geoscience*, *10*(6), 425–429.

Jehn, F.U., Breuer, L., Kraft, P., Bestian, K. and Houska, T., 2021. Simple catchments and where to find them: The storage–discharge relationship as a proxy for catchment complexity. *Frontiers in Water*, *3*, 631651.

Katsuyama, M., Tani, M. and Nishimoto, S., 2010. Connection between streamwater mean residence time and bedrock groundwater recharge/discharge dynamics in weathered granite catchments. *Hydrological Processes*, *24*(16), 2287–2299.

Keller, C., Van Der Kamp, G. and Cherry, J., 1988. Hydrogeology of two Saskatchewan tills, I. Fractures, bulk permeability, and spatial variability of downward flow. *Journal of Hydrology*, *101*(1–4), 97–121.

Kirchner, J.W., 2009. Catchments as simple dynamical systems: Catchment characterization, rainfall-runoff modeling, and doing hydrology backward. *Water Resources Research*, *45*(2).

Kirchner, J.W., 2016a. Aggregation in environmental systems – Part 1: Seasonal tracer cycles quantify young water fractions, but not mean transit times, in spatially heterogeneous catchments. *Hydrology and Earth System Sciences*, *20*(1), 279–297.

Kirchner, J.W., 2016b. Aggregation in environmental systems – Part 2: Catchment mean transit times and young water fractions under hydrologic nonstationarity. *Hydrology and Earth System Sciences*, *20*(1), 299–328.

Kirchner, J.W., 2019. Quantifying new water fractions and transit time distributions using ensemble hydrograph separation: theory and benchmark tests. *Hydrology and Earth System Sciences*, *23*(1), 303–349.

Kirchner, J.W., Feng, X. and Neal, C., 2001. Catchment-scale advection and dispersion as a mechanism for fractal scaling in stream tracer concentrations. *Journal of Hydrology*, *254*(1–4), 82–101.

Kirchner, J.W., Feng, X., Neal, C. and Robson, A.J., 2004. The fine structure of water-quality dynamics: the (high-frequency) wave of the future. *Hydrological Processes*, *18*(7), 1353–1359.

Kirchner, J.W. and Knapp, J.L.A., 2020. Technical note: Calculation scripts for ensemble hydrograph separation. *Hydrology and Earth System Sciences*, *24*(11), 5539–5558.

Klaus, J. and McDonnell, J., 2013. Hydrograph separation using stable isotopes: Review and evaluation. *Journal of Hydrology*, *505*, 47–64.

Knapp, J.L.A., Neal, C., Schlumpf, A., Neal, M. and Kirchner, J.W., 2019. New water fractions and transit time distributions at Plynlimon, Wales, estimated from stable water isotopes in precipitation and streamflow. *Hydrology and Earth System Sciences*, *23*(10), 4367–4388.

Lee, R.M., McGuire, K.J., Strahm, B.D., Knoepp, J.D., Jackson, C.R. and Stewart, R.D., 2020. Revisiting the Hewlett and Hibbert (1963) hillslope drainage experiment and modeling effects of decadal pedogenic processes and leaky soil boundary conditions. *Water Resources Research, 56*(1), e2019WR025090.

Li, Z. and Si, B., 2018. Reconstructed precipitation tritium leads to overestimated groundwater recharge. *Journal of Geophysical Research., 123*(17), 9858–9867.

Lyon, S.W., Laudon, H., Seibert, J., Mörth, M., Tetzlaff, D. and Bishop, K.H., 2010. Controls on snowmelt water mean transit times in northern boreal catchments. *Hydrological Processes, 24*(12), 1672–1684.

Marçais, J., Derry, L.A., Guillaumot, L., Aquilina, L. and de Dreuzy, J.R., 2022. Dynamic contributions of stratified groundwater to streams controls seasonal variations of streamwater transit times. *Water resources research, 58*(3), e2021WR029659.

Markovich, K.H., Manning, A.H., Condon, L.E. and McIntosh, J.C., 2019. Mountain-block recharge: a review of current understanding. *Water Resources Research, 55*(11), 8278–8304.

Martinec, J., Siegenthaler, U., Oeschger, H., and Tongiorgi, E., 1974. New insights into the run-off mechanism by environmental isotopes. In *Proceedings Symposium on Isotope Techniques in Groundwater Hydrology*, International Atomic Energy Agency, pp. 129–143.

McDonnell, J.J., 2017. Beyond the water balance. *Nature Geoscience, 10*(6), 396.

McGlynn, B., McDonnell, J., Stewart, M. and Seibert, J., 2003. On the relationships between catchment scale and streamwater mean residence time. *Hydrological Processes, 17*(1), 175–181.

McGuire, K.J. and McDonnell, J.J., 2006. A review and evaluation of catchment transit time modeling. *Journal of Hydrology, 330*(3–4), 543–563.

McGuire, K.J. and McDonnell, J.J., 2010. Hydrological connectivity of hillslopes and streams: Characteristic time scales and nonlinearities. *Water Resources Research, 46*(10), W10543.

McGuire, K.J., McDonnell, J.J., Weiler, M., Kendall, C., McGlynn, B.L., Welker, J.M. and Seibert, J., 2005. The role of topography on catchment-scale water residence time. *Water Resources Research, 41*(5), W05002.

McIntosh, J.C., Schlegel, M.E. and Person, M., 2012. Glacial impacts on hydrologic processes in sedimentary basins: evidence from natural tracer studies. *Geofluids, 12*(1), 7–21.

McNamara, J.P., Tetzlaff, D., Bishop, K., Soulsby, C., Seyfried, M., Peters, N.E., Aulenbach, B.T. and Hooper, R., 2011. Storage as a metric of catchment comparison. *Hydrological Processes, 25*(21), 3364–3371.

Muñoz-Villers, L.E. and McDonnell, J.J., 2012. Runoff generation in a steep, tropical montane cloud forest catchment on permeable volcanic substrate. *Water Resources Research, 48*(9).

National Research Council (NRC)., 2001. *Basic Research Opportunities in Earth Science*. The National Academies Press.

Pangle, L.A., Kim, M., Cardoso, C., Lora, M., Meira Neto, A.A., Volkmann, T.H.M., Wang, Y., Troch, P.A. and Harman, C.J., 2017. The mechanistic basis for storage-dependent age distributions of water discharged from an experimental hillslope. *Water Resources Research, 53*(4), 2733–2754.

Pearce, A.J., Stewart, M.K. and Sklash, M.G., 1986. Storm runoff generation in humid headwater catchments: 1. Where does the water come from? *Water Resources Research, 22*(8), 1263–1272.

Peters, N.E., Burns, D.A. and Aulenbach, B.T., 2014. Evaluation of high-frequency mean streamwater transit-time estimates using groundwater age and dissolved silica concentrations in a small forested watershed. *Aquatic Geochemistry, 20*(2–3), 183–202.

Pfister, L., Martínez-Carreras, N., Hissler, C., Klaus, J., Carrer, G.E., Stewart, M.K. and McDonnell, J.J., 2017. Bedrock geology controls on catchment storage, mixing, and release: A comparative analysis of 16 nested catchments. *Hydrological Processes, 31*(10), 1828–1845.

Rigon, R. and Bancheri, M., 2021. On the relations between the hydrological dynamical systems of water budget, travel time, response time and tracer concentrations. *Hydrological Processes, 35*(1).

Rigon, R., Bancheri, M. and Green, T.R., 2016. Age-ranked hydrological budgets and a travel time description of catchment hydrology. *Hydrology and Earth System Sciences, 20*(12), 4929–4947.

Rodhe, A., Nyberg, L. and Bishop, K., 1996. Transit times for water in a small till catchment from a step shift in the oxygen-18 content of the water input. *Water Resources Research, 32*(12), 3497–3511.

Sayama, T., McDonnell, J.J., Dhakal, A. and Sullivan, K., 2011. How much water can a watershed store? *Hydrological Processes, 25*(25), 3899–3908.

Soulsby, C. and Tetzlaff, D., 2008. Towards simple approaches for mean residence time estimation in ungauged basins using tracers and soil distributions. *Journal of Hydrology, 363*(1–4), 60–74.

Soulsby, C., Tetzlaff, D., Rodgers, P., Dunn, S. and Waldron, S., 2006. Runoff processes, stream water residence times and controlling landscape characteristics in a mesoscale catchment: An initial evaluation. *Journal of Hydrology, 325*(1–4), 197–221.

Spence, C., 2010. A paradigm shift in hydrology: Storage thresholds across scales influence catchment runoff generation. *Geography Compass, 4*(7), 819–833.

Sprenger, M., Stumpp, C., Weiler, M., Aeschbach, W., Allen, S.T., Benettin, P., Dubbert, M., Hartmann,

A., Hrachowitz, M., Kirchner, J.W., McDonnell, J.J., Orlowski, N., Penna, D., Pfahl, S., Rinderer, M., Rodriguez, N., Schmidt, M. and Werner, C., 2019. The demographics of water: a review of water ages in the critical zone. *Reviews of Geophysics.*, *57*(3), 800–834.

Stark, C.P. and Stieglitz, M., 2000. The sting in a fractal tail. *Nature*, *403*(6769), 493–495.

Stewart, M.K. and McDonnell, J.J., 1991. Modeling baseflow soil water residence times from deuterium concentrations. *Water Resources Research*, *27*(10), 2681–2694.

Tetzlaff, D., Seibert, J., McGuire, K.J., Laudon, H., Burns, D.A., Dunn, S.M. and Soulsby, C., 2009a. How does landscape structure influence catchment transit time across different geomorphic provinces? *Hydrological Processes: An International Journal*, *23*(6), 945–953.

Tetzlaff, D., Seibert, J. and Soulsby, C., 2009b. Inter-catchment comparison to assess the influence of topography and soils on catchment transit times in a geomorphic province; the Cairngorm mountains, Scotland. *Hydrological Processes: An International Journal*, *23*(13), 1874–1886.

Tetzlaff, D., Birkel, C., Dick, J., Geris, J. and Soulsby, C., 2014. Storage dynamics in hydropedological units control hillslope connectivity, runoff generation, and the evolution of catchment transit time distributions. *Water Resources Research*, *50*(2), 969–985.

Torres, M.A. and Baronas, J.J., 2021. Modulation of riverine concentration-discharge relationships by changes in the shape of the water transit time distribution. *Global Biogeochemical Cycles*, *35*(1), e2020GB006694.

Vaché, K.B. and McDonnell, J.J. 2006. A process-based rejectionist framework for evaluating catchment runoff model structure. *Water Resources Research*, *42*(2), W02409.

van der Velde, Y., Heidbüchel, I., Lyon, S., Nyberg, L., Rodhe, A., Bishop, K. and Troch, P., 2015. Consequences of mixing assumptions for time-variable travel time distributions. *Hydrological Processes*, 29, 3460–3474.

van der Velde, Y., Torfs, P.J.J.F., Van Der Zee, S.E.A.T.M. and Uijlenhoet, R., 2012. Quantifying catchment-scale mixing and its effect on time-varying travel time distributions. *Water Resources Research*, *48*(6), W06536.

van Meerveld, H.I., Kirchner, J.W., Vis, M.J., Assendelft, R.S. and Seibert, J., 2019. Expansion and contraction of the flowing stream network alter hillslope flowpath lengths and the shape of the travel time distribution. *Hydrology and Earth System Sciences*, *23*(11), 4825–4834.

Visser, A., Thaw, M., Deinhart, A., Bibby, R., Safeeq, M., Conklin, M., Esser, B. and van der Velde, Y., 2019. Cosmogenic isotopes unravel the hydrochronology and water storage dynamics of the Southern Sierra critical zone. *Water Resources Research*, *55*(2), 1429–1450.

Wang, H., Li, H., Xiang, W., Lu, Y., Wang, H., Hu, W., Si, B., Jasechko, S. and McDonnell, J.J., 2023. A 1000-year record of temperature from isotopic analysis of the deep critical zone in central China. *Geophysical Research Letters*, *50*(5), e2022GL101054.

Xie, J., Liu, X., Jasechko, S., Berghuijs, W.R., Wang, K., Liu, C., Reichstein, M., Jung, M. and Koirala, S., 2024. Majority of global river flow sustained by groundwater. *Nature Geoscience*, *17*(8), 770–777.

Xu, X. and Liu, W., 2017. The global distribution of Earth's critical zone and its controlling factors. *Geophysical Research Letters*, *44*(7), 3201–3208.

Chapter 8

Allen, S.T., Kirchner, J.W., Braun, S., Siegwolf, R.T.W. and Goldsmith, G.R., 2019. Seasonal origins of soil water used by trees. *Hydrology and Earth System Sciences*, *23*(2), 1199–1210.

Allison, G.B., Gat, J.R. and Leaney, F.W., 1985. The relationship between deuterium and oxygen-18 delta values in leaf water. *Chemical Geology: Isotope Geoscience Section*, *58*(1–2), 145–156.

Bachofen, C., Tumber-Dávila, S.J., Mackay, D.S., McDowell, N.G., Carminati, A., Klein, T., Stocker, B.D., Mencuccini, M. and Grossiord, C., 2024. Tree water uptake patterns across the globe. *The New Phytologist*, *242*(5), 1891–1910.

Barbeta, A., Burlett, R., Martín-Gómez, P., Fréjaville, B., Devert, N., Wingate, L., Domec, J. and Ogée, J., 2022. Evidence for distinct isotopic compositions of sap and tissue water in tree stems: consequences for plant water source identification. *The New Phytologist*, *233*(3), 1121–1132.

Bates, C.G. and A.J. Henry., 1928. Forest and streamflow experiment at Wagon Wheel Gap, Colorado. *Monthly Weather Review Supplement*, *30*, 192.

Benettin, P., Nehemy, M.F., Asadollahi, M., Pratt, D., Bensimon, M., McDonnell, J.J. and Rinaldo, A., 2021. Tracing and closing the water balance in a vegetated lysimeter. *Water Resources Research*, *57*(4), e2020WR029049.

Benettin, P., Volkmann, T.H.M., von Freyberg, J., Frentress, J., Penna, D., Dawson, T. E. and Kirchner, J.W., 2018. Effects of climatic seasonality on the isotopic composition of evaporating soil waters. *Hydrology and Earth System Sciences*, *22*(5), 2881–2890.

Berry, Z.C., Evaristo, J., Moore, G., Poca, M., Steppe, K., Verrot, L., Asbjornsen, H., Borma, L.S., Bretfeld, M., Hervé-Fernández, P., Seyfried, M., Schwendenmann, L., Sinacore, K., De Wispelaere, L. and McDonnell, J.J., 2018. The two water worlds hypothesis: Addressing multiple working hypotheses and proposing a way forward. *Ecohydrology, 11*(3), e1843.

Beyer, M., 2018. A brief on deep water uptake in water-limited environments: A conceptual explanation. *Isodrones*. https://www.isodrones.com/post/2018-08-31-isotopes-deep/

Bond, B.J., Jones, J.A., Moore, G., Phillips, N., Post, D. and McDonnell, J.J., 2002. The zone of vegetation influence on baseflow revealed by diel patterns of streamflow and vegetation water use in a headwater basin. *Hydrological Processes, 16*(8), 1671–1677.

Bosch, J.M. and Hewlett, J.D., 1982. A review of catchment experiments to determine the effect of vegetation changes on water yield and evapotranspiration. *Journal of Hydrology, 55*(1–4), 3–23.

Bowling, D.R., Schulze, E.S. and Hall, S.J., 2017. Revisiting streamside trees that do not use stream water: Can the two water worlds hypothesis and snowpack isotopic effects explain a missing water source? *Ecohydrology, 10*(1), e1771.

Brown, A.E., Zhang, L., McMahon, T.A., Western, A.W. and Vertessy, R.A., 2005. A review of paired catchment studies for determining changes in water yield resulting from alterations in vegetation. *Journal of Hydrology, 310*(1–4), 28–61.

Brooks, J.R., Barnard, H.R., Coulombe, R. and McDonnell, J.J., 2010. Ecohydrologic separation of water between trees and streams in a Mediterranean climate. *Nature Geoscience, 3*(2), 100–104.

Burvill, G.H., 1947. Soil salinity in the agricultural area of Western Australia. *Journal of the Australian Institute of Agricultural Science, 13*, 9–19.

Busch, D.E., Ingraham, N.L. and Smith, S.D., 1992. Water uptake in woody riparian phreatophytes of the southwestern United States: a stable isotope study. *Ecological Applications, 2*(4), 450–459.

Bush, S.A., Birch, A.L., Warix, S., Lininger, K.B., Musselman, K.N. and Barnard, H.R., 2025. Runoff composition is insensitive to summer rain contributions in a montane headwater stream. *Journal of Hydrology: Regional Studies, 61*.

Carrière, S.D., Martin-St Paul, N.K., Cakpo, C.B., Patris, N., Gillon, M., Chalikakis, K., Doussan, C., Olioso, A., Babic, M., Jouineau, A., Simioni, G. and Davi, H., 2020. The role of deep vadose zone water in tree transpiration during drought periods in karst settings – Insights from isotopic tracing and leaf water potential. *The Science of the Total Environment, 699*, 134332.

Ceperley, N., Gimeno, T.E., Jacobs, S.R., Beyer, M., Dubbert, M., Fischer, B., Geris, J., Holko, L., Kübert, A., Le Gall, S., Lehmann, M.M., Llorens, P., Millar, C., Penna, D., Prieto, I., Radolinski, J., Scandellari, F., Stockinger, M., Stumpp, C. and Tetzlaff, D., 2024. Toward a common methodological framework for the sampling, extraction, and isotopic analysis of water in the Critical Zone to study vegetation water use. *WIREs, 11*(4), e1727.

Craig, H., 1961. Isotopic variations in meteoric waters. *Science, 133*(3465), 1702–1703.

Dawson, T.E. and Ehleringer, J.R., 1991. Streamside trees that do not use stream water. *Nature, 350*(6316), 335–337.

Del Castillo, J., Comas, C., Voltas, J. and Ferrio, J.P., 2016. Dynamics of competition over water in a mixed oak–pine Mediterranean forest: Spatio-temporal and physiological components. *Forest Ecology and Management, 382*, 214–224.

Duvert, C., Barbeta, A., Hutley, L.B., Rodriguez, L., Irvine, D.J. and Taylor, A.R., 2024. Cavitron extraction of xylem water suggests cryogenic extraction biases vary across species but are independent of tree water stress. *Hydrological Processes, 38*(2), e15099.

Dye, P.J., 1996. Response of Eucalyptus grandis trees to soil water deficits. *Tree Physiology, 16*(1-2), 233–238.

Evaristo, J., Jasechko, S. and McDonnell, J.J., 2015. Global separation of plant transpiration from groundwater and streamflow. *Nature, 525*, 91–94.

Evaristo, J., Kim, M., van Haren, J., Pangle, L.A., Harman, C.J., Troch, P.A. and McDonnell, J.J., 2019. Characterizing the fluxes and age distribution of soil water, plant water, and deep percolation in a model tropical ecosystem. *Water Resources Research, 55*(4), 3307–3327.

Ezawa, T., Silvestri, A., Maruyama, H., Tawaraya, K., Suzuki, M., Duan, Y., Turina, M. and Lanfranco, L., 2023. Structurally distinct mitoviruses: are they an ancestral lineage of the Mitoviridae exclusive to arbuscular mycorrhizal fungi (Glomeromycotina)? *MBio, 14*(4), 00240–23.

Felton, A.J., Fisher, J.B., Hufkens, K., Purdy, A.J., Spawn-Lee, S.A., Duloisy, L.F. and Goldsmith, G.R., 2025. Global estimates of the storage and transit time of water through vegetation. *Nature Water, 3*(1), 59–69.

Gaines, K.P., Meinzer, F.C., Duffy, C.J., Thomas, E.M. and Eissenstat, D.M., 2016. Rapid tree water transport and residence times in a Pennsylvania catchment. *Ecohydrology, 9*(8), 1554–1565.

Gao, H., Fenicia, F. and Savenije, H.H., 2023. HESS Opinions: Are soils overrated in hydrology? *Hydrology and Earth System Sciences, 27*(14), 2607–2620.

Goeking, S.A. and Tarboton, D.G., 2020. Forests and water yield: A synthesis of disturbance effects on streamflow and snowpack in western coniferous forests. *Journal of Forestry, 118*(2), 172–192.

Goldsmith, G.R., Muñoz-Villers, L.E., Holwerda, F., McDonnell, J.J., Asbjornsen, H. and Dawson, T.E., 2012. Stable isotopes reveal linkages among ecohydrological processes in a seasonally dry tropical montane cloud forest. *Ecohydrology, 5*(6), 779–790.

Good, S.P., Noone, D. and Bowen, G., 2015. Hydrologic connectivity constrains partitioning of global terrestrial water fluxes. *Science, 349*(6244), 175–177.

Guntz, A.A., and Beltran, E., 1939. L'eau lourde et les végétaux. *Annales de l'Institut agricole, 1*, 12–32.

Hahm, W.J., Lapides, D.A., Rempe, D.M., McCormick, E.L. and Dralle, D.N., 2022. The age of evapotranspiration: Lower-bound constraints from distributed water fluxes across the continental United States. *Water Resources Research, 58*(10), e2022WR032961.

Hall, S.J., Weintraub, S.R. and Bowling, D.R., 2016. Scale-dependent linkages between nitrate isotopes and denitrification in surface soils: implications for isotope measurements and models. *Oecologia, 181*, 1221–1231.

Hamilton, I.L., 2019. *The interactions of jack pine trees (*Pinus banksiana*) and water.* Doctoral dissertation, University of Saskatchewan.

Harmon, R., Barnard, H.R. and Singha, K., 2020. Water table depth and bedrock permeability control magnitude and timing of transpiration-induced diel fluctuations in groundwater. *Water Resources Research, 56*(5), e2019WR025967.

Hartmann, A., Weiler, M. and Blume, T., 2020. The impact of landscape evolution on soil physics: Evolution of soil physical and hydraulic properties along two chronosequences of proglacial moraines. *Earth System Science Data, 12*(4), 3189–3204.

Hibbert, A.R., 1967. Forest treatment effects on water yield. In *Proceedings of International Symposium on Forest Hydrology*, Pergamon Press, pp. 527–543.

Hunt, A.G., Sahimi, M., Ghanbarian, B. and Poveda, G., 2024. Predicting ecosystem net primary productivity by percolation theory and optimality principle. *Water Resources Research, 60*(3), e2023WR036340.

Jones, J.A., Achterman, G.L., Augustine, L.A., Creed, I.F., Ffolliott, P.F., MacDonald, L. and Wemple, B.C., 2009. Hydrologic effects of a changing forested landscape—challenges for the hydrological sciences. *Hydrological Processes, 23*(18), 2699–2704.

Keim, R., Skaugset, A. and Weiler, M., 2005. Temporal persistence of spatial patterns in throughfall. *Journal of Hydrology, 314*(1–4), 263–274.

Klamerus-Iwan, A., Link, T.E., Keim, R.F. and Van Stan II, J.T., 2020. Storage and routing of precipitation through canopies. In *Precipitation partitioning by vegetation: A global synthesis* (pp. 17–34). Cham: Springer International Publishing.

Kline, J.R., Reed, K.L., Waring, R.H. and Stewart, M.L., 1976. Field measurement of transpiration in Douglas-fir. *Journal of Applied Ecology, 13*(1), 273–283.

Knighton, J., Kuppel, S., Smith, A., Soulsby, C., Sprenger, M. and Tetzlaff, D., 2020. Using isotopes to incorporate tree water storage and mixing dynamics into a distributed ecohydrologic modelling framework. *Ecohydrology, 13*(3), e2201.

Koeniger, P., Marshall, J.D., Link, T. and Mulch, A., 2011. An inexpensive, fast, and reliable method for vacuum extraction of soil and plant water for stable isotope analyses by mass spectrometry. *Rapid Communications in Mass Spectrometry, 25*(20), 3041–3048.

Li, H., Li, H., Wu, Q., Si, B., Jobbágy, E.G. and McDonnell, J.J., 2023. Afforestation triggers water mining and a single pulse of water for carbon trade-off in deep soil. *Agriculture, Ecosystems & Environment, 356*, 108655.

Liu, X., Chen, X., Zhang, Z., Liu, W., Gao, F., Cheng, Q., Chen, J. and Peng, T., 2025. *The role of rock fractures as a water source for trees growing in karst*. Authorea Preprints.

McCormick, E.L., Dralle, D.N., Hahm, W.J., Tune, A.K., Schmidt, L.M., Chadwick, K.D. and Rempe, D.M., 2021. Widespread woody plant use of water stored in bedrock. *Nature, 597*(7875), 225–229.

McDonnell, J.J., 2014. The two water worlds hypothesis: ecohydrological separation of water between streams and trees? *Wiley Interdisciplinary Reviews: Water, 1*(4), 323–329.

McDonnell, J.J., Evaristo, J., Bladon, K.D., Buttle, J., Creed, I.F., Dymond, S.F., Grant, G., Iroume, A., Jackson, C.R., Jones, J.A., Maness, T., McGuire, K.J., Scott, D.F., Segura, C., Sidle, R.C. and Tague, C. 2018. Water sustainability and watershed storage. *Nature Sustainability, 1*(8), 378–379.

Miguez-Macho, G. and Fan, Y., 2021. Spatiotemporal origin of soil water taken up by vegetation. *Nature, 598*(7882), 624–628.

Millar, C., Janzen, K., Nehemy, M. F., Koehler, G., Hervé-Fernández, P., Wang, H., Orlowski, N., Barbeta, A. and McDonnell, J.J., 2022. On the urgent need for standardization in isotope-based ecohydrological investigations. *Hydrological Processes, 36*(10), e14698.

Montgomery, D.R., Dietrich, W.E., Torres, R., Anderson, S.P., Heffner, J.T. and Loague, K., 1997. Hydrologic response of a steep, unchanneled valley to natural and applied rainfall. *Water Resources Research, 33*(1), 91–109.

Muñoz-Villers, L.E. and McDonnell, J.J., 2012. Runoff generation in a steep, tropical montane cloud forest catchment on permeable volcanic substrate. *Water Resources Research, 48*(9), W09528.

Najm, M.A., Lassabatere, L. and Stewart, R.D., 2019. Current insights into nonuniform flow across scales, processes, and applications. *Vadose Zone Journal, 18*(1), 190113.

Nehemy, M.F., Benettin, P., Asadollahi, M., Pratt, D., Rinaldo, A. and McDonnell, J.J., 2021. Tree water deficit and dynamic source water partitioning. *Hydrological Processes*, *35*(1), e14004.

Nehemy, M.F., Mattos, C.R.C., Oliveira, R.S., Hirota, M., Fan, Y., Schlickmann, M.B., Penha, D., Giacomin, L.L., Silva, J.S.G.M., Rocha, M., Rodrigues, G.A. and McDonnell, J.J., 2025. Embolism resistance supports the contribution of dry-season precipitation to transpiration in eastern Amazon forests. *Proceedings of the National Academy of Sciences of the United States of America*, *122*(33), e2501585122.

Nehemy, M.F., Maillet, J., Perron, N., Pappas, C., Sonnentag, O., Baltzer, J.L., Laroque, C.P. and McDonnell, J.J., 2022. Snowmelt water use at transpiration onset: phenology, isotope tracing, and tree water transit time. *Water Resources Research*, *58*(9), e2022WR032344.

Or, D., Lehmann, P., Shahraeeni, E. and Shokri, N., 2013. Advances in soil evaporation physics—A review. *Vadose Zone Journal*, *12*(4), vzj2012-0163.

Orlowski, N. and Breuer, L., 2020. Sampling soil water along the pF curve for δ^2H and δ^{18}O analysis. *Hydrological Processes*, *34*(25), 4959–4972.

Orlowski, N., Breuer, L., Angeli, N., Boeckx, P., Brumbt, C., Cook, C.S., Dubbert, M., Dyckmans, J., Gallagher, B., Gralher, B., Herbstritt, B., Hervé-Fernández, P., Hissler, C., Koeniger, P., Legout, A., Macdonald, C.J., Oyarzún, C., Redelstein, R., Seidler, C. and Siegwolf, R., 2018. Inter-laboratory comparison of cryogenic water extraction systems for stable isotope analysis of soil water. *Hydrology and Earth System Sciences*, *22*(7), 3619–3637.

Oshun, J., Dietrich, W.E., Dawson, T.E. and Fung, I., 2016. Dynamic, structured heterogeneity of water isotopes inside hillslopes. *Water Resources Research*, *52*(1), 164–189.

Peck, A.J. and Williamson, D.R., 1987. Effects of forest clearing on groundwater. In Peck. A.J. and Williamson, D.R., editors, Hydrology and Salinity in the Collie River Basin, Western Australia. *Journal of Hydrology*, *94*, 47–65.

Penna, D. and Van Meerveld, H.J., 2019. Spatial variability in the isotopic composition of water in small catchments and its effect on hydrograph separation. *Wiley Interdisciplinary Reviews: Water*, *6*(5), e1367.

Poca, M., Coomans, O., Urcelay, C., Zeballos, S.R., Bodé, S. and Boeckx, P., 2019. Isotope fractionation during root water uptake by *Acacia caven* is enhanced by arbuscular mycorrhizas. *Plant and Soil*, *441*(1–2), 485–497.

Radolinski, J., Vremec, M., Wachter, H., Birk, S., Brüggemann, N., Herndl, M., Kahmen, A., Nelson, D.B., Kübert, A., Schaumberger, A., Stumpp, C., Tissink, M., Werner, C. and Bahn, M., 2025. Drought in a warmer, CO_2-rich climate restricts grassland water use and soil water mixing. *Science*, *387*(6731), 290–296.

Rempe, D.M. and Dietrich, W.E., 2018. Direct observations of rock moisture, a hidden component of the hydrologic cycle. *Proceedings of the National Academy of Sciences*, *115*(11), 2664–2669.

Rodriguez-Iturbe, I., 2000. Ecohydrology: A hydrologic perspective of climate-soil-vegetation dynamics. *Water Resources Research*, *36*(1), 3–9.

Rothfuss, Y. and Javaux, M., 2017. Reviews and syntheses: Isotopic approaches to quantify root water uptake: a review and comparison of methods. *Biogeosciences*, *14*(8), 2199–2224.

Salve, R., Rempe, D.M. and Dietrich, W.E., 2012. Rain, rock moisture dynamics, and the rapid response of perched groundwater in weathered, fractured argillite underlying a steep hillslope. *Water Resources Research*, *48*, W11528.

Sohel, M.S.I., Grau, A.V., McDonnell, J.J. and Herbohn, J., 2021. Tropical forest water source patterns revealed by stable isotopes: A preliminary analysis of 46 neighboring species. *Forest Ecology and Management*, *494*, 119355.

Soulsby, C., Birkel, C. and Tetzlaff, D., 2016. Characterizing the age distribution of catchment evaporative losses. *Hydrological Processes*, *30*(8), 1308–1312.

Spencer, S.A. and van Meerveld, H.V., 2016. Double funnelling in a mature coastal British Columbia forest: spatial patterns of stemflow after infiltration. *Hydrological Processes*, *30*(22), 4185–4201.

Sprenger, M. and Allen, S.T., 2020. What ecohydrologic separation is and where we can go with it. *Water Resources Research*, *56*(7), e2020WR027238.

Sprenger, M., Herbstritt, B. and Weiler, M., 2015. Established methods and new opportunities for pore water stable isotope analysis. *Hydrological Processes*, *29*(25), 5174–5192.

Sprenger, M., Leistert, H., Gimbel, K. and Weiler, M., 2016. Illuminating hydrological processes at the soil–vegetation–atmosphere interface with water stable isotopes. *Reviews of Geophysics*, *54*(3), 674–704.

Sprenger, M., Stumpp, C., Weiler, M., Aeschbach, W., Allen, S.T., Benettin, P., Dubbert, M., Hartmann, A., Hrachowitz, M., Kirchner, J.W., McDonnell, J.J., Orlowski, N., Penna, D., Pfahl, S., Rinderer, M., Rodriguez, N., Schmidt, M. and Werner, C., 2019. The demographics of water: a review of water ages in the critical zone. *Reviews of Geophysics*, *57*(3), 800–834.

Stednick, J.D., 1996. Monitoring the effects of timber harvest on annual water yield. *Journal of Hydrology*, *176*(1–4), 79–95.

Taiz, L. and Zeiger, E., 2010. *Plants Physiology*, 5th edition. Sinauer Associates Inc.

Tian, F. and Xu, Z., 2024. Streamflow diurnal fluctuation and driving mechanism of headwater stream in a semi-humid mountainous region. *Journal of Hydrology, 641.*

Torres, R., Dietrich, W.E., Montgomery, D.R., Anderson, S.P. and Loague, K., 1998. Unsaturated zone processes and the hydrologic response of a steep, unchanneled catchment. *Water Resources Research., 34*(8), 1865–1879.

van Meerveld, H.J., Jones, J.P., Ghimire, C.P., Zwartendijk, B.W., Lahitiana, J., Ravelona, M. and Mulligan, M., 2021. Forest regeneration can positively contribute to local hydrological ecosystem services: Implications for forest landscape restoration. *Journal of Applied Ecology, 58*(4), 755–765.

van Meerveld, I. and Seibert, J., 2025. Reforestation effects on low flows: Review of public perceptions and scientific evidence. *WIREs, 12*(1), e1760.

von Freyberg, J., Allen, S.T., Grossiord, C. and Dawson, T.E., 2020. Plant and root-zone water isotopes are difficult to measure, explain, and predict: Some practical recommendations for determining plant water sources. *Methods in Ecology and Evolution, 11*(11), 1352–1367.

Washburn, E.W. and Smith, E.R., 1934. The isotopic fractionation of water by physiological processes. *Science, 79*(2043), 188–189.

Wershaw, R.L., Friedman, I., Heller, S.J. and Frank, P.A., 1966. Hydrogen isotopic fractionation of water passing through trees. In *Advances in Organic Geochemistry*, Pergamon Press, pp. 55–67.

White, J.W., Cook, E.R., Lawrence, J.R. and Wallace S.B., 1985. The D/H ratios of sap in trees: Implications for water sources and tree ring D/H ratios. *Geochimica et Cosmochimica Acta, 49*(1), 237–246.

Zhang, Z.Q., Evaristo, J., Li, Z., Si, B.C. and McDonnell, J.J., 2017. Tritium analysis shows apple trees may be transpiring water several decades old. *Hydrological Processes, 31*(5), 1196–1201.

Zhao, Y., Rahmati, M., Vereecken, H. and Or, D., 2024. Comment on "Are soils overrated in hydrology?" by Gao et al. (2023). *Hydrology and Earth System Sciences, 28*(17), 4059–4063.

Chapter 9

Allen, S.T. and Berghuijs, W.R., 2018. A need for incentivizing field hydrology, especially in an era of open data: discussion of "The role of experimental work in hydrological sciences–insights from a community survey." *Hydrological Sciences Journal, 63*(8), 1262–1265.

Anache, J.A., Wendland, E.C., Oliveira, P.T., Flanagan, D.C. and Nearing, M.A., 2017. Runoff and soil erosion plot-scale studies under natural rainfall: A meta-analysis of the Brazilian experience. *Catena, 152*, 29–39.

Arciniega-Esparza, S., Birkel, C., Durán-Quesada, A.M., Sánchez-Murillo, R., Moore, G.W., Maneta, M.P., Boll, J., Negri, L.B., Tetzlaff, D., Yoshimura, K. and Soulsby, C., 2023. Tracer-aided ecohydrological modelling across climate, land cover, and topographical gradients in the tropics. *Hydrological Processes, 37*(5), e14884.

Atwood, A., Hille, M., Clark, M.K., Rengers, F., Ntarlagiannis, D., Townsend, K. and West, A.J., 2023. Importance of subsurface water for hydrological response during storms in a post-wildfire bedrock landscape. *Nature Communications, 14*(1), 3814.

Auerswald, K., Fiener, P. and Dikau, R., 2009. Rates of sheet and rill erosion in Germany—A meta-analysis. *Geomorphology, 111*(3–4), 182–193.

Barthold, F.K. and Woods, R.A., 2015. Stormflow generation: A meta-analysis of field evidence from small, forested catchments. *Water Resources Research, 51*(5), 3730–3753.

Bejan, A., 2007. Constructal theory of pattern formation. *Hydrology and Earth System Sciences, 11*(2), 753–768.

Bejan, A., 2024. Vascular flow design and predicting evolution. *International Communications in Heat and Mass Transfer, 155*, 107517.

Bejan, A. and Lorente, S., 2011. The constructal law and the evolution of design in nature. *Physics of Life Reviews, 8*(3), 209–240.

Berghuijs, W.R., Woods, R.A., Hutton, C.J. and Sivapalan, M., 2016. Dominant flood generating mechanisms across the United States. *Geophysical Research Letters, 43*(9), 4382–4390.

Beven, K.J., 2006. *Streamflow Generation Processes*. International Association of Hydrological Sciences.

Beven, K.J., 2007. Working towards integrated environmental models of everywhere: uncertainty, data, and modelling as a learning process. *Hydrology and Earth System Science, 11*(1), 460–467.

Beven, K., 2021. The era of infiltration. *Hydrology and Earth System Sciences Discussions, 25*, 851–866.

Beven, K.J., 2025. A short history of philosophies of hydrological model evaluation and hypothesis testing. *WIREs, 12*(1), e1761.

Beven, K.J. and Alcock, R., 2012. Modelling everything everywhere: a new approach to decision making for water management under uncertainty. *Freshwater Biology, 56*, 124–132.

Beven, K.J. and Chappell, N.A., 2021. Perceptual perplexity and parameter parsimony. *Wiley Interdisciplinary Reviews: Water, 8*(4), e1530.

Beven, K. and Freer, J., 2001. A dynamic TOPMODEL. *Hydrological Processes, 15*(10), 1993–2011.

Beven, K.J. and Kirkby, M.J., 1979. A physically based, variable contributing area model of basin hydrology. *Hydrological Sciences Journal, 24*(1), 43–69.

Beveridge, W.I.B., 1950. *The Art of Scientific Investigation*. Blackburn Press.

Birkel, C., Miller, J., Watson, A., Anh Trinh, D., Durán-Quesada, A.M., Sánchez-Murillo, R., Soulsby, C., Terzer-Wassmuth, S., Tetzlaff, D., Uhlenbrook, S., Vystavna, Y. and Yoshimura, K., 2025. Demystifying the art of isotope-enabled hydrological and climate modelling. *The Science of the Total Environment*, *959*, 178242.

Blume, T., van Meerveld, I. and Weiler, M., 2017. The role of experimental work in hydrological sciences–insights from a community survey. *Hydrological Sciences Journal*, *62*(3), 334–337.

Brantley, S.L., Eissenstat, D.M., Marshall, J.A., Godsey, S.E., Balogh-Brunstad, Z., Karwan, D.L., Papuga, S.A., Roering, J., Dawson, T.E., Evaristo, J., Chadwick, O., McDonnell, J.J. and Weathers, K.C., 2017. Reviews and syntheses: on the roles trees play in building and plumbing the critical zone. *Biogeosciences*, *14*(22), 5115–5142

Brutsaert, W., 2005. *Hydrology: An Introduction*. Cambridge University Press.

Burt, T.P. and McDonnell, J.J., 2015. Whither field hydrology? The need for discovery science and outrageous hydrological hypotheses. *Water Resources Research*, *51*(8), 5919–5928.

Burt, T. and Thompson, D., 2020. *Curious About Nature: A Passion for Fieldwork*. Cambridge University Press.

Cain, M.R., Ward, A.S. and Hrachowitz, M., 2019. Ecohydrologic separation alters interpreted hydrologic stores and fluxes in a headwater mountain catchment. *Hydrological Processes*, *33*(20), 2658–2675.

Correa, A., Birkel, C., Gutierrez, J., Dehaspe, J., Durán-Quesada, A.M., Soulsby, C. and Sánchez-Murillo, R., 2020. Modelling non-stationary water ages in a tropical rainforest: A preliminary spatially distributed assessment. *Hydrological Processes*, *34*(25), 4776–4793.

Costa, J.T., 2017. *Darwin's Backyard: How Small Experiments Led to a Big Theory*. WW Norton & Company, New York.

Crawford, N.H. and Linsley, R.K., 1966. Digital Simulation in Hydrology: Stanford Watershed Model IV. Technical Report No. 39, Department of Civil Engineering, Stanford University, p. 210.

Datry, T., Truchy, A., Olden, J.D., Busch, M.H., Stubbington, R., Dodds, W.K., Zipper, S., Yu, S., Messager, M.L., Tonkin, J.D., Kaiser, K.E., Hammond, J.C., Moody, E.K., Burrows, R.M., Sarremejane, R., DelVecchia, A.G., Fork, M.L., Little, C.J., Walker, R.H. and Walters, A.W., 2023. Causes, responses, and implications of anthropogenic versus natural flow intermittence in river networks. *BioScience*, *73*(1), 9–22.

Davis, W.M., 1926. The value of outrageous geological hypotheses. *Science*, *63*(1636), 463–468.

Do Nascimento, T.V.M., Rudlang, J., Höge, M., van der Ent, R., Chappon, M., Seibert, J., Hrachowitz, M. and Fenicia, F., 2024. EStreams: An integrated dataset and catalogue of streamflow, hydro-climatic and landscape variables for Europe. *Scientific Data*, *11*(1), 879.

Dunne, T. and Leopold, L.B., 1978. *Water in Environmental Planning*. Freeman.

East, A.E., Logan, J.B., Dartnell, P., Lieber-Kotz, O., Cavagnaro, D.B., McCoy, S.W. and Lindsay, D.N., 2021. Watershed sediment yield following the 2018 Carr Fire, Whiskeytown National Recreation Area, Northern California. *Earth and Space Science*, *8*(9), e2021EA001828.

Ebel, B.A., Hammond, J.C., Walvoord, M.A., Partridge, T.F., Rey, D.M. and Murphy, S.F., 2025. A review and synthesis of post-wildfire shifts in hydrologic processes and streamflow generation mechanisms. *Environmental Research: Water*, *1*(4), 042001.

Eberhardt, L.L. and Thomas, J.M., 1991. Designing environmental field studies. *Ecological Monographs*, *61*(1), 53–73.

Engler, A., 1919. *Untersuchungen über den Einfluss des Waldes auf den Stand der Gewässer*, *12*. Kommissionsverlag von Beer & Cie.

Fan, Y., Clark, M., Lawrence, D.M., Swenson, S., Band, L.E., Brantley, S.L., Brooks, P.D., Dietrich, W.E., Flores, A., Grant, G., Kirchner, J.W., Mackay, D.S., McDonnell, J.J., Milly, P.C.D., Sullivan, P.L., Tague, C., Ajami, H., Chaney, N., Hartmann, A. and Hazenberg, P., 2019. Hillslope hydrology in global change research and earth system modeling. *Water Resources Research*, *55*(2), 1737–1772.

Fenicia, F., Kavetski, D. and Savenije, H.H.G., 2011. Elements of a flexible approach for conceptual hydrological modeling: 1. Motivation and theoretical development. *Water Resources Research*, *47*(11), W11510.

Finkenbiner, C.E., Good, S.P., Renée Brooks, J., Allen, S.T. and Sasidharan, S., 2022. The extent to which soil hydraulics can explain ecohydrological separation. *Nature Communications*, *13*(1), 6492.

Flanagan, D.C., Gilley, J.E. and Franti, T.G., 2007. Water Erosion Prediction Project (WEPP): Development history, model capabilities, and future enhancements. *Transactions of the ASABE*, *50*(5), 1603–1612.

Gao, H., Fenicia, F. and Savenije, H.H., 2023. HESS Opinions: Are soils overrated in hydrology? *Hydrology and Earth System Sciences*, *27*(14), 2607–2620.

Germann, P.F., 2017. Comment on "Understanding preferential flow in the vadose zone: Recent advances and future prospects" by N. Jarvis et al. *Vadose Zone Journal*, *16*(5), 1–2.

Getirana, A., Kumar, S., Bates, P., Boone, A., Lettenmaier, D. and Munier, S., 2024. The SWOT mission will reshape our understanding of the global terrestrial water cycle. *Nature Water*, *2*(12), 1139–1142.

Gharari, S., Hrachowitz, M., Fenicia, F. and Savenije, H.H.G., 2011. Hydrological landscape classification: investigating the performance of HAND based landscape classifications in a central European meso-scale catchment. *Hydrology and Earth System Sciences*, *15*(11), 3275–3291.

Gibson, J., Holmes, T., Stadnyk, T., Birks, S., Eby, P. and Pietroniro, A., 2021. Isotopic constraints on water balance and evapotranspiration partitioning in gauged watersheds across Canada. *Journal of Hydrology: Regional Studies*, *37*, 100878.

Gleason, C.J. and Brown, C.M., 2025. The once and future hydrology is Whole Earth hydrology. *Perspectives of Earth and Space Scientists*, *6*(1).

Golden, H.E., Christensen, J.R., McMillan, H.K., Kelleher, C.A., Lane, C.R., Husic, A., Li, L., Ward, A.S., Hammond, J., Seybold, E.C., Jaeger, K.L., Zimmer, M., Sando, R., Jones, C.N., Segura, C., Mahoney, D.T., Price, A.N. and Cheng, F., 2025. Advancing the science of headwater streamflow for global water protection. *Nature Water*, *3*(1), 16–26.

Gremillion, P., Gonyeau, A. and Wanielista, M., 2000. Application of alternative hydrograph separation models to detect changes in flow paths in a watershed undergoing urban development. *Hydrological Processes*, *14*(8), 1485–1501.

Guo, Q., Hao, Y. and Liu, B., 2015. Rates of soil erosion in China: A study based on runoff plot data. *Catena*, *124*, 68–76.

Hartmann, A., Weiler, M. and Blume, T., 2020. The impact of landscape evolution on soil physics: Evolution of soil physical and hydraulic properties along two chronosequences of proglacial moraines. *Earth System Science Data*, *12*(4), 3189–3204.

Heimsath, A.M., Dietrich, W.E., Nishiizumi, K. and Finkel, R.C., 1997. The soil production function and landscape equilibrium. *Nature*, *388*(6640), 358–361.

Hewlett, J.D. and Troendle, C.A., 1975. Non-point and diffused water sources: A variable source area problem. In *Watershed Management*, American Society of Civil Engineers (ASCE), pp. 21–46.

Hirmas, D.R., Giménez, D., Nemes, A., Kerry, R., Brunsell, N.A. and Wilson, C.J., 2018. Climate-induced changes in continental-scale soil macroporosity may intensify water cycle. *Nature*, *561*(7721), 100–103.

Hunt, A.G., Sahimi, M., Ghanbarian, B. and Poveda, G., 2024. Predicting ecosystem net primary productivity by percolation theory and optimality principle. *Water Resources Research*, *60*(3), e2023WR036340.

Imbeau, M.E., 1892. La Durance: Regime, crues et inundations. In *Annales des ponts et chaussées*. Carilian-Goeury.

Jarvis, N., Koestel, J. and Larsbo, M., 2016. Understanding preferential flow in the vadose zone: Recent advances and future prospects. *Vadose Zone Journal*, *15*(12), 1–11.

Kennedy, B.A., 1983. On outrageous hypotheses in geography. *Geography*, *68*(4), 326–330.

Kirchner, J.W., Benettin, P. and van Meerveld, I., 2023. Instructive surprises in the hydrological functioning of landscapes. *Annual Review of Earth and Planetary Sciences*, *51*(1), 277–299.

Knoben, W.J., Keshavarz, K., Torres-Rojas, L., Thébault, C., Chaney, N.W., Pietroniro, A. and Clark, M.P., 2025. Catchment Attributes and MEteorology for Large-Sample SPATially distributed analysis (CAMELS-SPAT): Streamflow observations, forcing data and geospatial data for hydrologic studies across North America. *EGUsphere*, 1–62.

Krabbenhoft, C.A., Allen, G.H., Lin, P., Godsey, S.E., Allen, D.C., Burrows, R.M., DelVecchia, A.G., Fritz, K.M., Shanafield, M., Burgin, A.J., Zimmer, M.A., Datry, T., Dodds, W.K., Jones, C.N., Mims, M.C., Franklin, C., Hammond, J.C., Zipper, S., Ward, A.S. and Costigan, K.H., 2022. Assessing placement bias of the global river gauge network. *Nature Sustainability*, *5*(7), 586–592.

Kuppel, S., Tetzlaff, D., Maneta, M.P. and Soulsby, C., 2018a. EcH 2 O-iso 1.0: Water isotopes and age tracking in a process-based, distributed ecohydrological model. *Geoscientific Model Development*, *11*(7), 3045–3069.

Kuppel, S., Tetzlaff, D., Maneta, M.P. and Soulsby, C., 2018b. What can we learn from multi-data calibration of a process-based ecohydrological model? *Environmental Modelling & Software: With Environment Data News*, *101*, 301–316.

Lane, C.R., D'Amico, E., Christensen, J.R., Fritz, K.M. and Golden, H.E., 2025. Linking wetlands to relatively permanent flowing waters: a conterminous United States geospatial analysis. *Wetlands Ecology and Management*, *33*(2).

Lapides, D.A., Hahm, W.J., Rempe, D.M., Dietrich, W.E. and Dralle, D.N., 2022. Controls on stream water age in a saturation overland flow-dominated catchment. *Water Resources Research*, *58*(4), e2021WR031665.

Leuteritz, A., Gauthier, V.A. and van Meerveld, I., 2025. Spatial and temporal variability in event water in near-surface flow pathways in a humid steep headwater catchment. *Hydrological Processes*, *39*(7), e70186.

Levia, D.F., Creed, I.F., Hannah, D.M., Nanko, K., Boyer, E.W., Carlyle-Moses, D.E., van de Giesen, N., Grasso, D., Guswa, A.J., Hudson, J.E., Hudson, S.A., Iida, S., Jackson, R.B., Katul, G.G., Kumagai, T., Llorens, P., Ribeiro, F.L., Pataki, D.E., Peters, C.A., Carretero, D.S., Selker, J.S., Tetzlaff, D., Zalewski, M. and Bruen, M., 2020.

Homogenization of the terrestrial water cycle. *Nature Geoscience*, *13*(10), 656–658

Li, L., Ni, J., Chang, F., Yue, Y., Frolova, N., Magritsky, D., Borthwick, A.G., Ciais, P., Wang, Y., Zheng, C. and Walling, D.E., 2020. Global trends in water and sediment fluxes of the world's large rivers. *Science Bulletin*, *65*(1), 62–69.

Maneta, M.P. and Silverman, N.L., 2013. A spatially distributed model to simulate water, energy, and vegetation dynamics using information from regional climate models. *Earth Interactions*, *17*(11), 1–44.

McDonnell, J., 2020. *Navigating an Academic Career: A Brief Guide for PhD Students, Postdocs, and New Faculty*. American Geophysical Union.

McDonnell, J., 2024. The ebbs and flows of creativity. *Science*, *383*(6688), 1262.

McDonnell, J.J., Sivapalan, M., Vaché, K., Dunn, S., Grant, G., Haggerty, R., Hinz, C., Hooper, R., Kirchner, J., Roderick, M.L., Selker, J. and Weiler, M., 2007. Moving beyond heterogeneity and process complexity: A new vision for watershed hydrology. *Water Resources Research*, *43*(7), W07301.

McDonnell, J.J. and Woods, R., 2004. On the need for catchment classification. *Journal of Hydrology*, *299*(1–2), 2–3.

McGuire, K.J., Klaus, J. and Jackson, C.R., 2024. Interflow, subsurface stormflow and throughflow: A synthesis of field work and modelling. *Hydrological Processes*, 38, e15263.

McMillan, H., Araki, R., Bolotin, L., Kim, D.-H., Coxon, G., Clark, M. and Seibert, J., 2025. Global patterns in observed hydrologic processes. *Nature Water*, *3*(4), 497–506.

McMillan, H., Araki, R., Gnann, S., Woods, R. and Wagener, T., 2023. How do hydrologists perceive watersheds? A survey and analysis of perceptual model figures for experimental watersheds. *Hydrological Processes*, *37*(3), e14845.

Milly, P.C.D., Betancourt, J., Falkenmark, M., Hirsch, R.M., Kundzewicz, Z.W., Lettenmaier, D. P. and Stouffer, R.J., 2008. Stationarity is dead: whither water management? *Science*, 319(5863), 573–574.

Mulvaney, T.J., 1851. On the use of self-registering rain and flood gauges in making observations of the relations of rainfall and flood discharges in a given catchment. *Proceedings of the Institution of Civil Engineers of Ireland*, *4*(2), 18–33.

Najm, M.A., Lassabatere, L. and Stewart, R.D., 2019. Current insights into nonuniform flow across scales, processes, and applications. *Vadose Zone Journal*, *18*(1), 190113.

Nasta, P., Blöschl, G., Bogena, H.R., Zacharias, S., Baatz, R., De Lannoy, G., Jensen, K.H., Manfreda, S., Pfister, L., Tarquis, A.M., van Meerveld, I., Voltz, M., Zeng, Y., Kustas, W., Li, X., Vereecken, H. and Romano, N., 2025. HESS Opinions: Towards a common vision for the future of hydrological observatories. *Hydrology and Earth System Sciences*, *29*(2), 465–483.

Pearce, A.J., 1990. Streamflow generation processes: An Austral view. *Water Resources Research*, *26*(12), 3037–3047.

Penna, D., 2024. A recipe for why and how to set up and sustain an experimental catchment. *Hydrological Processes*, *38*(5), e15163.

Pfister, L. and Kirchner, J.W., 2017. Debates—Hypothesis testing in hydrology: Theory and practice. *Water Resources Research*, *53*(3), 1792–1798.

Porkka, M., Virkki, V., Wang-Erlandsson, L., Gerten, D., Gleeson, T., Mohan, C., Fetzer, I., Jaramillo, F., Staal, A., te Wierik, S., Tobian, A., van der Ent, R., Döll, P., Flörke, M., Gosling, S.N., Hanasaki, N., Satoh, Y., Müller Schmied, H., Wanders, N. and Famiglietti, J.S., 2024. Notable shifts beyond pre-industrial streamflow and soil moisture conditions transgress the planetary boundary for freshwater change. *Nature Water*, *2*(3), 262–273.

Radolinski, J., Pangle, L., Klaus, J. and Stewart, R.D., 2021. Testing the 'two water worlds' hypothesis under variable preferential flow conditions. *Hydrological Processes*, *35*(6), e14252.

Radolinski, J., Vremec, M., Wachter, H., Birk, S., Brüggemann, N., Herndl, M., Kahmen, A., Nelson, D.B., Kübert, A., Schaumberger, A., Stumpp, C., Tissink, M., Werner, C. and Bahn, M., 2025. Drought in a warmer, CO_2-rich climate restricts grassland water use and soil water mixing. *Science*, *387*(6731), 290–296.

Riebe, C.S., Hahm, W.J. and Brantley, S.L., 2017. Controls on deep critical zone architecture: a historical review and four testable hypotheses. *Earth Surface Processes and Landforms*, *42*(1), 128–156.

Scherrer, S., Naef, F., Faeh, A.O. and Cordery, I., 2007. Formation of runoff at the hillslope scale during intense precipitation. *Hydrology and Earth System Sciences*, *11*(2), 907–922.

Seibert, J. and McDonnell, J.J., 2002. On the dialog between experimentalist and modeler in catchment hydrology: Use of soft data for multicriteria model calibration. *Water Resources Research*, *38*(11), 1241.

Seibert, J. and McDonnell, J.J., 2010. Land-cover impacts on streamflow: a change-detection modelling approach that incorporates parameter uncertainty. *Hydrological Sciences Journal*, *55*(3), 316–332.

Seibert, J., McDonnell, J.J. and Woodsmith, R.D., 2010. Effects of wildfire on catchment runoff response: a modelling approach to detect changes in snow-dominated forested catchments. *Hydrology Research*, *41*(5), 378–390.

Seo, K.-W., Ryu, D., Jeon, T., Youm, K., Kim, J.-S., Oh, E.H., Chen, J., Famiglietti, J.S. and Wilson, C.R., 2025. Abrupt sea level rise and Earth's gradual pole shift reveal permanent hydrological regime changes in the 21st century. *Science*, *387*(6741), 1408–1413.

Sharif, H. and Ameli, A.A., 2025. Searching for functional simplicity of stormflow generation. *Water Resources Research*, *61*(3), e2024WR037179.

Sidle, R.C., Tsuboyama, Y., Noguchi, S., Hosoda, I., Fujieda, M. and Shimizu, T., 2000. Stormflow generation in steep forested headwaters: a linked hydrogeomorphic paradigm. *Hydrological Processes*, *14*(3), 369–385.

Sivapalan, M., 2003. Process complexity at hillslope scale, process simplicity at watershed scale: Is there a connection? *Hydrological Processes*, *17*, 1037–1041

Tauro, F., Selker, J., van de Giesen, N., Abrate, T., Uijlenhoet, R., Porfiri, M., Manfreda, S., Caylor, K., Moramarco, T., Benveniste, J., Ciraolo, G., Estes, L., Domeneghetti, A., Perks, M.T., Corbari, C., Rabiei, E., Ravazzani, G., Bogena, H., Harfouche, A. and Brocca, L., 2018. Measurements and Observations in the XXI century (MOXXI): innovation and multi-disciplinarity to sense the hydrological cycle. *Hydrological Sciences Journal*, *63*(2), 169–196.

Tetzlaff, D., Birkel, C., Dick, J., Geris, J. and Soulsby, C., 2014. Storage dynamics in hydropedological units control hillslope connectivity, runoff generation, and the evolution of catchment transit time distributions. *Water Resources Research*, *50*(2), 969–985.

Tetzlaff, D., Smith, A., Luo, S. and Soulsby, C., 2024. Integrated monitoring and isotope-aided modelling to assess ecohydrological fluxes and storage dynamics in a drought sensitive lowland catchment, Germany. In *EGU General Assembly Conference Abstracts*, p. 4416.

Tetzlaff, D., Soulsby, C., Waldron, S., Malcolm, I.A., Bacon, P.J., Dunn, S. M., Lilly, A. and Youngson, A.F., 2007. Conceptualization of runoff processes using a geographical information system and tracers in a nested mesoscale catchment. *Hydrological Processes*, *21*(10), 1289–1307.

Vaché, K.B. and McDonnell, J.J., 2006. A process-based rejectionist framework for evaluating catchment runoff model structure. *Water Resources Research*, *42*(2), W02409.

Van Stan, J.T., Allen, S.T., Aubrey, D.P., Berry, Z.C., Biddick, M., Coenders-Gerrits, M.A., Giordani, P., Gotsch, S.G., Gutmann, E.D., Kuzyakov, Y. and Magyar, D., 2023. Shower thoughts: why scientists should spend more time in the rain. *Bioscience*, *73*(6), 441–452.

Wagener, T., Dadson, S.J., Hannah, D.M., Coxon, G., Beven, K., Bloomfield, J.P., Buytaert, W., Cloke, H., Bates, P., Holden, J., Parry, L., Lamb, R., Chappell, N.A., Fry, M. and Old, G., 2021. Knowledge gaps in our perceptual model of Great Britain's hydrology. *Hydrological Processes*, *35*(7), e14288.

Wagener, T., Sivapalan, M., Troch, P. and Woods, R., 2007. Catchment classification and hydrologic similarity. *Geography Compass*, *1*(4), 901–931.

Wang, H., Liu, J., Klaar, M., Chen, A., Gudmundsson, L. and Holden, J., 2024. Anthropogenic climate change has influenced global river flow seasonality. *Science*, *383*(6686), 1009–1014.

Weiler, M. and McDonnell, J.J., 2007. Conceptualizing lateral preferential flow and flow networks and simulating the effects on gauged and ungauged hillslopes. *Water Resources Research*, *43*(3), W03403.

Wischmeier, W.H. and Smith, D.D., 1978. *Predicting rainfall erosion losses: a guide to conservation planning* (No. 537). Department of Agriculture, Science and Education Administration.

Wrede, S., Fenicia, F., Martínez-Carreras, N., Juilleret, J., Hissler, C., Krein, A., Savenije, H.H.G., Uhlenbrook, S., Kavetski, D. and Pfister, L., 2015. Towards more systematic perceptual model development: a case study using 3 Luxembourgish catchments. *Hydrological Processes*, *29*(12), 2731–2750.

Wu, S., Zhao, J., Wang, H. and Sivapalan, M., 2021. Regional patterns and physical controls of streamflow generation across the conterminous United States. *Water Resources Research*, *57*(6), e2020WR028086.

Yu, G., Liu, T., McGuire, L.A., Wright, D.B., Hatchett, B.J., Miller, J.J., Berli, M., Giovando, J., Bartles, M. and Floyd, I.E., 2023. Process-based quantification of the role of wildfire in shaping flood frequency. *Water Resources Research*, *59*(12), e2023WR035013.

Zhao, Y., Rahmati, M., Vereecken, H. and Or, D., 2024. Comment on "Are soils overrated in hydrology?" by Gao et al. (2023). *Hydrology and Earth System Sciences*, *28*(17), 4059–4063.

Zimmer, M.A., Kaiser, K.E., Blaszczak, J.R., Zipper, S.C., Hammond, J.C., Fritz, K.M., Costigan, K.H., Hosen, J., Godsey, S.E., Allen, G.H., Kampf, S., Burrows, R.M., Krabbenhoft, C.A., Dodds, W., Hale, R., Olden, J.D., Shanafield, M., DelVecchia, A.G., Ward, A.S. and Mims, M.C., 2020. Zero or not? Causes and consequences of zero-flow stream gage readings. *WIREs*, *7*(3), e1436.

Index